Bird–Parasite Interactions

Oxford Ornithology Series

Edited by C. M. Perrins

1 *Bird population studies: relevance to conservation and management* (1991). Edited by C. M. Perrins, J.-D. Lebreton, and G. J. M. Hirons

2 *Bird–parasite interactions: ecology, evolution, and behaviour* (1991). Edited by J. E. Loye and M. Zuk

Bird–Parasite Interactions

ECOLOGY, EVOLUTION, AND BEHAVIOUR

Edited by

J. E. LOYE
Department of Biology, University of Utah

M. ZUK
Department of Biology, University of California, Riverside

Oxford New York Tokyo
OXFORD UNIVERSITY PRESS
1991

This book has been printed digitally in order to ensure its continuing availability

OXFORD
UNIVERSITY PRESS

Great Clarendon Street, Oxford OX2 6DP

Oxford University Press is a department of the University of Oxford.
It furthers the University's objective of excellence in research, scholarship,
and education by publishing worldwide in

Oxford New York

Auckland Bangkok Buenos Aires Cape Town Chennai
Dar es Salaam Delhi Hong Kong Istanbul Karachi Kolkata
Kuala Lumpur Madrid Melbourne Mexico City Mumbai Nairobi
São Paulo Shanghai Singapore Taipei Tokyo Toronto

with an associated company in Berlin

Oxford is a registered trade mark of Oxford University Press
in the UK and in certain other countries

Published in the United States
by Oxford University Press Inc., New York

Reprinted 2002

A catalogue record for this book is available from the British Library

Library of Congress Cataloging in Publication Data
Bird-parasite interactions : ecology, evolution, and behaviour /
Edited by J. E. Loye and M. Zuk.
p. cm. — (Oxford ornithology series; 2)
1. Birds—Parasites. I. Loye, J. E. (Jenella E.) II. Zuk, M.
(Marlene) III. Series.
QL698.95.B57 1991 598.25'249—dc20 90-42349

ISBN 0-19-857738-9

Foreword

PETER PRICE

With this volume a paradigm is born: parasites are likely to play a role in practically every aspect of the evolutionary biology of birds, and probably vertebrates in general. Such a view has been a long time in gestation, probably because in a fetal condition it was roundly thwarted by two eminent ecologists. Early researchers were obviously convinced that bird parasites must play a role in their host's ecology, but experienced difficulties in establishing this as fact. Even in 1980 in preparation for my book *Evolutionary biology of parasites,* I was continually frustrated with a literature on parasites of birds that was speculative and uncompelling. As a result avian parasites received a low profile in my book, somewhat paradoxically for such an enthusiastically studied host species. This volume on the ecology, behaviour, and evolution of bird–parasite interactions changes all this radically and permanently.

In 1927 Charles Elton, in his classic book *Animal ecology,* devoted a whole chapter to parasites, but none to predators or mutualists or competitors. Did this bode well for the launching of a focus on host–parasite relationships in the ecological literature? Hardly. To give Elton credit, he was developing a generality, not particulars, but the generality was 'that it is best to treat parasites as being essentially the same as carnivores . . . the resemblances between the two classes of animals are more important than the differences' (p. 75). This was because 'If parasites are to occupy a special place in the scheme we must, to be consistent, accuse cows of petty larceny against grasses, and cactuses of cruelty to the sun' (p. 75). Cute phrases but misleading, for the complexity of many parasite life cycles and the details of the host–parasite interaction do indeed result in special roles needing special attention from evolutionary biologists.

Had Edward A. Wilson not died with Scott in the Antarctic, would avian diseases have received emphasis rather than epitaph from David Lack in 1954? In his chapter on *The influence of disease on bird numbers*, Lack used strongylosis of Red Grouse, caused by *Trichostrongylus tenuis*, as the premier example of parasite epidemiology in avian populations. Wilson, in 1911, had been a strong proponent of disease as a critical factor in grouse population ecology, and no doubt would have persisted in his case had he returned from the Antarctic. He had also focused on the role of physical appearance of pheasants with parasites in sexual selection among birds. Speculation aside, Lack concluded: 'While further evidence is needed, it seems unlikely that disease is an important factor regulating the numbers of most wild birds.' (p. 169, Lack 1954).

The 'further evidence' was a long time coming, for in his influential treatise on bird populations in 1966 Lack continued to downplay the role of parasites. This position was reinforced even in 1976 by Bennett *et al.* who argued that insufficient data were available, and more critical studies were needed. Kluyver, throughout his research career, was intrigued by the parasitic larvae of the fly *Protocalliphora* in the nests of great tits, but even in 1970 the convincing evidence was unpublished. In Belgium and England such parasites were reported to play little role in fledgling viability (as observed by den Boer and Gradwell). Yet the tremble-thrust behaviour of stabbing into the nest material, described by Hartshorne in 1966, must surely be involved with *Protocalliphora* attack, suggesting evolved and specialized behaviours in response to selection by parasites. But measuring the impact of *Protocalliphora*, one of the more conspicuous nest parasites, has continued to foil investigators (e.g. Gold and Dahlsten 1983). Warner's arguments for extinction of endemic Hawaiian birds, published in *Condor* in 1969, marked a turning-point in the literature, at least for me, but a monumental shift in viewpoint was needed for the general field of avian biology to become sensitized to the role of parasites.

It has been symptomatic of the broader perspective on parasite–host interactions that the literature on bird diseases and their epidemiology has remained almost isolated from developments in avian ecology, behaviour, and evolution. Considerable understanding of poultry diseases exists, but without extrapolation to the field. Extensive surveys of bird populations as potential natural reservoirs of human diseases have been conducted without the conviction that the epidemiology of diseases in the bird populations themselves must be important and worthy of study.

Sensitivity among biologists to the role of parasites in bird populations will be greatly increased by this book. Avian populations and communities have always held a central role in the development of ecology. Moreover, the juxtaposition of ecology, behaviour, and evolution sends a compelling message to the community of biologists that all aspects of natural populations are potentially impacted by parasites. This book is clearly a landmark in the study of bird–parasite relationships, which I expect will establish a paradigm for all vertebrates. It sets the stage for rapid advances in understanding the complexity and subtlety of host–parasite interactions, and will stimulate great interest among young scientists, and recruit many into this fascinating field.

References

Den Boer, P.J. and Gradwell, G.R. (1970). Discussion on the papers of Kluyver, Perrins and Dhondt. In *Dynamics of populations* (ed. P.J. den Boer and G.R. Gradwell), pp. 544–7. Centre for Agricultural Publishing and Documentation, Wageningen.

Elton, C. (1927). *Animal ecology*. Macmillan, New York.

Gold, C.S. and Dahlsten D.L. (1983). Effects of parasitic flies (*Protocalliphora* spp.) on nestlings of mountain and chestnut-backed chickadees. *Wilson Bulletin* **95**, 560–72.

Haftorn, S. (1966). Efflegging og ruging hos meiser basert pa temperaturmalinger og direkte iakttagelser. *Sterna* **7**, 49–102.

Kluyver, H.N. (1970). Regulation of numbers in populations of Great Tits (*Parus m. major*). In *Dynamics of populations* (ed. P.J. den Boer and G.R. Gradwell), pp. 507–23. Centre for Agricultural Publishing and Documentation, Wageningen.

Lack, D. (1954). *The natural regulation of animal numbers*. Oxford University Press, London.

Lack, D. (1966). *Population studies of birds*. Oxford University Press, London.

Price, P.W. (1980). *Evolutionary biology of parasites*. Princeton University Press, Princeton.

Warner, R.E. (1968). The role of introduced diseases in the extinction of the endemic Hawaiian avifauna. *Condor* **70**, 101–20.

Wilson, E.A. (1911). Grouse disease. In *The grouse in health and disease* (ed. Lord Lovat), pp. 185–206. Smith Elder, London.

Preface

> It became evident . . . that the diversity of insects and life in general may be at least partly understood by gaining knowledge of the parasitic way of life. . . . Parasites affect the life and death of practically every other living organism.
>
> *The evolutionary biology of parasites*
> Peter Price (1980)

The effects of parasites on their avian hosts have always intrigued natural historians. Every person who handles birds has a riveting, often gruesome, story of parasite pathology in their study subject. When Dame Miriam Rothschild and T. Clay wrote *Flies, fleas and cuckoos* (1952, Literary Press, London) the available information was mainly anecdotal, but their work had a broad impact on generations of workers. When, as graduate students in the 1980s, we enthusiastically delved into the literature on avian parasitism, we were frustrated both by the difficulty of locating primary sources and by the paucity of empirical data that could be used for hypothesis testing. In 1982 R. M. Anderson and R. M. May stated, 'The accumulation of well-understood case studies from a wide variety of animals, both vertebrate and invertebrate, is clearly crucial. These are the studies that restrain mathematical flights of fancy and guide experimental work.' (p. 22, *Population biology of infectious diseases*, Springer-Verlag).

Now, in 1990, 10 years after Peter Price piqued the conscience of our generation of parasite watchers with *The evolutionary biology of parasites* (1980, Princeton Monograph in Population Biology, Princeton University Press) we have those data and are developing theory. Avian biology is unique in the extent of its data on population trends and on the reproductive success of individuals in long-term studies. Thus avian hosts offer a most fruitful area for testing host–parasite theory.

The ecology of parasitism in vertebrate biology has aroused much interest in topics such as the effects of parasites on the individual fitness, population regulation, and behaviour of their hosts, including mate choice and the resulting opportunity for sexual selection. Information about relevant parasite–host systems is, however, still diffusely distributed in specialized and multilingual journals of medicine, entomology, parasitology, and vertebrate biology. The research on bird–parasite systems presented here seeks to draw together this field, combining new theoretical concepts, current literature, and real data. Such a compilation has not been available before and we hope this volume will prove to be useful as a reference and seminar text in many disciplines.

The chapters in this book present data collected both in the laboratory and in the field. With the exception of some introductory review chapters that generalize the theory of the effects of parasites on host ecology and behaviour, these data are here published for the first time. We collected the most current research together because the diverse nature of workers in evolutionary ecology of host–parasite interactions means that we seldom meet in person or overlap in journals. We feel that the contributors to this volume are in the process of originating a field of evolutionary ecology that is of increasing interest and importance to other workers. We limit our attention to avian hosts, not to slight the excellent work being done with mammals, fishes, and herptiles but in order to use the resources with which the editors are most familiar. As Peter Price points out in his Foreword, the history of bird–parasite interactions is both long and idiosyncratic. We hope to smooth out some of the less desirable idiosyncrasies while acknowledging our debt to that history.

In Parts II and III on 'Ecology' and 'Behaviour', respectively, the chapters examine a representative selection of current findings on endo- and ectoparasite infections infecting birds both colonial and solitary, migratory and resident, in tropical and temperate environments.

Each part begins with an introductory chapter that reviews a topic in that area, taking a theoretical approach with reference to empirical data. Subsequent chapters examine hypotheses and report primary data that quantify the latest findings in the subject discussed, and speculate on the directions of future enquiry.

In addition we endeavour to provide a survey of current techniques of data collection and analysis for the testing of hypotheses involving parasites. Note that this volume is not a textbook of unifying explanations or interpretations of results. This is deliberate, reflecting the infancy of the field. A truly comprehensive work awaits the arrival of both more data and the successful marriage of those data with more sophisticated theory. Perhaps this book will inspire talented theoreticians to apply their skills to the unique problems posed by parasite–host interactions.

Ten of these chapters were presented as papers at the 1988 meeting of the American Ornithologists' Unions in a symposium organized by Jenella Loye and Dale Clayton and entitled: 'Avian responses to parasitism'.

We have thoroughly enjoyed getting to know the work and workers in our areas of common interest and are bound forever by shared enthusiasm and frustration at the 'how did I get myself into this' response of all writers to deadlines. Our group suffered a terrible loss with the death of Carl Kirkpatrick this year. Carl was a man who begat infectious enthusiasm even over a telephone line, making the listener feel hopeful and ready to do research. This volume is the result of a common desire to have this information on our own bookshelves.

Salt Lake City J. L.
Riverside M. Z.
1991

Acknowledgements

In addition to our contributors and reviewers we gratefully acknowledge discussions with Charles Brown, Larry Clark, Alex Cruz, Janis Dickinson, Nigella Hillgarth, Patricia Gowaty, Ray Gill, Peter Frederick, Carl Kirkpatrick, Walt Koenig, David Ligon, Tom Scott, Marilyn Spaulding, Randy Thornhill, and Cathy Toft. J. Loye thanks the Department of Biology and the University of Utah for providing the resources that made this volume possible, Janice Moore and Marlene Zuk for stimulating her efforts in this direction by way of their excellent research, and David Loye, Riane Eisler, Scott Carroll, Vicky Prager, Michael Salamon, and Buck and Minnow for creating a psychological climate conducive to creative effort.

Contents

Part III Behaviour

Part IV Sexual selection

Associate editors and contributors

Associate editors

EDWARD H. BURTT, JR, Department of Zoology, Ohio Wesleyan University, Delaware, OH 43015, USA

SCOTT P. CARROLL, Department of Biology, University of Utah, Salt Lake City, UT 84112, USA

BRIAN R. CHAPMAN, Department of Zoology, 730 Van Vleet Oval, Room 314, University of Oklahoma, Norman, OK 73019, USA

JOSEPH J. SCHALL, Department of Zoology, Marsh Life Sciences Building, University of Vermont, Burlington, VT 05405-0086, USA

CHARLES VAN RIPER III, Department of Biological Sciences and Coop. Park Unit, Northern Arizona University, Flagstaff, AZ 86011, USA

THOMAS Q. VU, Department of Engineering, University of Utah, Salt Lake City, UT 84112, USA

Contributors

CARTER T. ATKINSON, Institute of Pathology, Case Western Reserve University, School of Medicine, Cleveland, OH 44106, USA

G. A. BABBITT, Department of Zoology, Ohio Wesleyan University, Delaware, OH 43015, USA

DAVID M. BIRD, Department of Renewable Resources, McGill University, MacDonald College, 21, 111 Lakeshore Rd, Ste. Anne de Bellevue PQ H9X 1CO, Canada

MARK S. BOYCE, Department of Zoology and Physiology, University of Wyoming, Box 3166, University Station, Laramie, WY 82071-3166, USA

NANCY BURLEY, Department of Ecology, Ethology and Evolution, University of Illinois, 606 E. Healey, Champaign, IL 61820, USA

EDWARD H. BURTT, JR, Department of Zoology, Ohio Wesleyan University, Delaware, OH 43015, USA

SCOTT P. CARROLL, Department of Biology, University of Utah, Salt Lake City, UT 84112, USA

BRIAN R. CHAPMAN, Department of Zoology, 730 Van Vleet Oval, Room 314, University of Oklahoma, Norman, OK 73019, USA

W. CHOW, School of Biological Sciences, Univ. Sains Malaysia, 11800 USM, Pulau Pinang, Malaysia

LARRY CLARK, Monell Chemical Senses Center, 3500 Market Street, Philadelphia, PA 19104, USA

DALE H. CLAYTON, Committee on Evolutionary Biology, University of Chicago, 915 East 57th Street, Chicago, IL 60637, USA

ALEXANDER CRUZ, Environmental, Population and Organismic Biology Department, University of Colorado, Boulder, CO 80309-0334, USA

CARLOS A. DELANNOY, Department of Biology, University of Puerto Rico, Mayaquez, PR 00708, USA

ANDREW P. DOBSON, Department of Biology, Princeton University, Princeton, NJ 08544, USA

DAVID C. DUFFY, The Seatuck Foundation, Box 31, Islip, Long Island, NY 11751, USA

JOHN D. EDMAN, Department of Entomology, University of Massachusetts, Amherst, MA 01003, USA

JOHN E. GEORGE, Knipling-Bushland US Livestock Insects Research Laboratory, P.O. Box 232, Kerrville, TX 78029, USA

KARL HALUPKA, Department of Ecology, Ethology and Evolution, University of Illinois, 515 Morrill Hall, Urbana, IL 61801, USA

PETER J. HUDSON, Upland Research Unit, The Game Conservancy, Crubenmore Newtonmore, Inverness-shire PH20 1BE, Scotland

LINDA L. JOHNSON, Department of Zoology and Physiology, University of Wyoming, Box 3166, University Station, Laramie, WY 82071-3166, USA

CARL E. KIRKPATRICK, Department of Veterinary Pathobiology, School of Veterinary Medicine, University of Illinois, Urbana, IL 61801, USA

URIEL D. KITRON, Department of Veterinary Pathobiology, School of Veterinary Medicine, University of Illinois, Urbana, IL 61801, USA

JENELLA E. LOYE, Department of Biology, University of Utah, Salt Lake City, UT 84112, USA

BRYAN F. J. MANLY, University of Otago, Dunedin, New Zealand

ANDERS PAPE MØLLER, Department of Zoology, Uppsala University, Box 561, S-751 22 Uppsala, Sweden

PETER W. PRICE, Department of Biology, N. Arizona University, Flagstaff, AZ 86001, USA

MELINDA PRUETT-JONES, Field Museum of Natural History, Roosevelt Road at Lake Shore Drive, Chicago, IL 60605, USA

STEPHEN PRUETT-JONES, Department of Ecology and Evolution, University of Chicago, Chicago, IL 60637, USA

MANFRED E. RAU, Institute of Parasitology, McGill University, MacDonald College, 21, 111 Lakeshore Rd, Ste. Anne de Bellevue, PQ H9X 1CO, Canada

RALEIGH J. ROBERTSON, Department of Biology, Queens University, Kingston, Ontario K7L 3N6, Canada

SCOTT K. ROBINSON, Illinois Natural History Survey, Peabody Drive, Champaign, IL 61820, USA

CATHERINE A. ROGERS, Department of Biology, Queens University, Kingston, Ontario K7L 3N6, Canada

THOMAS W. SCOTT, Department of Entomology and Center for Agricultural Biotechnology, University of Maryland, College Park, MD 20742, USA

MICHELE D. SAUMIER, Institute of Parasitology, McGill University, MacDonald College, 21, 111 Lakeshore Rd, Ste. Anne de Bellevue, PQ H9X 1CO, Canada

MARGO F. SPURRIER, Department of Zoology and Physiology, University of Wyoming, Laramie, WY 82071-3166, USA

BRIDGET J. STUTCHBURY, Department of Biology, Yale University, P.O. Box 6666, New Haven, CT 06511, USA

CATHERINE A. TOFT, Department of Zoology, University of California, Davis, CA 95616, USA

SONIA C. TIDEMANN, Conservation Commission of the Northern Territory, P.O. Box 496, Palmerston, NT 0831, Australia

CHARLES VAN RIPER III, Department of Biological Sciences and Coop. Park Unit, Northern Arizona University, Flagstaff, AZ 86011, USA

MARLENE ZUK, Department of Biology, University of California, Riverside, CA 92521, USA

Part I
Introduction

1 *Current theory of host–parasite interactions*

CATHERINE A. TOFT

Introduction

As Peter Price emphasizes in his foreword, ecologists and evolutionary biologists virtually ignored parasites until recently, even as a source of mortality for host species of primary interest. Ecologists have now, finally, turned their attention to parasites. As if in compensation, much of the recent, seminal work on parasite–host interactions in ecology and evolutionary biology has given hosts an ancillary role, focusing instead mainly on the parasites (e.g. Anderson and May 1978, 1979; May and Anderson 1978, 1979; Price 1980; Marshall 1981). The chapters in this book focus on the interaction between parasites and hosts but emphasize the host species—the book itself is delimited to a certain host taxon (i.e. birds).

The paucity of good, empirical work on parasites and their interactions with hosts in Nature has inhibited conceptual progress on this exiciting and important topic (Anderson and May 1982); the studies reported in this book seek to remedy this situation. In particular, they address, for some avian systems, the questions: (1) whether parasites regulate host populations; and (2) whether parasites act as selective agents that result in adaptive responses in their hosts, including the amazing sexually selected traits often exhibited by birds.

I wish to provide here a conceptual framework (only one of several possible) linking the chapters in this book. I will discuss the contents of these chapters; at the same time, I will point to areas in the study of parasite–host interactions that still need strengthening. In this connection I will develop three points. First, although the following chapters represent some of the finest ecological and evolutionary work to date on parasite–host interactions, we nevertheless have a long way to go before these interactions will be fully integrated into the study of the host's biology. Second, the study of parasite–host interactions will require more integration between empirical work and theory, for reasons I detail below. Most importantly, we need the tool of theory to gain intuition about any population-level phenomenon, including that of the interaction between parasite and host populations. Third, in all areas of ecology and evolutionary biology, conceptual progress would benefit from marriage between types of theory. In particular, development of theory on the role of parasites in sexual selection of hosts requires more integration with theory on the population dynamics of parasite–host interactions.

Pathogenicity: the parasites' effect on their hosts

Conventional wisdom among parasitologists and until recently among evolutionary biologists says that the 'good' parasite evolves to do little or no harm to the host. We see that assumption operating in this volume (e.g. Delannoy and Cruz, Chapter 5; Rogers *et al.*, Chapter 7). Nevertheless, some ancient parasites remain measurably pathogenic to their long-standing hosts; examples in this volume include the ectoparasites in cliff swallows (Loye and Carroll, Chapter 12 and ticks of nesting seabirds Duffy, Chapter 13). Recent coevolutionary theory also challenges the view that commensalism is the only evolutionary outcome (Anderson and May 1982, 1983; Ewald 1983; May and Anderson 1983; review by Toft and Karter 1990). Moreover, famous examples of the evolution of lowered pathogenicity do not always stand up to close scrutiny (Toft and Karter 1990). Thus, as population theory emphasizes (Anderson and May 1978, 1979; May and Anderson 1978, 1979), 'successful' parasites potentially can have a strong impact on hosts—on both their survivorship and reproduction and therefore on their population dynamics and evolution (Anderson and May 1978, 1979; May and Anderson 1978, 1979; Levin 1983). Because bird populations have long been the subject of intensive study (relative to many other groups), they offer a good system for investigating how parasites affect host populations.

Words like pathogenicity and virulence are shorthand terms for the effect that parasites have on the host; unlike predators, parasites do not necessarily kill the host outright. It is useful to define such intuitive terms more precisely. The impact of parasites on their hosts can be measured in any or all of the following ways (Bremmermann and Pickering 1983) among others.

1. Parasite reproduction, measured as either the number of new parasite individuals or the number of new hosts infected, surely affects hosts and host populations, minimally by drawing upon limited host resources and in causing the host to mount possibly costly immune responses.
2. Parasite infectiousness i.e., rate at which new host individuals become infected, is a somewhat less direct measure; it correlates with the impact of parasites on the host population, assuming that some

probability of host mortality or loss of reproduction accrues when new host individuals are infected.

3. At the extreme, the population impact of parasites on hosts can be felt directly through parasite-induced host mortality.

Any of these three measures have been variously termed 'pathogenicity' and 'virulence' (Bremmermann and Pickering 1983); some authors prefer to define specific and different attributes for these respective terms. Here I will use pathogenicity and virulence interchangeably to refer generally to the impact that parasites have on hosts.

However we measure it, the impact of parasites on their host populations is poorly known. Until recently, many ecologists did not bother to look for any impact, partly because of the assumption that good parasites do not harm the host or because few parasites or diseased host individuals are discovered in field studies of hosts (more below). Yet studies focusing on parasites often reveal strong, present-day effects of parasites on their hosts, as do many of the chapters in this book (Atkinson and van Riper, Chapter 2; Hudson and Dobson, Chapter 3; Chapman and George, Chapter 4; Delannoy and Cruz, Chapter 5; Scott and Edman, Chapter 10; Loye and Carroll, Chapter 12; Duffy, Chapter 13; Clayton, Chapter 14; Saumier *et al.*, Chapter 15; Johnson and Boyce, Chapter 20; Spurrier *et al.*, Chapter 21). The challenge of future work will be to take these efforts even further, to understand more fully the multitude of effects that parasites may have on hosts in nature.

Part of this understanding will include the recognition that pathogenicity, or virulence, is not simply an attribute of the parasite. Rather, the impact that parasites have on hosts is the net result of the parasite's attack and the host's counterattack. Resistance in hosts is therefore the 'flip-side' of parasite virulence (Toft and Karter 1990), and both involve physiological, morphological, and behavioural characteristics of the two participant organisms. As chapters in this book reveal, the host's immune system is not the only line of defence against parasites. Hosts may employ a variety of behavioural tactics against parasite attack, which may be just as costly as physiological defences (in this volume: Delannoy and Cruz, Chapter 5; Scott and Edman, Chapter 10; Clark, Chapter 11; Loye and Carroll, Chapter 12; Duffy, Chapter 13; Clayton, Chapter 14; Saumier *et al.*, Chapter 15).

An aspect of parasite–host interactions we know little about is the exact relationship between disease, or the total harm that parasites cause the host, and the benefits that parasites accrue from the host, in particular the parasite's reproduction. The two are unlikely to map one-on-one, and in this way parasites may be especially different from predators. We assume that the resources parasites use from the host for reproduction are thus unavailable to the host (e.g. in this volume: Atkinson and van Riper, Chapter 2; discussion in Clark, Chapter 11). In addition, mounting an immune response to fight the parasite is costly and stressful for the host, and the host can suffer mechanical damage from parasites' movement in and on the host and physiological stress from the parasites' waste products. The latter do not contribute directly to parasite reproduction and even detract from it. Most theory does not yet link such parameters explicitly, i.e. the parasite's rate of egg production, the parasite-induced host mortality, and the transmission rate between infected and susceptible hosts, and so on (see May and Anderson 1979, 1982, 1983; Endler and Lyles 1989). Until such linkages are explicitly made, we will not understand fully the evolutionary interactions between parasite and host.

Extrapolation between individual- and population-level phenomena

The role of population models

The title of this book includes the phrase 'bird–parasite interactions'. Such interactions can occur both on the level of individual hosts and parasites, i.e. their physiological, morphological, and behavioural interfaces, and on the level of populations, i.e. their population dynamics, coexistence, and coevolution. Unfortunately perhaps, extrapolation between the two levels is anything but straightforward, and there is a danger in being misled if we think that it is. For understanding population-level phenomena, theory must be used to develop intuition. As Roughgarden (1979, 1983) eloquently points out, we do not have natural intuition about population behaviour because it occurs on a temporal and spatial scale that we cannot observe. One does not go into the field with a pair of binoculars and pad of paper to observe populations grow and interact. Population models as oversimplified metaphors can isolate the qualities of populations that lead us to understand their behaviour, i.e. how the numbers and qualities of individuals in the population change with time and in space, and compress the temporal and spatial scale on which populations behave.

Do parasites regulate host populations?

One pressing issue of both ecological and evolutionary importance is whether parasites regulate host populations. Here, 'regulation' means that host populations with the parasite are lower at equilibrium than those without. Or, if the interaction lacks an equilibrium, we need to know the difference in the growth rates of host populations with and without the parasite, respectively. This information is central to understanding host population dynamics, host community structure, and host evolution, to name a few topics. Because parasites were often ignored in earlier studies, we do not yet know the extent to which such patterns are modified by parasites. For example (to mention just one), none of the food webs in Cohen's (1978) and Pimm's (1982) analyses includes endoparasites, which are surely present; potentially parasites can have as much impact as the predators in those food webs.

Importantly, we cannot tell whether parasites regulate populations by simply looking at collections of individuals and the degree to which they are parasitized—a common practice. Rather, we need a model framework in which various parameters are explicitly linked. For example, many parasites (such as endoparasitic worms and arthropod ectoparasites) are aggregated in the host population, with most hosts having a few or no parasites and a few hosts being heavily infected (e.g. in this volume: Hudson and Dobson, Chapter 3; van Riper, Chapter 8; Pruett-Jones and Pruett-Jones, Chapter 9). Simple intuition might say that such parasites would therefore be unlikely to regulate the host population and, indeed, population models show that parasite aggregation in the host population makes regulation by parasites less likely. However, this tendency toward aggregation often makes a stable equilibrium possible for the parasite–host interaction (Anderson and May 1978). Furthermore, the parasite's aggregation in the host population and virulence interact to determine whether regulation occurs. Even highly aggregated parasites can regulate the host population if mortality from parasites is not too high or too low (Anderson and May 1978).

Because parasites are contained in or on hosts (by definition, parasites are not free-living), they remove their own effects when they cause host death. That is, when hosts die, so do the parasites. (This depends somewhat on the kind of parasite; this is true of all endoparasites—see below. And, here I am considering the definitive host, in which the adult parasite resides, as is usually the case in birds.) Intuitively, would you say that virulent or relatively avirulent parasites would depress host population size more? Most people would probably answer 'Virulent!' at first, but the patterns (as revealed by population models) are not straightforward. Highly virulent parasites, i.e. ones that cause high host mortality, may have greater impact on their own population than they do the host's, especially if they are aggregated in the host population. Thus, in the event that the host population is regulated solely by a parasite, the equilibrium size of the host population, H^*, is *directly* related to parasite-induced host mortality (i.e. virulence; Fig. 3.1 in Hudson and Dobson, Chapter 3, this volume). That is, the more virulent the parasite,

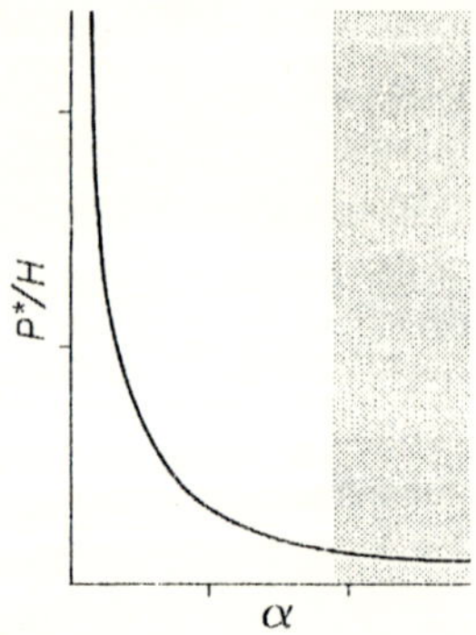

Fig. 1.1. The mean parasite burden, P^*/H, for various values of α, the parasite-induced host mortality. The shaded area denotes regions in which parameter values lead to unregulated population growth (there is no stable equilibrium for the parasite–host interaction). (Adapted from Anderson and May 1978, Fig. 5e; see that paper for parameter values used in their model.)

the larger the host's equilibrium population size with the parasite present; again, if the parasite is sufficiently virulent, there may be no parasite-regulated host equilibrium (Anderson and May 1978, 1979).

Perhaps ironically, the mean parasite burden (P^*/H or number of parasites per host) is for the same reason *inversely* related to the parasite-induced host mortality α (Fig. 1.1). As parasites kill hosts, they also kill themselves, and this is true whether parasites are aggregated in the host population or not (Anderson and May 1978). Thus, the more virulent the parasite, the lower the mean parasite burden. There is nothing suspicious or new about this counterintuitive result; an inverse relation between predator population size at equilibrium and prey capture rate arises in the Volterra predator–prey model (Volterra 1926). However, for field workers the implications are great. If you sample a population of hosts and find few parasites or diseased individuals, what can you conclude? Most would conclude that parasites are not an important factor in contributing to host mortality, and this conclusion might be true. Or, the exact opposite might be true, with the parasites being highly virulent and removing hosts and themselves from the population. Such parasites, if highly aggregated or too virulent, might indeed not be able to regulate the host population (above). However, virulent parasites can still slow the population growth rate (Anderson and May 1979), and this effect could influence the host's ability to compete or coexist with their predators and other parasites. Moreover, such parasites might be extremely important in the host's evolution. In addition, a parasite-regulated host equilibrium can exist under a relatively wide range of parameter values (Anderson and May 1978, 1979) even when parasites cause appreciable host mortality.

These diverse possibilities mean that we can say little about the effect of parasites on their host when we know only the mean parasite burden or proportion of infected individuals. Models tell us that we must estimate a number of other parameters to understand what is going on. Very few studies have provided estimates of these essential parameters. Some parameters are easy to measure, but most present field workers with considerable challenges (Table 1.1). Nevertheless, we must attempt to

Table 1.1. Parameters from models of parasite–host interactions. Few of these have been estimated for real populations (from Anderson and May 1978, 1979)

Symbol	Meaning	Difficulty/frequency of measurement
α	Rate of host mortality due to the influence of the parasite	High; rarely measured
λ	Birth rate of parasite transmission stages, produced by adult	Low; often estimated
μ	Death rate of parasites within the host, due to natural or host-induced (immunological) causes	?; rarely estimated
c	Death rate of transmission stages	Probably high; rarely measured
β	Transmission rate, the probability that a transmission stage will successfully establish itself in a host (also the rate that susceptible hosts become infected upon contact with infectious hosts)	?; usually indirectly measured
ν	Rate at which infected hosts recover and become immune to new infections	Low in clinical studies—high in nature; estimated from clinical studies
γ	Rate at which immunity to that parasite is lost	Low in clinical studies—high in nature; estimated from clinical studies
k, k'	Measures of over- and underdispersion; moments of the negative and positive binomial distribution which describes the distribution of parasites in the host population	Low; often reported

measure these parameters for real populations in order to understand the effects of parasites on hosts. The chapters included herein have accomplished part of this task but still more work and more collaboration among scientists with different skills (Dobson and Merenlender 1991) is required.

The most important lesson provided by models of parasite–host interactions is that our intuition in extrapolating from individual- to population-level phenomena is untrustworthy. Population models, while still imperfect, provide an agenda for empirical work and, with feedback between empirical and theoretical studies, we can make real progress. Chapter 3 by Hudson and Dobson is a particularly good example; the effect of parasites on host fecundity, and its relation to population cycles, could not have been detected without the aid of simplifying models.

Selected issues in parasite–host interactions

This volume contributes to a number of areas that I consider to be the 'cutting-edge' of research in parasite–host interactions, which I will now outline. The chapters themselves are organized into three parts: 'Ecology', including population-level interactions; 'Behaviour', which examines the host's behavioural response to parasitism; and 'Sexual selection', which examines whether females select healthier males using some cue that might indicate parasite load. I now turn to some areas of research in which theory has provided special insights and inspiration for further progress and discuss how the chapters herein have provided a contribution to them.

Interactions between parasitism and other demographic factors

Most models and demographic treatments of cause of death assume that factors contributing to mortality are independent (Carey 1989), i.e. they assume additive mortality. Yet, interactions between types of mortality have been considered in ecology for a long time (e.g. Errington 1963), and parasitism in particular raises this possibility. For example, individuals weakened by disease may be more vulnerable to predation. Some studies in this volume provide evidence for these interactions in mortality factors (e.g. Atkinson and van Riper, Chapter 2; Hudson and Dobson, Chapter 3) and others for additive mortality (e.g. Delannoy and Cruz, Chapter 5). Clearly, the pattern of parasite-induced host mortality depends on the species involved.

Other factors slowing population growth, but falling short of outright mortality, can interact with parasitism. Examples include the relation between nutritional status of the host and infection (Hudson and Dobson, Chapter 3, this volume) and the relation between behaviour (i.e. territoriality, feeding, parental care, nest-site and habitat selection) and infection (in this volume: Hudson and Dobson, Chapter 3; Delannoy and Cruz, Chapter 5; Burtt *et al.*, Chapter 6; Pruett-Jones and Pruett-

Jones, Chapter 9; Scott and Edman, Chapter 10; Clark, Chapter 11; Loye and Carroll, Chapter 12; Duffy, Chapter 13; Saumier *et al.*, Chapter 15; Møller, Chapter 17; Johnson and Boyce, Chapter 20; Spurrier *et al.*, Chapter 21.) As yet, few model systems (but see May and Anderson 1979) incorporate such interactions in rates of mortality.

Host fecundity versus mortality

Some have suggested that vertebrates mainly suffer mortality and invertebrates mainly loss of reproduction from parasites (e.g. May and Anderson 1978). However, chapters herein show clearly that parasites can affect birds' reproductive success by causing nestling morbidity and clutch abandonment and by changing parental behaviour during pair-bonding, breeding, and nesting (Atkinson and van Riper, Chapter 2; Hudson and Dobson, Chapter 3; Chapman and George, Chapter 4; Delannoy and Cruz, Chapter 5; Burtt *et al.*, Chapter 6; Scott and Edman, Chapter 10; Clark, Chapter 11; Loye and Carroll, Chapter 12; Duffy, Chapter 13; Saumier *et al.*, Chapter 15; Zuk, Chapter 16; Møller, Chapter 17; Burley *et al.*, Chapter 19; Johnson and Boyce, Chapter 20; Spurrier *et al.*, Chapter 21). Models tell us that the relative importance of parasites in increasing mortality versus decreasing reproduction can have a strong effect on population dynamics. For example, if the parasite's effect on host fecundity is much greater than that on host mortality, host populations can cycle (Hudson and Dobson, Chapter 3, this volume) or even go extinct (May and Anderson 1978).

Types of parasites

Chapters in this volume deal with a whole spectrum of parasite types, particularly those of importance to birds: ectoparasites on the birds and in nests (nine chapters); haematozoa (four chapters); myiasis (two chapters); worms (two chapters); and viruses (one chapter). Much of the previous work on bird parasites concentrates primarily on worms and haematozoa (Toft and Dobson, in preparation; A. O. Bush, in preparation). Elsewhere (Toft 1986), I emphasize that various types of parasites could have different types of population dynamics, i.e. in interactions with the host based on a review of model systems. The chapters herein show that parasites other than worms, bacteria, and viruses can have just as profound an effect on the mortality or reproductive success of their hosts.

Sexual selection in hosts

Hamilton and Zuk (1982) hypothesized that bright plumage in male birds might have evolved from females' preference for males with a healthy appearance. This recent debate has stimulated much exciting new experimental and observational work on parasites.

One element of this debate is over how to test Hamilton and Zuk's hypothesis (see reviews by Read 1988; Endler and Lyles 1989), a topic which all of the chapters in Part IV also discuss. For example, the degree to which parasites affect sexual selection surely depends on the type of parasite, just as different types of parasite may variously affect the population behaviour of the host (above). Of these, Read (1988) has ventured that pathogenic protozoans and helminths would be the most likely candidates for Hamilton and Zuk's scenario. Certainly the chapters in Part IV yield different results depending on the general type of parasite—sometimes, opposite trends occur in different parasite groups (Burley *et al.*, Chapter 19). Also significant are such variables as geographic differences in parasitism (Zuk, Chapter 16; Burley *et al.*, Chapter 19), temporal and age-specific characteristics of hosts and transmission dynamics (Kirkpatrick *et al.*, Chapter 18; Johnson and Boyce, Chapter 20), and host mating system (Møller, Chapter 17; Johnson and Boyce, Chapter 20; Spurrier *et al.*, Chapter 21). Exploration of such confounding factors is only now beginning to give this area the strength of accumulated data necessary to test alternative hypotheses about parasites and sexual selection. As yet, these variables have not yet been incorporated explicitly into models.

More controversial are interspecific tests and the role of correlational data (see Zuk, Chapter 16, this volume); one of Hamilton and Zuk's novel predictions (Read 1988) is that certain trends in the correlation of brightness and parasite load will occur across species. The majority of the chapters herein examine single species, and a developing consensus (Read 1988; Endler and Lyles 1989) is that the fate of Hamilton and Zuk's hypothesis will lie with experimental intraspecific tests—some of the best examples of which are in this volume. Such tests so far have provided some support for Hamilton and Zuk's hypothesis, although most authors remain cautious at this juncture because of the confounding factors mentioned above (see also Read 1988; Endler and Lyles 1989).

This introductory chapter focuses on theory, and I would like to emphasize here the need to combine epidemiological (that is, demographic) models with those developing Hamilton and Zuk's hypothesis, such as May and Anderson (1982) have begun to do. In my opinion, demographic models clearly show that prevalence and occurrence data taken from a static population sample (mean number of parasites per host or mean number of infected individuals) are unsuitable for tests of Hamilton and Zuk's hypothesis. As I explained above, we cannot distinguish from such data whether parasites are rare or are common and virulent. That is, we don't know whether the hypothesis to be tested is a negative or positive correlation of sexually selected traits with parasite load (see also Harvey *et al.* 1991). In reality, both situations might occur, so that correlations comparing many static population samples might be statistically non-significant because of opposing effects. I know of only one paper on this subject (Kirkpatrick *et al.*, Chapter 18, this volume) that has addressed this problem.

While others have debated the impact of taxonomic effects (Read 1989) or the problem of assigning causation in correlation studies (Burley *et al.*, Chapter 19, this volume), these considerations are relevant only once a proper test for Hamilton and Zuk's hypothesis has been established. Interspecific and interpopulational correlations of brightness and parasite load, when they occur (Zuk, Chapter 16, this volume), are interesting and require explanation. Not all interspecific tests (or comparative data) are subject to this problem, and some are quite appropriate for Hamilton and Zuk's hypothesis. For example, Ward's (1988) correlation of the number of parasite genera and sexual dichromatism in fish is a valid test and one that provides support for Hamilton and Zuk's hypothesis. By using number of genera, he is asking a different question, one that focuses more directly on coevolution between host and parasite taxa.

The ideal study testing these exciting ideas on the role of parasites in the sexual selection of host species will examine whether sexually selected ornaments, or behaviours, are specifically responsive to parasites and will include experimental manipulation of marked host individuals within a single population (e.g., Johnson and Boyce, Chapter 20, this volume; Spurrier *et al.*, Chapter 21, this volume; Clayton, 1990; Zuk *et al.* 1990).

Conclusion

I originally began this chapter with a remark that ecological study of parasites had almost become a bandwagon; I decided instead to end with this idea in order to develop it more completely. Recently, Abrahamson *et al.* (1989) published a commentary on fads in ecology, defining a fad (i.e. bandwagon) as a concept in science that receives more attention than it deserves. I would like to conclude by making two points about their article and how it applies to the ecological and evolutionary study of parasites.

First, although parasites are receiving more attention than ever before, with study of parasites perhaps diverting ecologists and evolutionary biologists from other topics, parasites can hardly be receiving more attention than they deserve. We know that a considerable proportion of described species are parasites (Price 1980; Toft 1991) and that they occur in nearly all natural ecosystems and food webs. Even if it turns out that parasites rarely regulate their host populations or influence either natural or sexual selection of hosts, *we need to know whether they do;* at present we don't. In addition, of course, many of us would say that parasites are fascinating in their own right, whether they influence the biologies of their hosts or not.

Second, and a more general point, is that even if an idea is eventually shown to be wrong, does that mean that it has got more attention than it deserved? To paraphrase Thomas Huxley's summary of the Popperian method, a powerful, and time-honoured, way to do science is by making mistakes as fast as possible. Certainly scientific progress can be made by testing as many ideas as we can about Nature and rejecting those that don't work out. While I strongly recommend reading Abrahamson *et al.* (1989) because of its many valuable ideas, I would also say that there is nothing wrong with bandwagons. And, I am confident that the study of parasites will not be a vacuous undertaking. For those reluctant to step on to a bandwagon, much, much more empirical work needs to be done on parasites.

Acknowledgements

I thank J. Loye, R. May, P. Price, J. Theis, and M. Zuk for helpful discussion, comments, and suggestions on this chapter.

References

Abrahamson, W.G., Whitham, T.G., and Price, P.W. (1989). Fads in ecology. *Bioscience,* **39**, 321–5.

Anderson, R.M. and May, R.M. (1978). Regulation and stability of host–parasite population interactions. I. Regulatory processes. *Journal of Animal Ecology*, **47**, 219–47.

Anderson, R.M. and May, R.M. (1979). Population biology of infectious diseases: Part I. *Nature,* **280**, 361–7.

Anderson, R.M. and May, R.M. (1982). Coevolution of hosts and parasites. *Parasitology*, **85**, 411–26.

Bremmermann, H.J. and Pickering, J. (1983). A game-theoretical model of parasite virulence. *Journal of Theoretical Biology*, **100**, 411–26.

Carey, J.R. (1989). The multiple decrement table: a unifying framework for cause-of-death analysis in ecology. *Oecologia*, **78**, 131–7.

Clayton, D.H. (1990). Mate choice in experimentally parasitized Rock Doves: lousy males lose. *American Zoologist*, **30**, 251–62.

Cohen, J.E. (1978). *Food webs and niche space*. Princeton University Press, Princeton.

Dobson, A and Merenlender, A. (1991). Coevolution of macroparasites and their hosts. In *Parasite-host associations: coexistence or conflict?* C.A. Toft, A. Aeschlimann, and L. Bolis (ed.). Oxford University Press, Oxford.

Endler, J.A. and Lyles, A.M. (1989). Bright ideas about parasites. *Trends in Ecology and Evolution*, **4**, 246–8.

Errington, P.L. (1963). Muskrat populations. Iowa State University Press, Ames.

Ewald, P.W. (1983). Host–parasite relations, vectors, and the evolution of disease severity. *Annual Review of Ecological Systematics*, **14**, 465–85.

Hamilton, W.D. and Zuk, M. (1982). Heritable true fitness and bright birds: a role for parasites? *Science*, **218**, 384–7.

Harvey, P.H., Read, A.F., John, J.L., Gregory, R.D., and Keymer, A.E. (1991). An evolutionary perspective: using the comparative method. In *Parasite-host associations: coexistence or conflict?* (ed. C.A. Toft, A. Aeschlimann, and L. Bolis), Oxford University Press.

Levin, S.A. (1983). Coevolution. In *Population biology*, Lecture Notes in Biomathematics, no. 52 (ed. H.I. Freedman and C. Strobeck), pp. 328–34. Springer-Verlag, Berlin.

Marshall, A.G. (1981). *The ecology of ectoparasitic insects*. Academic Press, London.

May, R.M. and Anderson, R.M. (1978). Regulation and stability of host–parasite interactions. II. Destabilizing processes. *Journal of Animal Ecology*, **47**, 249–67.

May, R.M. and Anderson, R.M. (1979). Population biology of infectious diseases: Part II. *Nature*, **2800**, 455–61.

May, R.M. and Anderson, R.M. (1982). Epidemiology and genetics in the coevolution of parasites and hosts. *Proceedings of the Royal Society of London*, **B219**, 281–313.

May, R.M. and Anderson, R.M. (1983). Parasite–host coevolution. In *Coevolution* (ed. D. J. Futuyma and M. Slatkin) pp. 186–206. Sinaur Associates, Sunderland, Massachusetts.

Pimm, S.L. (1982). *Food webs*. Chapman and Hall, London.

Price, P.W. (1980). *Evolutionary biology of parasites*. Princeton University Press, Princeton.

Read, A.F. (1988). Sexual selection and the role of parasites. *Trends in Ecology and Evolution*, **3**, 97–102.

Read, A.F. (1989). Sexual selection and parasites: comparative tests of the Hamilton and Zuk hypothesis. *American Naturalist*. (In press.)

Roughgarden, J. (1979). *Theory of population genetics and evolutionary ecology: an introduction*. MacMillan, New York.

Roughgarden, J. (1983). The theory of coevolution. In *Coevolution* (ed. D.J. Futuyma and M. Slatkin), pp. 33–64. Sinauer Associates, Sunderland, Massachusetts.

Toft, C.A. (1986). Communities of species with parasitic life-styles. In *Community ecology* (ed. J. Diamond and T.J. Case), pp. 445–64. Harper and Row, New York.

Toft, C.A. (1991). An ecological perspective: The population and community consequences of parasitism. In *Parasite-host associations: coexistence or conflict?* (ed. C.A. Toft, A. Aeschlimann, and L. Bolis). Oxford University Press.

Toft, C.A. and Karter, A.J. (1990). Parasite–host coevolution. *Trends in Ecology and Evolution*, **5**, 326–9.

Volterra, V. (1926). Fluctuations in the abundance of species, considered mathematically. *Nature*, **118**, 558–60.

Ward, P.I. (1988). Sexual dichromatism and parasitism in British and Irish freshwater fish. *Animal Behaviour*, **36**, 1210–15.

Zuk, M., Thornhill, R., Ligon, J.D. Johnson, J.E., and Johnson, K. (1990). Parasites and mate choice in red jungle fowl. *American Zoologist*, **30**, 235–44.

Part II
Ecology

2 *Pathogenicity and epizootiology of avian haematozoa:* Plasmodium, Leucocytozoon, *and* Haemoproteus

CARTER T. ATKINSON and CHARLES VAN RIPER III

Introduction

A diverse group of parasitic helminths and protozoans use the avian bloodstream as a habitat for growth and reproduction. These parasites include the filarial nematodes which live in the tissues or peritoneal cavity of their hosts and release larval microfilariae into the circulation, flagellated trypanosomes which divide and circulate in the bloodstream, and intracellular sporozoan parasites (phylum: Apicomplexa) which complete part of their complex life cycles within circulating blood cells. Within this last group, the haemosporidian parasites of birds, i.e. species of *Plasmodium, Leucocytozoon,* and *Haemoproteus,* have been studied intensely by parasitologists in all of the major zoogeographic regions of the world. An extensive body of literature has accumulated on the biology of these organisms during the past century which makes them exceptional candidates for studies of the evolution of bird–parasite interactions and outstanding reference systems for studies of other vertebrate–parasite interactions.

It has only been recently that scientists have recognized the benefits of using data on the prevalence and distribution of avian haemosporidians to answer ecological, evolutionary, and behavioural questions. Its use does, however, present potential pitfalls that might lead to spurious conclusions if a worker is not aware of them. This chapter is designed to provide background on the natural history of avian haemosporidians and to describe some of the potential problems that may await future researchers who wish to work in this area. It also provides the reader with a brief outline of haemosporidian life cycles so that this information can act as a springboard for chapters in this book that deal with the ecology, behaviour, and evolution of avian hosts and their blood parasites.

This chapter will begin with brief descriptions of the general characteristics, taxonomy, host distribution, and life cycles of avian haemosporidians. These will be followed by major sections on the pathogenicity of *Plasmodium, Leucocytozoon,* and *Haemoproteus* spp., how it correlates with life cycles, and ultimately, how it affects individual host fitness. The final section of this chapter will discuss the impact of epizootiological factors on the transmission and prevalence of haemosporidians in avian populations.

General characteristics and life history

Haemosporidian parasites are traditionally classified within the suborder Haemospororina in the protozoan phylum Apicomplexa. Members of this phylum are intracellular parasites which have a number of common morphological and developmental characteristics (Levine 1985). The invasive stages of these protozoans, i.e. merozoites and sporozoites, have an apical complex composed of specialized organelles that are important in host cell penetration, while their developmental characteristics are related to a fundamental life cycle based on alternating cycles of sexual (gametogony) and asexual (schizogony and sporogony) reproduction (Fig. 2.1).

Haemosporidians are common blood parasites of reptiles, birds, and mammals which undergo some stage of development in both the tissues and the circulating blood cells of infected hosts (Garnham 1966). *Plasmodium, Haemoproteus,* and *Leucocytozoon* are the three genera found in birds. They are transmitted from host to host by the bite of bloodsucking flies from the order Diptera, such as mosquitoes.

Avian haemosporidians have a cosmopolitan distribution and have been recorded in about 68 per cent of the avian species that have been examined. This wide distribution of blood parasites in a broad spectrum of host species makes haemosporidians excellent models for testing host–parasite evolutionary theories. *Haemoproteus* is the most common genus encountered and has been reported in about 67 per cent of infected bird species. It is followed by *Plasmodium* (41.5 per cent) and *Leucocytozoon* (39 per cent) (Bennett *et al.* 1982*a*). However, certain host families (e.g. Fringillidae, Columbidae, and Phasianidae) appear to be much more susceptible to infection and have high prevalences of infection, while other families (e.g. Laridae, Scolopacidae, Charadriidae) are only rarely infected with blood parasites. The reasons for these differences are not completely understood, but probably involve behaviour and habitat, presence or absence of suitable vectors, climate and other environmental conditions, and host specificity as a result of host–parasite coevolutionary patterns over time. Most species of *Haemoproteus* and *Leucocytozoon* are relatively host-specific and restricted to closely related species within the same host family (Fallis *et al.* 1974; Atkinson 1986). By contrast, species of avian *Plasmodium* have much broader host specificity and may occur in several avian families (Bennett *et al.* 1982*a*).

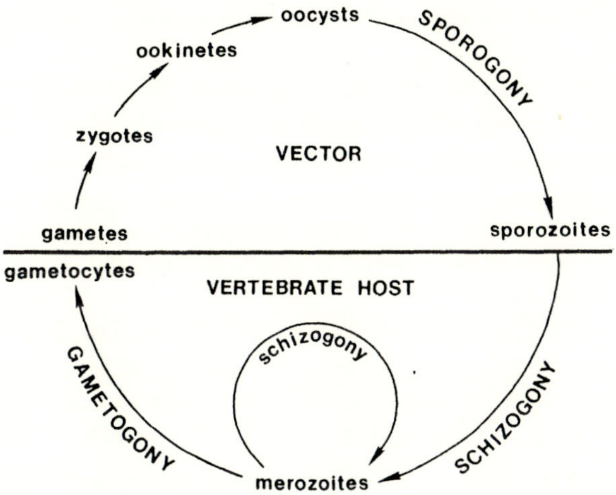

Fig. 2.1. Schematic diagram of a typical haemosporidian life cycle. Transmission occurs when infective sporozoites are inoculated into a suitable host by a bloodsucking vector. Sporozoites invade tissues of the host, grow, and undergo asexual reproduction (schizogony) to produce merozoites. These may invade more tissues and undergo one or more additional cycles of asexual schizogony, invade circulating erythrocytes and continue schizogony (*Plasmodium*), or develop directly into gametocytes (*Leucocytozoon* and *Haemoproteus*). All three genera are transmitted to new hosts when gametocytes are ingested by a suitable bloodsucking vector. Gametogenesis occurs within the midgut of the vector to form gametes which fuse to form fertile zygotes. Zygotes transform into elongate, invasive ookinetes which penetrate the midgut epithelium, round up and develop into oocysts under the basal lamina of midgut epithelial cells. Asexual reproduction (sporogony) occurs in the oocyst to yield numerous infective sporozoites. These rupture from mature oocysts and move through the haemocoel of the vector to the salivary glands where they rest until the next blood meal.

The distribution of avian haemosporidians by host, habitat, and zoogeographical region has been summarized in detail by several extensive reviews. Prevalence data from a number of these compilations have been used to test evolutionary hypotheses of host–parasite systems (Hamilton and Zuk 1982; Borgia 1986; Read 1987; Zuk, Chapter 16, this volume). Bennett and colleagues have produced the most current host–parasite index of avian haemosporidians (Bennett *et al.* 1982*a*) as well as several thorough bibliographies of the literature on avian species of *Plasmodium, Leucocytozoon,* and *Haemoproteus* (Herman *et al.* 1976; Bennett *et al.* 1981). Major surveys of avifauna for blood parasites have been conducted in most parts of the world, including North America (Greiner *et al.* 1975*a*; Barnard and Bair 1986; Kirkpatrick and Suthers 1988), South America (White *et al.* 1978), Asia and Indo-Pacific regions (McClure *et al.* 1978), Africa (Bennett and Herman 1976; Peirce 1984) and Europe (Peirce and Mead 1978*a*; Bennett *et al.* 1982*b*; Peirce *et al.* 1983).

The first avian haemosporidians to be observed in smears of peripheral blood were probably species of *Plasmodium* and *Haemoproteus* from crows, owls, and magpies in central Europe (Danilewsky 1885). Most early observations were made from unstained blood smears or preparations that were treated with carmine-, picric acid-, or methylene blue-based stains (Garnham 1966; Smith and Sanford 1985). In spite of technical difficulties in clearly distinguishing these parasites within blood cells, it was quickly determined that haemosporidians were common in a wide variety of avian species. Nothing was known, however, about their host specificity, taxonomy, and life cycles. Major breakthroughs included development of Romanovsky (Giemsa) stains for clear demonstration of the parasites in methanol-fixed blood films and the discovery by Ross (1898) that *Plasmodium relictum* can be transmitted by mosquitoes (Garnham 1966).

Discovery of the vectors of *Plasmodium* was followed in the early 1900s by the demonstration that *Haemoproteus* of pigeons and doves could be transmitted by ectoparasitic hippoboscid flies (Sergent and Sergent 1906). It was not until the late 1950s that ceratopogonid flies in the genus *Culicoides* were also identified as important vectors (Fallis and Wood 1957). The complete life cycle of *Leucocytozoon* was not established until the early 1930s by O'Roke (1930*a*, 1931) and Skidmore (1931) who demonstrated transmission of *L. simondi* of ducks and *L. smithi* of turkeys by black flies.

Haemosporidian infections are usually diagnosed by examination of Giemsa-stained blood films and quantified as number of parasites or infected erythrocytes per 1000–10 000 uninfected erythrocytes. Detailed techniques for quantifying and analysing intensity data from blood smears have been described (Godfrey *et al.* 1987). Inoculation of blood into one or more uninfected susceptible hosts (Sergent 1920; Herman *et al.* 1966) is the most sensitive method for identification of chronic infections of *Plasmodium* that may be virtually impossible to detect by examination of blood films. This technique will not diagnose *Leucocytozoon* and *Haemoproteus* infections, since only gametocytes are found in the circulating blood cells. It can be used, however, to definitively determine whether pigmented erythrocytic parasites are species of *Plasmodium* or *Haemoproteus*.

Life cycle and development

Species of *Plasmodium, Haemoproteus,* and *Leucocytozoon* have similar life cycles that involve cycles of asexual reproduction in both the avian host and the arthropod vector (Fallis and Desser 1977; Seed and Manwell 1977). Transmission occurs when the worm-like infective stage, the sporozoite, is introduced into the host with salivary gland secretions from the bloodsucking vector. After gaining entry to the bloodstream, sporozoites invade the tissues and undergo intracellular growth, development, and asexual reproduction (schizogony) to form numerous merozoites. The merozoites invade other cells and initiate new generations of schizogony in fixed tissues of the host or invade the circulating blood cells. Merozoites that invade circulating blood cells may continue schizogony (e.g. *Plasmodium*) or develop (gametogony) into male (micro-) or female (macro-) gametocytes that are infective to another bloodsucking arthropod (Fig. 2.1).

When mature micro- and macrogametocytes are ingested by a suitable vector, development continues within the blood meal in the midgut of the insect. The gametocytes round up, and leave their host cells in a process called gametogenesis. Microgametocytes produce threadlike, flagellated microgametes which fertilize larger spherical macrogametes in the midgut of the insect. This process of sexual reproduction produces a fertile zygote which quickly differentiates into an elongate, motile form called the ookinete. Ookinetes move to the periphery of the blood meal, and penetrate the midgut epithelial cells. They move to the basal lamina of the epithelium, round up, and develop into oocysts. Oocysts undergo asexual reproduction (sporogony) to produce numerous elongate sporozoites. These rupture from mature oocysts into the haemocoele of the vector and penetrate the salivary glands of the insect where they can gain entry into a new host during the next blood meal (Fig. 2.1).

While the life cycles of *Plasmodium, Haemoproteus,* and *Leucocytozoon* are similar, they differ in the types of tissues where asexual schizogony occurs. *Leucocytozoon* and *Haemoproteus* undergo schizogony only in fixed, non-circulating cells in the host, e.g. hepatocytes and vascular endothelial cells, and the only stages of development found in the circulating blood cells are gametocytes. By contrast, *Plasmodium* undergoes schizogony in fixed non-circulating cells as well as in circulating blood cells and both gametocytes and schizonts may be found in erythrocytes. The presence of both erythrocytic schizonts and gametocytes in infections with *Plasmodium* is a key difference from *Leucocytozoon* and *Haemoproteus* which is important in the identification, pathogenicity, and experimental transmission of *Plasmodium.*

Since erythrocytic schizonts of *Plasmodium* contain asexually reproducing stages of the parasite, they can be used to transmit the infection to new hosts by the intraperitoneal, intravenous, or intramuscular inoculation of infected blood. Gametocytes of haemosporidian parasites, however, are only infective to a suitable arthropod vector and will not infect new hosts if transferred by blood inoculation. Thus, species of *Plasmodium* can be transmitted from infected to uninfected hosts by simple blood inoculation, while *Leucocytozoon* and *Haemoproteus* cannot.

Other differences in life cycles among *Plasmodium*, *Leucocytozoon*, and *Haemoproteus* are related to development of schizonts in tissues of the host. Some species of *Haemoproteus* and *Leucocytozoon* form large, macroscopic megaloschizonts in host tissues that yield millions of merozoites. By contrast, species of *Plasmodium* form smaller tissue schizonts that produce tens to hundreds of merozoites.

The current taxonomy of avian species of *Plasmodium*, *Leucocytozoon*, and *Haemoproteus* is based on host family or species and the morphology of parasites that are found in circulating blood cells. While this approach ignores important phylogenetic characteristics that may be associated with type of vector or site and mode of pre-erythrocytic development in fixed tissues in the host, it is currently the only practical means for distinguishing these parasites. Characteristics that are used to separate *Plasmodium*, *Haemoproteus*, and *Leucocytozoon* on blood smears include the presence or absence of pigment granules in erythrocytic parasites, the size and shape of the intraerythrocytic parasites, and the presence or absence of asexual schizogony in the circulating blood cells (Fig. 2.2; Greiner and Bennett 1975; Greiner *et al.* 1975*b*).

Pigment granules of haemozoin are insoluble residues produced during digestion and metabolism of host haemoglobin (Yamada and Sherman 1979). They appear as small golden-brown or black deposits which are highly refractive when examined by dark-field or polarized light microscopy in both stained and unstained blood smears (Kirkpatrick 1989). Pigment is produced by both *Plasmodium* and *Haemoproteus*, but not *Leucocytozoon*. Many *Leucocytozoon* spp. may be further distinguished by the extensive changes in host cell morphology they may produce (Fig. 2.2(c)) and by their ability to develop in circulating lymphocytes (Greiner and Bennett 1975). As mentioned earlier, *Plasmodium* and *Haemoproteus* are distinguished by the presence or absence of erythrocytic schizogony.

Mortality and effects on individual fitness

It is likely that all species of *Plasmodium*, *Leucocytozoon*, and *Haemoproteus* are pathogenic in their natural hosts, although little is known about the physiological, behavioural, and ecological costs of haemosporidian infections in wild avian hosts. Most of what is known about potential impact of these parasites is based on laboratory studies of

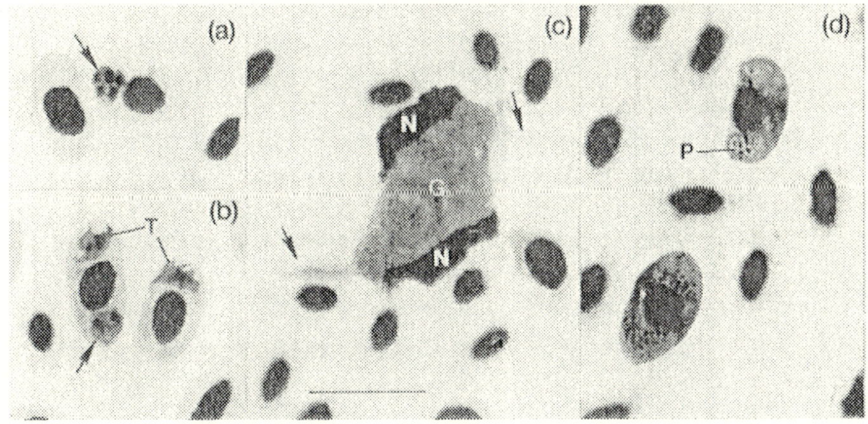

Fig. 2.2. Haemosporidian parasites of the wild turkey (*Meleagris gallopavo*). Scale bar, 20 μm. (a) Multinucleated erythrocytic schizont (arrow) of *Plasmodium hermani*. Each nucleus of the schizont will eventually be incorporated into a budding merozoite which will be capable of infecting additional erythrocytes after the host cell ruptures. (b) Early uninucleated trophozoites (T) of *Plasmodium hermani* which develop from merozoites after host cell invasion. These grow and eventually undergo nuclear division to form early binucleated schizonts (arrow). Pigment granules are not visible in these micrographs. (c) Mature gametocyte (G) of *Leucocytozoon smithi*. The host erythrocyte (arrows) has undergone extensive hypertrophy and elongation to form two horn-like extensions from either end of the host cell. The nucleus (N) of the infected erythrocyte is hypertrophied and divided into two halves on either side of the parasite. (d) Mature (lower left) and immature (upper right) gametocytes of *Haemoproteus meleagridis*. The mature gametocyte completely fills the host cell cytoplasm. Note dark pigment granules (P) which are characteristic of *Plasmodium* and *Haemoproteus* spp.

domesticated birds (e.g. ducks, chickens, and turkeys) that have been used as experimental models. Since a variety of complex factors, including environmental stress, genetics, age, nutrition, and immunological status, may determine the outcome and impact of infection, extrapolation of laboratory results to field conditions and wild individuals is difficult. However, reports of dead or moribund field-collected wild birds with intense haemosporidian infections have generally supported laboratory findings about the potential pathogenicity of these parasites (Manwell 1951; Herman *et al.* 1975; Bierer *et al.* 1981; Atkinson and Forrester 1987).

Pathological effects of infection with *Plasmodium*, *Leucocytozoon*, and *Haemoproteus* can be correlated with the development and life cycle of these protozoans in the fixed tissues and circulating blood cells of their hosts. Infections with parasites from all three genera, undergo similar clinical phases: (1) prepatent phase, when parasites undergo initial schizogonic cycles in tissues and cannot be found in circulating blood cells; (2) acute phase, when parasites begin to appear in the circulation and rapidly increase in numbers; (3) crisis phase, when parasitaemia in circulating blood cells and physiological stresses reach a peak; and (4) latent or chronic phase, when immune responses have reduced parasitaemia to low levels and surviving hosts show little or no overt signs of infection.

Infections with species of *Plasmodium, Leucocytozoon,* and *Haemoproteus* may persist for years and perhaps for the lifetime of infected birds (Bishop *et al.* 1938; Sergent and Sergent 1952; Ahmed and Mohammed 1978). Infected hosts develop some degree of resistance to reinfection with the homologous strain of parasite and often exhibit chronic or latent infections that may not be detectable in blood films until they are reactivated by hormones (spring relapse) or environmental and physiological stresses.

Acute haemosporidian infections in non-immune hosts are generally more severe, although a number of factors may affect innate or non-specific resistance to infection, including age, genetic background, or host species. Young birds are most susceptible and the most serious mortality occurs in this age group. Epizootics in nestling birds have been documented for *Plasmodium* (Gabaldon and Ulloa 1980) and *Leucocytozoon* (Herman *et al.* 1975). Gabaldon and Ulloa (1980) studied transmission of *Plasmodium* in nesting Ciconiiformes in Venezuela and documented extensive mortality in young nestlings. Transmission was most intense during the rainy season when birds were nesting and was similar in many respects to holoendemic human malaria where rates of transmission are high throughout the year, where most of the adult population carries chronic infections, and where significant mortality occurs in the young non-immune cohort. Herman *et al.* (1975) and other workers have described epizootics of *Leucocytozoon* in nesting geese in Canada, where intense transmission coincided with the springtime appearance of large populations of black fly vectors and susceptible host goslings. Epizootics appeared to run in 4-year cycles.

Epizootics and severe mortality attributable to acute haemosporidian infections may also occur when non-immune individuals are translocated into an endemic area or when a parasite is introduced into a population that had no previous history of exposure. Either of these scenarios has serious implications for the management of critical populations of endangered or threatened species, including reintroductions of captive-reared individuals into former ranges. Extensive malaria-induced morbidity and mortality have been documented in zoos when non-immune birds are moved to endemic areas, especially when recently relocated individuals under the stress of adapting to new surroundings are exposed to local vector populations. Localized epizootics of *Plasmodium* have been described in captive penguins (Stoskopf and Beier 1979; Laird and van Riper 1981; Fix *et al.* 1988). Similar epizootics have been documented in domestic turkeys when large poultry farms were established in habitats endemic for *L. smithi* in South Carolina (Kissam *et al.* 1975). Epizootics of *L. simondi* occurred in subarctic Quebec when the Canada Department of Agriculture attempted to establish flocks of domestic geese in endemic habitats (Laird and Bennett 1970). The most detailed studies of how avian populations as a whole adapt when a new parasite is introduced to a geographic region have been done by Warner (1968) and van Riper *et al.* (1986) on endemic Hawaiian birds.

Relatively little is known about the behavioural effects of haemosporidian infections on wild birds. Overt behavioural changes may only be evident during prepatent, acute, and crisis phases of infection when physiological stress reaches a peak. In very intense infections, these typically manifest as signs of discomfort and include changes in posture and decreases in physical activity. Behavioural changes may not be noticeable in infections of low intensity, although behavioural time budgets may reveal shifts in activity (Schall and Sarni 1987; Saumier *et al.*, Chapter 15, this volume). In lizard populations, where the behavioural effects of malarial infections have been easier to quantify, fence lizards, *Sceloporus occidentalis*, infected with *Plasmodium mexicanum* engage in fewer social interactions and may have territories of inferior size or quality. This change in social activity could be correlated with physiological effects of the parasite infection, including decreases in size of testes, disruption of lipid storage, and decreased aerobic metabolic capacity (Schall *et al.* 1982; Schall and Sarni 1987). Experimental studies of mate choice using the same host–parasite model revealed similar differences between infected and uninfected lizards, with uninfected males dominant in social encounters more often than infected ones (Schall and Dearing 1987).

Little is known about the long-term effects of chronic or latent infections on individual fitness. Most studies that have been done suggest that stress (either nutritional or environmental) is an important factor in inducing relapses (Ben-Harel 1923; Manwell 1929; Garnham 1966). Physiological studies of haemosporidian infections have shown that acute infections with *Plasmodium* may have significant effects on oxygen transport and ability to thermoregulate (Schall *et al.* 1982; Hayworth *et al.* 1987). Among lizards, where these effects have been studied in greater detail, malarial infection can have significant impact on host physiology, behaviour, breeding success, and coloration (Schall *et al.* 1982; Schall 1983; Schall and Sarni 1987; Ressel and Schall 1989). Few corresponding studies have been done using birds (see in this volume: Kirkpatrick *et al.*, Chapter 18; Burley *et al.*, Chapter 19; Johnson and Boyce, Chapter 20). Hayworth *et al.* (1987) found that canaries infected with *P. relictum* have decreased ability to thermoregulate and lowered oxygen capacity at the crisis stage when peripheral parasitaemia reaches a peak. Effects at other stages of infection were less significant. A number of workers have suggested that unseasonal extremes in environmental conditions may induce relapses in individuals with chronic or latent infections (Bennett *et al.* 1976). Surveys of breeding populations have found little correlation between infection with haemosporidians and body mass and fat levels, suggesting that hosts are able to tolerate infection under the physiological stresses of reproduction (Bennett *et al.* 1988). Surveys of migrating populations have shown a slight negative correlation between body mass and haemosporidian infection (Ashford 1971). Peirce and Mead (1978*b*) have suggested that the stress of migration may be an important factor in inducing relapses, but critical studies of the impact of relapsing infections on host survival, competition for nesting territory and mates, and reproductive success have not been done.

Pathogenicity

Plasmodium

Laboratory studies of species of *Plasmodium* in canaries, chickens, ducks, and turkeys have shown that significant pathological effects are associated with both tissue and erythrocytic stages of infection. The onset of signs in acute infections may be very rapid, and infected birds may be lethargic, refuse to eat or drink, and appear to be in distress, with ruffled feathers, drooped heads, and closed eyes. Neurological symptoms may be evident with infections of some species, e.g. *P. gallinaceum*, where development occurs in the capillary endothelium of the brain (Garnham 1966). Danilewsky (1891) was probably the first to describe gross pathological changes in birds with malarial infections and observed severe anaemia, enlargement and discoloration of the spleen and liver, and the presence of pigment deposits in phagocytic cells of these organs.

Tissue stages

After sporozoites gain access to a suitable host, species of avian malaria develop in tissues of mesodermal origin (i.e. endothelial cells lining capillaries; cells of the haemopoietic system; cells of the lymphoid–macrophage system). In heavy infections, developing schizonts may restrict or completely block blood flow to vital organs such as the brain and spleen (e.g. *P. cathemerium* and *P. durae*), or preferentially develop in cells of the haemopoietic system and cause severe anaemia (e.g. *P. elongatum*). Thus, serious disease and death may occur prior to the appearance of parasites in the circulating blood cells. This can make diagnosis difficult and dependent on histopathological examination of Giemsa-stained impression smears of fresh tissues or paraffin sections of fixed and embedded organs. Characteristic gross pathological signs of malarial infection, e.g. enlargement and discoloration of the liver and spleen, may not be evident at this stage of infection because the parasites have not yet invaded circulating blood cells. Significant pathological effects have been attributed to pre-erythrocytic schizonts of *P. cathemerium, P. gallinaceum, P. fallax, P. durae, P. lophurae, P. juxtanucleare,* and *P. elongatum*. Many of these effects may be secondary to blockage of blood flow and related to tissue necrosis (Garnham 1966).

Erythrocytic stages

Red blood cell destruction and anaemia associated with erythrocytic schizogony are the most serious pathological consequences of infection with species of *Plasmodium*. Red blood cells are destroyed by two mechanisms: (1) rupture of infected erythrocytes during release of merozoites from mature intracellular schizonts; and (2) removal of infected erythrocytes by cells of the reticulo-endothelial system, particularly macrophages of the spleen, liver, and bone marrow. Severe anaemia may result when destruction and removal of infected erythrocytes is not balanced by the synthesis and release of immature erythroblasts. Changes in plasma chemistry may accompany destruction of red blood cells, resulting in decreases in plasma pH and increases in plasma proteins such as gamma globulins and fibrin (Sherman and Hull 1960). Such changes may compound the effects of erythrocyte destruction by reducing the oxygen binding capacity of haemoglobin and slowing blood flow in capillary beds of major organs (Seed and Manwell 1977). Enlargement and discoloration of the spleen is a characteristic sign of *Plasmodium* infection and is associated with hyperplasia of lymphoid–macrophage cells. In heavy infections, the accumulation of insoluble malarial pigment in macrophages of the spleen and liver after phagocytosis of parasites may turn these organs black.

Leucocytozoon

This blood parasite is a cause of serious disease in domestic poultry and waterfowl in many parts of the world and has been documented as a cause of localized epizootics in Canadian geese and other wild waterfowl populations. It is a common parasite of many other families of birds, but its impact is not as well known. Much of what is known about the pathological effects of *Leucocytozoon* infection is based on laboratory studies of chickens, turkeys, ducks, and geese and post-mortem examinations of wild birds.

Tissue stages

After sporozoites of *Leucocytozoon* spp. gain access to a suitable host via the bite of an infected blackfly, the parasites invade hepatocytes of the bird and undergo schizogony. In heavy infections, numerous hepatic schizonts may cause dilation and congestion of the sinusoids, although no accompanying inflammatory host responses have been described around these early schizonts (Newberne 1957; Wehr 1962; Desser 1967). Hepatic schizonts eventually rupture, releasing merozoites or 'syncytia' composed of cytomeres with incompletely differentiated merozoites. In infections of *L. simondi, L. caulleryi,* and *L. sakharoffi,* released syncytia are ingested by phagocytic cells in most of the major organs, including spleen, liver, lung, heart, and brain, and grow into megaloschizonts. Pathogenicity of most *Leucocytozoon* species seems to correlate with the presence or absence of these megaloschizonts which grow up to 200 μm in diameter and contain millions of merozoites. Significant host inflammatory responses occur around megaloschizonts, and parasites are often encapsulated by fibrotic tissue and surrounded by mixed inflammatory infiltrates of red blood cells, plasma cells, macrophages, and heterophils. Megaloschizonts may eventually become necrotic and calcified (Cowan 1957; Newberne 1957; Desser 1967; Desser and Fallis 1967; Commichau and Jonas 1977; Miller *et al.* 1983).

Among *Leucocytozoon* spp. that do not appear to form megaloschizonts, e.g. *L. dubreuili, L. fringillinarum, L. berestneffi,* and *L. smithi,* pathological changes associated with tissue stages do not appear to be severe. Many of these species do not appear to have significant effects on their hosts, although this question has not been examined carefully.

Erythrocytic stages

Like *Plasmodium,* infection with *Leucocytozoon* causes severe anaemia in infected birds. Anaemia is caused by destruction of infected blood cells by cells of the reticulo-endothelial system in the spleen (Fallis *et al.* 1951; Kocan and Clark 1966*a,b;* Maley and Desser 1977) and also by the destruction of unparasitized erythrocytes by a putative anti-erythrocyte factor that appears in the serum during acute infections (Kocan 1968).

Significant pathology is also caused by the circulating gametocytes themselves. Studies of *L. smithi* in turkeys have shown that gametocytes can cause pneumonia, congestion of the lungs, and have significant effects on respiratory function by blockage of alveolar capillaries (Siccardi *et al.* 1974; Kohne and Jones 1975). Numerous other pathological changes may occur in organs throughout the body in association with infection, including necrosis and lymphocytic infiltration of the liver, enlargement of the spleen, lymphocytic infiltration of the heart, and haemosiderosis (Newberne 1957; Commichau and Jonas 1977).

Haemoproteus

Species of *Haemoproteus* are the most common haemosporidians encountered in wild birds and, historically, have been believed to be the least pathogenic. Numerous anecdotal descriptions of moribund birds with high peripheral parasitaemias have been published suggesting that haemoproteids from pigeons and doves (Coatney 1933; Markus 1972) and California quail (O'Roke 1930*b*) might be pathogenic. However, until recently, there have been no controlled experimental studies. This is largely because *Haemoproteus* spp. are host-specific and few suitable experimental models are available. These parasites are also difficult to transmit experimentally because few vectors are known and passage by blood and tissue inoculation is only rarely successful (Lastra and Coatney 1950).

Tissue stages

The pre-erythrocytic development of *Haemoproteus* spp. is still poorly understood. Most work has been based on histological examination of tissues from naturally infected birds (O'Roke 1930*b*; Mohammed 1965; Khan and Fallis 1969; Miltgen *et al.* 1981). Most studies of development have focused on *H. columbae* of Columbiformes (Mohammed 1965; Bradbury and Gallucci 1972; Ahmed and Mohammed 1977, 1978). *H. columbae* undergoes asexual schizogony in endothelial cells of the lungs where it may cause congestion and pneumonia-like symptoms in pigeons with heavy infections (Garnham 1966). The prepatent period of columbiform haemoproteids is 21–32 days, suggesting that a number of asexual generations occur before gametocytes appear in the circulating blood cells (Ahmed and Mohammed 1977). Most other species of *Haemoproteus* are traditionally believed to undergo asexual schizogony in lung tissue as well, with few significant pathological effects. Recent studies of natural infections of *H. handai* (= *H. desseri*) of South-east Asian parakeets (Miltgen *et al.* 1981) and experimental infections of *H. meleagridis* in domestic turkeys (Atkinson *et al.* 1986, 1988*a*) demonstrated development in capillary endothelial cells and myofibroblasts in skeletal muscle of infected birds. Atkinson *et al.* (1986, 1988*a*) found two generations of schizogony before merozoites invaded erythrocytes and developed into gametocytes. Both generations of schizogony caused necrosis of surrounding muscle tissue and lameness in experimental birds. Intense inflammatory infiltrates composed of lymphocytes and red blood cells surrounded or filled the megaloschizonts, as occurs in *Leucocytozoon* infections (Fig. 2.3). Surrounding muscle tissue became necrotic and calcified as the infections progressed. Similar host responses were also reported by Miltgen *et al.* in infections of *H. handai* and by Atkinson and Forrester (1987) in a naturally infected wild turkey. Atkinson *et al.* (1988*a*) observed significant declines in growth and weight gain in experimentally infected turkey poults that coincided with development of megaloschizonts in muscle tissue.

Whether similar megaloschizonts occur in other species of *Haemoproteus* is an important question for future research. There have been numerous reports during the past 80 years of unidentified protozoan megaloschizonts in muscle tissue of a wide variety of birds, including California quail, grass parakeets, chickens, pied currawongs, and psittacine birds (Graham-Smith 1907; Levine *et al.* 1970; Walker and Garnham 1972; Hartley *et al.* 1981; Opitz *et al.* 1982; Gardiner *et al.* (1984). In most cases, significant myopathy was associated with these megaloschizonts. Because of their morphological similarities to the large megaloschizonts of *Leucocytozoon* spp. and the absence of circulating gametocytes in most cases, most workers have identified them as aberrant *Leucocytozoon* infections in abnormal hosts. Observations by Atkinson *et al.* (1986, 1988*a*) and Miltgen *et al.* (1981) now suggest that many of these organisms may actually be species of *Haemoproteus*. Interestingly, Farmer (1964) and Becker *et al.* (1956) reported unusual thick-walled megaloschizonts and 'granular tissue' in gizzard muscle from pigeons with natural infections of *H. sacharovi* and *H. columbae* and suggested that these may be related to *Haemoproteus* infection. This work was largely ignored because it did not fit well with current ideas about the pre-erythrocytic stages of avian haemoproteids.

Erythrocytic stages

Most reports of pathogenic *Haemoproteus* infections have suggested that anaemia was a serious consequence of infection since parasitaemias may be very intense and involve more than half of the circulating erythrocytes (Garnham 1966). O'Roke (1930*b*) in studies of *H. lophortyx* in naturally infected California quail suggested that gametocytes made infected erythrocytes more fragile and susceptible to rupture during movement through the capillary beds. Atkinson *et al.* (1988*a*) measured

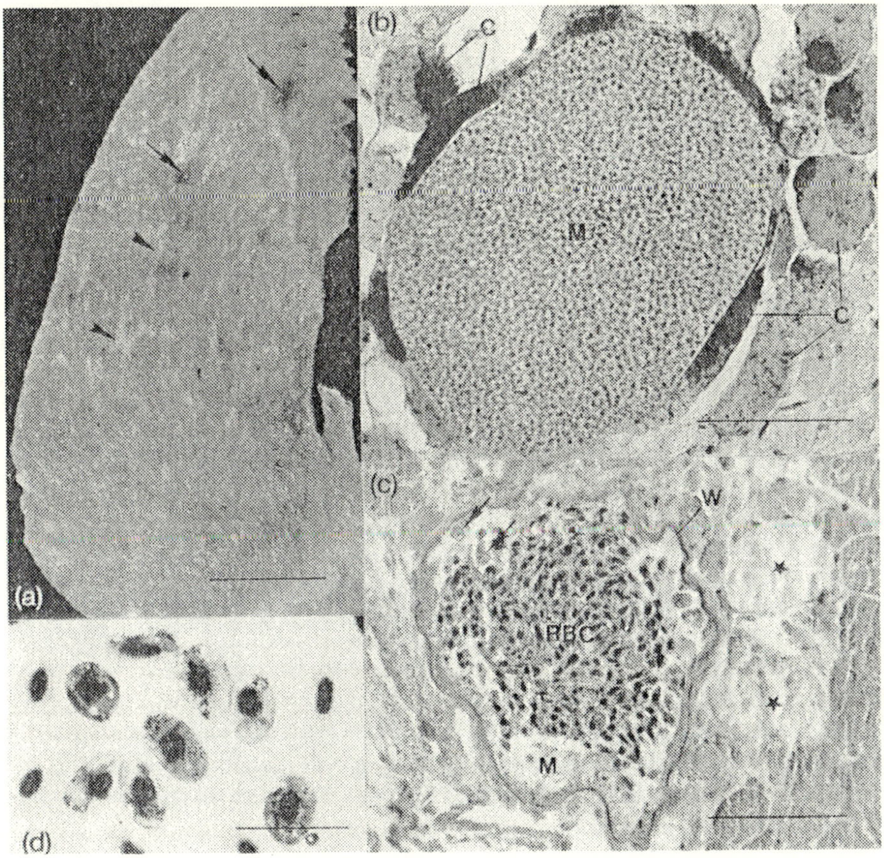

Fig. 2.3. Pathological changes associated with the schizogony of *Haemoproteus meleagridis* in muscle tissue of domestic turkeys. Fig. 3.3(a)–(c) are reproduced with permission of the *Journal of Parasitology* (Atkinson *et al.* 1988a). (a) Formalin-fixed pectoral muscle from a turkey that was inoculated with sporozoites of *H. meleagridis*. The scattered white streaks (arrowheads), dark flecks, and darkened areas (arrows) are associated with calcified muscle fibres and haemorrhagic megaloschizonts in the tissue of the host. This bird died 19 days after infection. Scale bar, 0.5 cm (b) Histological sections of the tissue in (a) revealed numerous megaloschizonts (M) that were packed with millions of tiny spherical merozoites. Each dot in the megaloschizont is the nucleus of an individual merozoite. Muscle fibres (C) surrounding the megaloschizont are necrotic and contain prominent, darkly-staining deposits of calcium. Scale bar, 50 µm. (c) Ruptured and haemorrhagic megaloschizont (M) that is filled with red blood cells (RBC) and macrophages (arrows). A thick, hyaline wall (W) surrounds the megaloschizont. Merozoites have been liberated from the megaloschizont and are no longer visible. Note pale-staining necrotic muscle fibres (*) adjacent to the megaloschizont. Scale bar 50 µm. (d) Blood smear from a domestic turkey that received a comparable experimental infection with sporozoites of *H. meleagridis*. Merozoites that rupture from megaloschizonts invade the circulating erythrocytes and develop into gametocytes. More than 50 per cent of the circulating erythrocytes contain immature pigmented gametocytes. Scale bar, 20 µm.

haemoglobin levels and packed cell volume in domestic turkeys with experimental infections of *H. meleagridis* and did not find evidence of anaemia, in spite of intense parasitaemias that often exceeded 70 per cent of the circulating erythrocytes. In their study, removal of infected cells by cells of the reticuloendothelial system was balanced by the development and release of erythroblasts. It is likely, however, that weakened hosts under nutritional and environmental stresses may not be able to replace gametocyte-infected erythrocytes that are removed from the circulation. These conditions may lead to the development of severe anaemia.

Like *Plasmodium*, enlargement and discoloration of the spleen is a characteristic pathological change of *Haemoproteus* infections (Becker *et al.* 1956). Atkinson *et al.* (1988*a*) observed extensive deposition of pigment in macrophages of lungs, liver, and spleen in turkeys during acute and crisis phases of experimental infections of *H. meleagridis*. The spleens from infected birds were significantly larger than those of controls and the degree of hyperplasia was correlated with parasitaemia.

Epizootiology

Since nestling and juvenile birds are most susceptible to the effects of haemosporidian infections, timing of transmission of *Plasmodium*, *Leucocytozoon*, and *Haemoproteus* and the relative pathogenicity of each species may be important in determining morbidity and mortality in juvenile birds. Unusual environmental conditions and changes in vector populations may delay, advance, or affect the intensity of transmission and create conditions suitable for epizootics.

Studies of *Plasmodium*, *Leucocytozoon*, and *Haemoproteus* in temperate climates in the northern hemisphere have shown that most transmission occurs during the breeding season when vector populations are increasing with the onset of warmer weather, when adult birds with chronic, relapsing infections are available as sources of infection, and when non-immune juvenile birds are hatching and leaving the nest (Herman 1938; Bennett and Fallis 1960; Janovy 1966; Bennett and Cameron 1974; Baker 1975). Studies by Janovy (1966), Herman (1938), Herman *et al.* (1954), Bennett (1960), and Bennett *et al.* (1975) have shown that breeding behaviour, habitat, vector bionomics, and host susceptibility may all strongly influence the epizootiology of haemosporidian infections and determine the age at which nestlings are exposed to infection and the overall incidence and prevalence of infection in a population. Studies of *Leucocytozoon simondi* infections in waterfowl in the northern USA and Canada have suggested that epizootiological factors, especially vector distribution and numbers, may be important in limiting waterfowl populations (Herman *et al.* 1975; Herman and Bennett 1976). More recently, work by van Riper *et al.* (1986) has shown that introduction of mosquito vectors and *Plasmodium relictum* to the Hawaiian Islands may be an important limiting factor in the distribution of Hawaiian land birds. In most of these studies, interactions between avian host and invertebrate vector have been of key importance in understanding the epizootiology of disease in individuals and populations.

The wild turkey (*Meleagris gallopavo*) is a good host for studying haemosporidian infections because domestic turkeys can be used as a convenient experimental model for studies of pathogenicity, vectors, and epizootiology and because extensive field studies have been conducted of the wild host. In eastern North America, *Leucocytozoon smithi, Haemoproteus meleagridis,* and *Plasmodium hermani* occur in wild turkeys (Forrester *et al.* 1974; Telford and Forrester 1975). Sentinel studies with uninfected domestic turkeys have been used to monitor natural transmission of the parasites in wild turkey habitat in the south-eastern USA where *P. hermani, L. smithi,* and *H. meleagridis* occur together. Unlike the intense bursts of springtime transmission of haemosporidians that occur in Canada (Bennett and Cameron 1974), transmission of *L. smithi, P. hermani,* and *H. meleagridis* may occur either throughout the year or in discrete seasons, depending on climate, region, and vector distribution. This was true during a 2-year period in northern Florida when transmission of all three parasites was monitored continuously with domestic turkeys that were exposed to vectors for 2-week intervals in a wild turkey habitat (Fig. 2.4). *Haemoproteus meleagridis* was transmitted continually between April and December when the primary vector, *Culicoides edeni* was most active (Atkinson *et al.* 1988*b*). By contrast, transmissions of *Leucocytozoon* and *Plasmodium* were segregated into two discrete seasons. *Leucocytozoon* was transmitted during an 8-week period during both years of the study between the middle of March and the middle of May while *Plasmodium* was transmitted only between August and December.

Bennett *et al.* (1975) studied the distribution and prevalence of haemosporidian parasites of waterfowl in different locations in maritime Canada and found that prevalence varied with host species and geographic area. They noted that sweeping generalizations about the prevalence and transmission of avian haemosporidians should not be made because each habitat has unique environmental characteristics which may ultimately affect vector populations and the transmission and

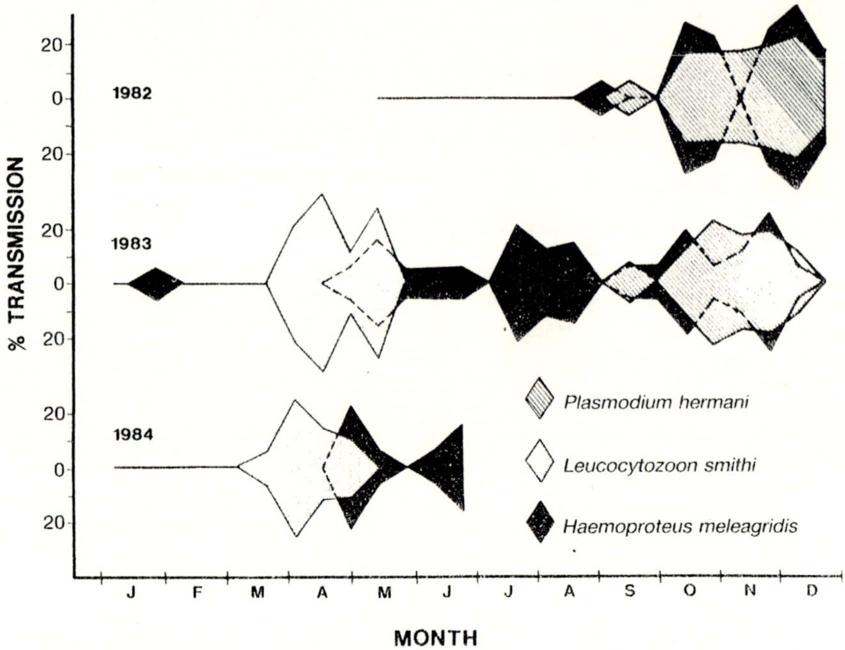

Fig. 2.4. Kite diagram showing seasonal transmission of *Plasmodium hermani, Leucocytozoon smithi,* and *Haemoproteus meleagridis* to domestic turkeys that were exposed in wild turkey habitat in Paynes Prairie State Preserve, Alachua County, northern Florida. Successive groups of nine, 2-week-old domestic turkeys were exposed for 2-week intervals in groups of three at two study sites at Paynes Prairie as described by Atkinson *et al.* (1988*b*). The birds were housed in plywood and hardware cloth cages that allowed passage of vectors but prevented access by predators. The *y*-axis refers to the percentage of birds that became infected with blood parasites. Data on *H. meleagridis* is redrawn from the study by Atkinson *et al.* (1988*b*). Data on transmission of all three parasites is based on combined totals from the two study areas. Transmission of *Plasmodium* and *Leucocytozoon* is segregated into two discrete seasons in the fall and spring, respectively. By contrast, transmission of *Haemoproteus* was continual throughout the warmer months of the year between April and December.

prevalence of haemosporidian parasites. Comparisons between transmission of *H. meleagridis* at two study sites at Paynes Prairie, Florida less than 1 km apart revealed differences in the time of onset of transmission, duration, and intensity, in spite of similar vector faunas (Atkinson 1988; Atkinson *et al.* 1988*b*). Atkinson *et al.* (1988*b*) suggested that this might be due to the limited dispersal and flight range of vectors at each site and local differences in habitat utilization and movement of turkeys in the two study areas. Differences in seasonal transmission of *H. meleagridis* were also evident when comparisons were made between study areas in northern and southern Florida. *H. meleagridis* was transmitted throughout the year in southern Florida where the semitropical climate and larger vector populations probably contributed to a much higher prevalence of infection in wild birds (Forrester *et al.* 1974; Atkinson *et al.* 1988*b*).

Similar studies of *L. smithi* in South Carolina and Florida have found regional differences in transmission of the parasite that are related to distribution of the blackfly vectors (Noblet *et al.* 1972, 1975; Pinkovsky *et al.* 1981). In both states, two species of blackflies, *Simulium slossonae* and *S. congareenarum,* are important vectors. *Simulium slossonae* produces several generations per year, is active year round, and present in areas of South Carolina where transmission occurs throughout the year (Noblet *et al.* 1975. By contrast, *S. congareenarum* produces a single generation each year and is active only during the springtime. Sentinel findings at Paynes Prairie suggest that *S. slossonae* may be absent at this study area, thus limiting transmission of *L. smithi* to early spring (Fig. 2.4).

Detailed studies of vector competence and behaviour may also reveal important aspects of parasite transmission that determine when and where hosts become infected. Studies of the epizootiology of *H. meleagridis* in Florida have shown that at least five species of *Culicoides* can support development of the parasite (Atkinson *et al.* 1983; Atkinson 1988). Only two of these species, *C. edeni* and *C. hinmani,* have been collected in large enough numbers in northern and southern Florida from turkey-baited traps to be implicated as major vectors. Of these two species, *C. edeni* is most susceptible to development of the parasite and about 60 per cent of engorged flies (versus 20 per cent of *C. hinmani*) eventually develop salivary gland infections with sporozoites of *H. meleagridis*.

Studies of the activity cycles and vertical distribution of *C. edeni* and *C. hinmani* in southern Florida indicate that both species are more active in the forest canopy, with major peaks in flight/host-seeking activity at sunset and minor peaks at sunrise (Fig. 2.5). Since wild turkeys normally fly up into the forest canopy at sunset to roost (Williams and Austin 1988), these findings suggest that exposure to *H. meleagridis* most likely occurs in roosting habitat. Careful evaluation of capture times of *C. edeni* and *C. hinmani* on bait turkeys that were exposed in the forest canopy at sunset demonstrated a clear difference in time and peak biting activity for these two species (Fig. 2.6). *Culicoides hinmani* reached a peak in numbers about 40 min prior to sunset, before most turkeys move to their roosts, while *C. edeni* peaked shortly after sunset. These subtle differences in vector behaviour illustrate how one species of ornithophilic

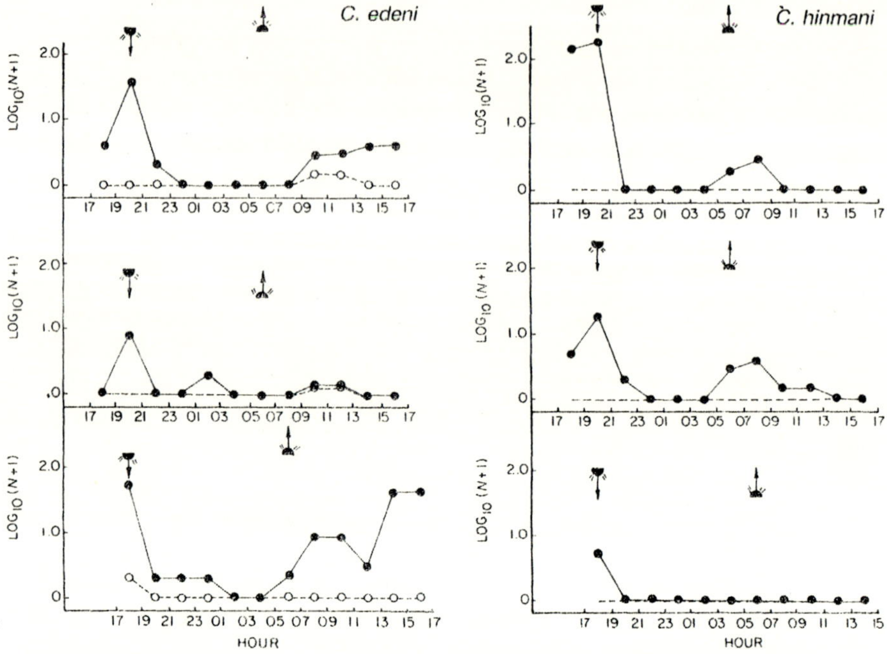

Fig. 2.5. Activity cycles of *Culicoides edeni* and *Culicoides hinmani* at Lykes Fisheating Creek Wildlife Refuge, Glades County, Florida. Data were collected by suspending a pair of modified New Jersey light traps in a live oak hammock that was surrounded by a cypress creek swamp. One trap was suspended 1 m above the ground and the second was suspended 7 m above it in the middle canopy. Traps were operated and modified as described by Atkinson (1988) for the collection of *Culicoides*. In this study, light bulbs were replaced with a paper envelope containing 2–3 kg of dry ice as a CO_2 attractant to minimize diurnal and nocturnal differences in their attractiveness to biting arthropods. The traps were run continuously for single 24–36 hour periods during May, April, and March 1983 (top to bottom on the graph) at the study site described by Atkinson (1988). Sample bottles from each trap were changed every 2 hours between 17.00 and 09.00 hours and every 4 hours between 09.00 and 17.00 hours. Dry ice was replenished every 4 hours. Data are plotted as the average number of individuals captured per hour of trap time for each of the three sampling periods as canopy (●—●) versus ground (○—○) catches. Rising and setting suns indicate the 2-hour sampling period that included dawn or dusk. Major peaks in host-seeking activity for *C. edeni* and *C. hinmani* occurred at sunset with a second smaller peak at sunrise. Both species were most active in the forest canopy where turkeys roost. *Culicoides himani* was never collected near the ground.

biting fly may be favoured over another in selection for efficient vector–parasite relationships.

These types of observations also open interesting questions about the possible role of vector distribution and host-seeking behaviour as selective pressures in modifying host behaviour and habitat utilization.

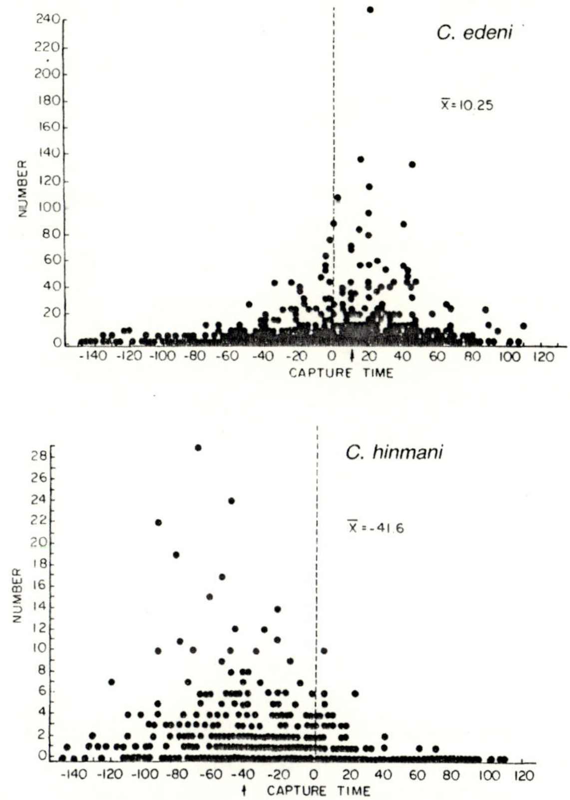

Fig. 2.6. Numbers of *Culicoides edeni* and *Culicoides hinmani* collected from exposed domestic turkeys from 2 hours prior to 2 hours after sunset (dotted vertical line). Specimens were collected in Bennett traps as described by Atkinson (1988) in the middle level of the forest canopy. Scatter plots represent collections of *Culicoides* that were made in northern and southern Florida at study sites described by Atkinson (1988) between May, 1982 and October, 1984. Sunset times for each collection data were calculated from standard tables (Nautical Almanac, US Naval Observatory) and the capture time for each *Culicoides* was graphed as number of minutes (x-axis) before or after sunset. Scatter plots for each species were calculated by collection site, year, and quarter (January–March, April–June, July–September, October–December). Mean capture times were tested by an analysis of variance for significant differences before data were combined. Both species exhibited unimodal peaks in host-seeking activity that were significantly different ($p < 0.05$). *Culicoides hinmani* reached a peak in host-seeking activity 41.6 min before sunset, over 50 min earlier than *C. edeni*. These data indicate that *C. hinmani* may be of little epizootiological significance in transmission of *H. meleagridis* because peaks in biting activity are over before turkeys move to roosts in the forest canopy.

In Hawaii, daily round-trip movement of some species of native birds between high elevation forests and lowland areas with high mosquito populations appears to have evolved in response to selective pressures exerted by *Plasmodium relictum* at lower elevations (van Riper *et al.* 1986). These movements to lower elevations occur at times of the day when the chief malaria vector, *Culex quinquefasciatus*, is inactive and probably evolved initially from the gradual downslope movements which followed the flowering of nectar-producing trees. Since *C. quinquefasciatus* is primarily a noctural biter, most birds are exposed to malaria on their roosts. Van Riper *et al.* (1986) noted that forest birds slept with bills and faces tucked into back feathers and one leg raised into abdominal feathers, effectively protecting soft, exposed skin of the head and legs from mosquito bites. These observations differed from earlier findings of Warner (1968) who did not find evidence of anti-insect behaviour in the native birds he examined. They suggest that selective pressures may favour modification of sleeping behaviour as well. Other studies have quantified anti-insect behaviour in ciconiiform birds, but its impact in interrupting transmission of haemosporidian parasites is still poorly known (Maxwell and Kale 1977; Webber and Edman 1972).

In Florida, wild turkey poults normally hatch between mid-April and mid-June, with a peak in mid-May (Williams and Austin, 1988). Thus, yearly variations in timing of breeding season, environmental conditions, and vector populations may determine when poults are exposed to avian haemosporidians. At the study sites at Paynes Prairie, transmission of *L. smithi* was over by the time of peak hatching, indicating that poults would be exposed to *H. meleagridis* within their first few weeks of life, followed by *P. hermani* between August and December. Exposure to *L. smithi* may not occur until the following spring. In years in which transmission of *Leucocytozoon* is extended because of abnormally wet environmental conditions or large vector populations, a significant proportion of broods may be infected during their first month of life, perhaps increasing pathological interactions between *L. smithi* and *H. meleagridis*.

Multiple infections of *Plasmodium*, *Leucocytozoon*, and *Haemoproteus* are common in Florida turkeys as well as in other species of birds and are the rule rather then the exception. The potential impact of combined infections on mortality has not been studied. Richie (1988) has suggested that interactions between different species of malarial parasites may be important in the coevolution of host and parasites. Thus, it would be particularly interesting to determine the effects of multiple infections on tissue specificity, parasitaemia, and efficiency of transmission in host species that share more than one species of haemosporidian. The simultaneous existence of species of *Plasmodium*, *Leucocytozoon*, and *Haemoproteus* within the same host suggests that all are able to simultaneously share available physiological resources. This suggests that each species has evolved distinctive antigenic repertoires that avoid immunological cross-reactions. Thus, initial infection and development of immunity to one parasite will not prevent infection with another species. This question deserves further study because of the potential impact multiple haemosporidian infections may have on the pathogenic effects, transmission, and epizootiology of the parasites.

Conclusions

Avian haemosporidians have complex life cycles involving vertebrate and invertebrate hosts that are influenced by host immunology, physiology, and behaviour, as well as a variety of local and regional environmental factors. Research on the pathogenicity and epizootiology of these parasites indicates that they probably have their greatest impact on non-immune, juvenile birds at times when abnormal environmental conditions, e.g. high rainfall, and large vector populations favour high rates of transmission. Unfortunately, problems in capturing wild hosts and monitoring haemosporidian infections within individual birds over a long period of time have made it difficult to precisely quantify the impact of these parasites on behaviour, reproductive success, and habitat selection. As a result, most work on wild populations has been based on surveys of large numbers of migrating or resident birds by blood smears to determine the incidence of haemosporidian infection. These techniques are invaluable in studies of the taxonomy and distribution of avian haemosporidians and have been useful in testing a number of important hypotheses concerning host behavioural ecology, but need to be combined with other disciplines to fully measure the impact of these parasites on host evolution. Future studies need to rely on the careful coordination of field and laboratory studies of hosts, vectors, and parasites.

References

Ahmed, F.E. and Mohammed, A.H.H. (1977). Schizogony in *Haemoproteus columbae* Kruse. *Journal of Protozoology*, **24**, 389–93.

Ahmed, F.E. and Mohammed, A.H.H. (1978). *Haemoproteus columbae:* course of infection, relapse and immunity to reinfection in the pigeon. *Zeitschrift für Parasitenkunde*, **57**, 229–36.

Ashford, R.W. (1971). Blood parasites and migratory fat at Lake Chad. *Ibis*, **224**, 100–1.

Atkinson, C.T. (1986). Host specificity and morphometric variation of *Haemoproteus meleagridis* Levine, 1961 (Protozoa: Haemosporina) in gallinaceous birds. *Canadian Journal of Zoology*, **64**, 2634–8.

Atkinson, C.T. (1988). Epizootiology of *Haemoproteus meleagridis* (Protozoa: Haemosporina) in Florida: potential vectors and prevalence in naturally infected

Culicoides (Diptera: Ceratopogonidae). *Journal of Medical Entomology*, **25**, 39–44.

Atkinson, C.T. and Forrester, D.J. (1987). Myopathy associated with megaloschizonts of *Haemoproteus meleagridis* in a wild turkey from Florida. *Journal of Wildlife Diseases*, **23**, 495–8.

Atkinson, C.T., Greiner, E.C., and Forrester, D.J. (1983). Experimental vectors of *Haemoproteus meleagridis* Levine from wild turkeys in Florida. *Journal of Wildlife Diseases*, **19**, 366–8.

Atkinson, C.T., Greiner, E.C., and Forrester, D.J. (1986). Pre-erythrocytic development and associated host responses to *Haemoproteus meleagridis* (Haemosporina: Haemoproteidae) in experimentally infected domestic turkeys. *Journal of Protozoology*, **33**, 375–81.

Atkinson, C.T., Forrester, D.J., and Greiner, E.C. (1988*a*). Pathogenicity of *Haemoproteus meleagridis* (Haemosporina: Haemoproteidae) in experimentally infected domestic turkeys. *Journal of Parasitology*, **74**, 228–39.

Atkinson, C.T., Forester, D.J. and Greiner, E.C. (1988*b*). Epizootiology of *Haemoproteus meleagridis* (Protozoa: Haemosporina) in Florida: seasonal transmission and vector abundance. *Journal of Medical Entomology*, **25**, 45–51.

Baker, J.R. (1975). Epizootiology of some haematozoic protozoa of English birds. *Journal of Natural History*, **9**, 601–9.

Barnard, W.H. and Bair, R.D. (1986). Prevalence of avian hematozoa in central Vermont. *Journal of Wildlife Diseases*, **22**, 365–74.

Becker, E.R., Hollander, W.F., and Pattillo, W.H. (1956). Naturally occurring *Plasmodium* and *Haemoproteus* infection in the common pigeon. *Journal of Parasitology*, **42**, 474–8.

Ben-Harel, S. (1923). Studies on bird malaria in relation to the mechanism of relapse. *American Journal of Hygiene*, **3**, 652–85.

Bennett, G.F. (1960). On some ornithophilic blood-sucking Diptera in Algonquin Park, Ontario, Canada. *Canadian Journal of Zoology*, **38**, 377–89.

Bennett, G.F. and Cameron, M. (1974). Seasonal prevalence of avian hematozoa in passeriform birds of Atlantic Canada. *Canadian Journal of Zoology*, **52**, 1259–64.

Bennett, G.F. and Fallis, A.M. (1960). Blood parasites of birds in Algonquin Park, Canada, and a discussion of their transmission. *Canadian Journal of Zoology*, **38**, 261–73.

Bennett, G.F. and Herman, C.M. (1976). Blood parasites of some birds from Kenya, Tanzania and Zaire. *Journal of Wildlife Diseases*, **12**, 59–65.

Bennett, G.F., Smith, A.D., Whitman, W., and Cameron, M. (1975). Hematozoa of the Anatidae of the Atlantic flyway. II. The maritime provinces of Canada. *Journal of Wildlife Diseases*, **11**, 280–9.

Bennett, G.F., Greiner, E.C., Cameron, M.F., and Herman, C.M. (1976). Hematozoos de aves paserifirmes silvestres muestreadas en anos sucesivos. *Boletin de la Direccion de Marlariologia y Saneamiento Ambiental*, **16**, 313–19.

Bennett, G.F., Kucera, J., Woodworth-Lynas, C., and Whiteway, M. (1981). *Bibliography of the avian blood-inhabiting protozoa*. Supplement 1, Memorial University of Newfoundland Occasional Papers in Biology, no. 4. St. John's, Newfoundland.

Bennett, G.F., Whiteway, M., and Woodworth-Lynas, C. (1982*a*). *A host–parasite catalogue of the avian haematozoa*. Memorial University of Newfoundland Occasional Papers in Biology, no. 5, St. John's, Newfoundland.

Bennett, G.F., Thommes, F., Blancou, J., and Artois, M. (1982*b*). Blood parasites of some birds from the Lorraine Region, France. *Journal of Wildlife Diseases*, **18**, 81–8.

Bennett, G.F., Caines, J.R., and Bishop, M.A. (1988). Influence of blood parasites on the body mass of passeriform birds. *Journal of Wildlife Diseases*, **24**, 339–43.

Bierer, J.C., Strandberg, J., Stoskopf, M.K., and Craft, C. (1981). Mortality in robins (*Turdus miqratorius*) due to avian malaria. *Journal of Wildlife Diseases*, **17**, 247–50.

Bishop, A., Tate, P., and Thorpe, M. (1938). The duration of *Plasmodium relictum* infection in canaries. *Parasitology*, **30**, 388–91.

Borgia, G. (1986). Satin Bowerbird parasites: a test of the bright male hypothesis. *Behavioural Ecology and Sociobiology*, **19**, 355–8.

Bradbury, P.C. and Gallucci, B.B. (1972). Observations on the fine structure of the schizonts of *Haemoproteus columbae* Kruse. *Journal of Protozoology*, **19**, 43–9.

Coatney, G.R. (1933). Relapse and associated phenomena in the *Haemoproteus* infection of the pigeon. *American Journal of Hygiene*, **18**, 133–60.

Commichau, C. and Jonas, D. (1977). Eine durch *Leucocytozoon simondi* verursachte Erkrankung bei Entenkuken unter besonderer Berucksichtigung des histologischen Nachweises. *Zentralblatt für Veterinärmedizin*, **B24**, 662–7.

Cowan, A.B. (1957). Reactions against the megaloschizonts of *Leucocytozoon simondi* Mathis and Leger in ducks. *Journal of Infectious Diseases*, **100**, 82–7.

Danilewsky, B. (1885). Zur parasitologie des blutes. *Biologisches Zentralblatt*, **5**, 529–37.

Danilewsky, B. (1891). Contribution a l'étude de la microbiose malarique. *Annales de l'Institut Pasteur*, **5**, 758–82.

Desser, S.S. (1967). Schizogony and gametogony of *Leucocytozoon simondi* and associated reactions in the avian host. *Journal of Protozoology*, **14**, 244–54.

Desser, S.S. and Fallis, A.M. (1967). The cytological development and encapsulation of megaloschizonts of *Leucocytozoon simondi*. *Canadian Journal of Zoology*, **45**, 1061–5.

Fallis, A.M. and Desser, S.S. (1977). On species of *Leucocytozoon*, *Haemoproteus*, and *Hepatocystis*. In *Parasitic protozoa*, Vol. 3 (ed. J.P. Kreier), pp. 239–66. Academic Press, New York.

Fallis, A.M. and Wood, D.M. (1957). Biting midges (Diptera: Ceratopogonidae) as intermediate hosts for *Haemoproteus* of ducks. *Canadian Journal of Zoology*, **35**, 425–35.

Fallis, A.M., Davies, D.M., and Vickers, M.A. (1951). Life history of *Leucocytozoon simondi* Mathis and Leger in natural and experimental infections and blood changes produced in the avian host. *Canadian Journal of Zoology*, **29**, 305–28.

Fallis, A.M., Desser, S.S., and Khan, R.A. (1974). On species of *Leucocytozoon*. *Advances in Parasitology*, **12**, 1–67.

Farmer, J.N. (1964). Gizzard lesions associated with *Haemoproteus sacharovi* infections of pigeons. *Proceedings of the Iowa Academy of Sciences*, **71**, 537–42.

Fix, A.S., Waterhouse, C., Greiner, E.C., and Stoskopf, M.K. (1988). *Plasmodium relictum* as a cause of avian malaria in wild-caught Magellanic penguins (*Spheniscus magellanicus*). *Journal of Wildlife Diseases*, **24**, 610–19.

Forrester, D.J., Hon, L.T., Williams, L.E. Jr., and Austin, D.H. (1974). Blood protozoa of wild turkeys in Florida. *Journal of Protozoology*, **21**, 494–7.

Gabaldon, A. and Ulloa, G. (1980). Holoendemicity of malaria: an avian model. *Transactions of the Royal Society of Tropical Medicine and Hygiene*, **74**, 501–7.

Gardiner, C.H., Jenkins, H.J., and Mahoney, K.S. (1984). Myositis and death in bobwhites, *Colinus virginianus* (L.), due to hemorrhagic cysts of a haemosporozoan of undermined taxonomic status. *Journal of Wildlife Diseases*, **20**, 308–18.

Garnham, P.C.C. (1966). *Malaria parasites and other haemosporidia*. Blackwell Scientific Publishers, Oxford.

Godfrey, R.D. Jr., Fedynich, A.M., and Pence, D.B. (1987). Quantification of hematozoa in blood smears. *Journal of Wildlife Diseases*, **23**, 558–65.

Graham-Smith, G.S. (1907). A cystic disease of the heart, gizzard and muscles of young grass parakeets (*Psittacus undulatus*) due to a protozoan parasite. *Journal of Hygiene*, **7**, 552–7.

Greiner, E.C. and Bennett, G.F. (1975). A color pictorial guide to some species of *Haemoproteus, Leucocytozoon*, and *Trypanosoma. Journal of Wildlife Diseases*, Supplement on microfiche.

Greiner, E.C., Bennett, G.F., White, E.M., and Coombs, R.F. (1975*a*). Distribution of the avian hematozoa of North America. *Canadian Journal of Zoology*, **53**, 1762–87.

Greiner, E.C., Bennett, G.F., Laird, M., and Herman, C.M. (1975*b*). Avian haematozoa 2. Taxonomic keys and color pictorial guide to species of *Plasmodium. Journal of Wildlife Diseases*, Supplement on microfiche.

Hamilton, W.D. and Zuk, M. (1982). Heritable true fitness and bright birds: a role for parasites? *Science*, **218**, 384–7.

Hartley, W.J., Reddacliff, G.L., Finnie, E.P., and Turner, D. (1981). Suspected lethal *Leucocytozoon* infection in the Pied Currawong (*Strepera graculina*). In *Wildlife diseases of the Pacific Basin and other countries* (ed. M. Fowler), pp. 95–7. Wildlife Diseases Association, Lawrence, Kansas.

Hayworth, A.M., van Riper III, C., and Weathers, W.W. (1987). Effects of *Plasmodium relictum* on the metabolic rate and body temperature in canaries (*Serinus canarius*). *Journal of Parasitology*, **73**, 850–3.

Herman, C.M. (1938). Epidemiology of malaria in eastern red-wings (*Agelaius P. Phoneiceus*). *American Journal of Hygiene*, **28**, 232–43.

Herman, C.M. and Bennett, G.F. (1976). Use of sentinel ducks in epizootiological studies of anatid blood protozoa. *Canadian Journal of Zoology*, **54**, 1038–43.

Herman, C.M., Reeves, W.C., McClure, H.E., French, E.M., and Hammon, W.McD. (1954). Studies on avian malaria in vectors and hosts of encephalitis in Kern County, California. I. Infections in avian hosts. *American Journal of Tropical Medicine and Hygiene*, **3**, 676–95.

Herman, C.M., Knisley, J.O. Jr., and Snyder, E.L. (1966). Subinoculation as a technique in the diagnosis of avian *Plasmodium. Avian Diseases*, **10**, 541–7.

Herman, C.M., Barrow, J.H. Jr., and Tarshis, I.B. (1975). Leucocytozoonosis in Canada geese at the Seney National Wildlife Refuge. *Journal of Wildlife Diseases*, **11**, 404–11.

Herman, C.M., Greiner, E.C., Bennett, G.F., and Laird, M. (1976). *Bibliography of the avian blood-inhabiting protozoa*. International Reference Centre for Avian Haematoza, St. John's, Newfoundland.

Janovy, J. Jr. (1966). Epidemiology of *Plasmodium hexamerium* Huff, 1935, in meadowlarks and starlings of the Cheyenne bottoms, Barton County, Kansas. *Journal of Parasitology*, **52**, 573–8.

Khan, R.A. and Fallis, A.M. (1969). Endogenous stages of *Parahaemoproteus fringillae* (Labbe, 1894) and *Leucocytozoon fringillinarum* Woodcock 1910. *Canadian Journal of Zoology*, **47**, 37–9.

Kirkpatrick, C.E. (1989). Dark-field microscopy for detection of some hemotropic parasites in blood smears of mammals and birds. *Transactions of the American Microscopical Society,* **108**, 190–6.

Kirkpatrick, C.E. and Suthers, H.B. (1988). Epizootiology of blood parasite infections in passerine birds from central New Jersey. *Canadian Journal of Zoology,* **66**, 2374–82.

Kissam, J.B., Noblet, R., and Garris, G.I. (1975) Large-scale aerial, treatment of an endemic area with abate granular lavicide to control black flies (Diptera: Simuliidae) and suppress *Leucocytozoon smithi* of turkeys. *Journal of Medical Entomology,* **12**, 359–62.

Kocan, R.M. (1968). Anemia and mechanism of erythrocyte destruction in ducks with acute *Leucocytozoon* infections. *Journal of Protozoology,* **15**, 455–62.

Kocan, R.M. and Clark, D.T. (1966*a*). Anemia in ducks infected with *Leucocytozoon simondi. Journal of Protozoology,* **13**, 465–8.

Kocan, R.M. and Clark, D.T. (1966*b*). Prepatent period and parasitemia in *Leucocytozoon simondi* infections resulting from short exposures to sporozoites. *Journal of Parasitology,* **52**, 962–5.

Kohne, H.J. and Jones, J.E. (1975). Blood parameters in *Leucocytozoon smithi* infected poults fed Clopidol. *Poultry Science,* **54**, 472–4.

Laird, M. and Bennett, G.F. (1970). The subarctic epizootiology of *Leucocytozoon simondi. Journal of Parasitology,* **56**, (II), 198.

Laird, M. and van Riper, C. III. (1981). Questionable reports of *Plasmodium* from birds in Hawaii, with the recognition of *P. relictum* ssp. *capistranoae* (Russell, 1932) as the avian malaria parasite there. In *Parasitological topics,* Society of Protozoology Special Publication, no. 1. (ed. E.V. Cunning), pp. 159–65. Society of Protozoology, Lawrence, Kansas.

Lastra, I. and Coatney, G.R. (1950). Transmission of *Haemoproteus columbae* by blood inoculation and tissue transplants. *Journal of the National Malaria Society,* **9**, 151–2.

Levine, N.D. (1985). Phylum Apicomplexa. In *Illustrated guide to the protozoa* (ed. J.J. Lee, S.H. Hunter, and E.C. Bovee), pp. 322–74. Society of Protozoology, Lawrence, Kansas.

Levine, N.D., Beamer, P.D., and Simon, J. (1970). A disease of chickens associated with *Arthrocystis galli* n.g., n.sp., an organism of uncertain taxonomic position. In *H.D. Srivastava commemoration volume* (ed. K.W. Singh and B.K. Tardon), pp. 429–34. Indian Veterinary Research Institute, Izatnagar.

Maley, G.J.M. and Desser, S.S. (1977). Anemia in *Leucocytozoon simondi* infections. I. Quantification of anemia, gametocytemia, and osmotic fragility of erythrocytes in naturally infected Pekin ducklings. *Canadian Journal of Zoology,* **55**, 355–8.

Manwell, R.D. (1929). Relapse in bird malaria. *American Journal of Hygiene,* **9**, 308–45.

Manwell, R.D. (1951). Acute malaria in a Canada jay of the high Rockies. *Journal of Parasitology,* **37**, 322.

Markus, M.B. (1972). Pathogenicity of *Haemoproteus columbae. Transactions of the Royal Society of Tropical Medicine and Hygiene,* **66**, 186–7.

Maxwell, G.R. II and Kale, H.W. II (1977). Maintenance and anti-insect behaviour of six species of ciconiiform birds in south Florida. *Condor,* **79**, 51–5.

McClure, H.E., Poonswad, P., Greiner, E.C., and Laird, M. (1978). *Haematozoa in the birds of eastern and southern Asia.* Memorial University of Newfoundland, St. John's, Newfoundland.

Miller, R.E., Trampel, D.W., Dresser, S.S., and Boever, W.J. (1983). *Leucocytozoon simondi* infection in European and American eiders. Journal of the American Veterinary Medecine Association, **183**, 1241–4.

Miltgen, F., Landau, I., Ratanaworabhan, N., and Yenbutra, S. (1981). *Parahaemoproteus desseri* n. sp.; gametogonie et schizogonie chez l'hôte naturel: *Psittacula roseata* de Thailande, et sporogonie experimentale chez *Culicoides nubeculosus*. *Annales de Parasitologie Humaine et Comparée*, **56**, 123–30.

Mohammed, A.H.H. (1965). Studies on the schizogony of *Haemoproteus columbae* Kruse 1890. *Proceedings of the Egyptian Academy of Science*, **19**, 37–46.

Newberne, J.W. (1957). Studies on the histopathology of *Leucocytozoon simondi* infection. *American Journal of Veterinary Research*, **18**, 191–9.

Noblet, R., Adkins, T.R., and Kissam, J.B. (1972). *Simulium congareenarum* (Diptera: Simuliidae), a new vector of *Leucocytozoon smithi* (Sporozoa: Leucocytozoidae) in domestic turkeys. *Journal of Medical Entomology*, **9**, 580.

Noblet, R., Kissam, J.B., and Adkins, T.R. Jr. (1975). *Leucocytozoon smithi*: incidence of transmission by black flies in South Carolina (Diptera: Simuliidae). *Journal of Medical Entomology*, **12**, 111–14.

Opitz, H.M., Jakob, H.J., Wiensenhuetter, E., and Vasandra Devi, V. (1982). A myopathy associated with protozoan schizonts in chickens in commercial farms in peninsular Malaysia. *Avian Pathology*, **11**, 527–34.

O'Roke, E.C. (1930*a*). The incidence, pathogenicity and transmission of *Leucocytozoon anatis* of ducks. *Journal of Parasitology*, **17**, 112.

O'Roke, E.C. (1930*b*). The morphology, transmission, and life history of *Haemoproteus lophortyx* O'Roke, a blood parasite of the California valley quail. *University of California Publications in Zoology*, **36**, 1–50.

O'Roke, E.C. (1931). Life history of *Leucocytozoon anatis* Wickware. *Journal of Parasitology*, **18**, 127–33.

Peirce, M.A. (1984). Haematozoa of Zambian birds. I. General survey. *Journal of National History*, **18**, 105–22.

Peirce, M.A. and Mead, C.J. (1978*a*). Haematozoa of British birds IV. Blood parasites of birds from Wales. *Journal of Natural History*, **12**, 361–3.

Peirce, M.A. and Mead, C.J. (1978*b*). Haematozoa of British birds III. Spring incidence of blood parasites of birds from Hertfordshire, especially returning migrants. *Journal of Natural History*, **12**, 337–40.

Pierce, M.A., Greenwood, A.G., and Cooper, J.E. (1983). Haematozoa of raptors and other birds from Britain, Spain and the United Arab Emirates. *Avian Pathology*, **12**, 443–6.

Pinkovsky, D.D., Forrester, D.J., and Butler, J.F. (1981). Investigations on black fly vectors (Diptera: Simuliidae) of *Leucocytozoon smithi* (Sporozoa: Leucocytozoidae) in Florida. *Journal of Medical Entomology*, **18**, 153–7.

Read, A.F. (1987). Comparative evidence supports the Hamilton and Zuk hypothesis on parasites and sexual selection. *Nature*, 327, 68–70.

Ressel, S. and Schall, J.J. (1989). Parasites and showy males—malarial infection and color variation in fence lizards. *Oecologia*, **78**, 158–64.

Richie, T.L. (1988). Interactions between malaria parasites infecting the same vertebrate host. *Parasitology*, **96**, 607–39.

Ross, R. (1898). Report on the cultivation of Proteosoma, Labbe, in grey mosquitoes. *Indian Medical Gazette*, **33**, 401–8, 448–51.

Schall, J.J. (1983). Lizard malaria: cost to vertebrate host's reproductive success. *Parasitology*, **87**, 1–6.

Schall, J.J. and Dearing, M.D. (1987). Malarial parasitism and male competition for mates in the western fence lizard, *Sceloporus occidentalis*. *Oecologia*, **73**, 389–92.

Schall, J.J. and Sarni, G.A. (1987). Malarial parasitism and the behavior of the lizard *Sceloporus occidentalis*. *Copeia*, **8**, 84–93.

Schall, J.J., Bennett, A.F., and Putman, R.W. (1982). Lizards infected with malaria: physiological and behavioral consequences. *Science*, **217**, 1057–9.

Seed, T.M. and Manwell, R.D. (1977). Plasmodia of birds. In *Parasitic protozoa*, Vol. 3 (ed. J.P. Kreier), pp. 311–57. Academic Press, New York.

Sergent, Et. (1920). Le diagnostic de l'infection latente dans le paludisme des oiseaux (*Plasmodium relictum*). *Comptes Rendus des Séances de la Société de Biologie et de ses Filiales*, **83**, 1063–4.

Sergent, Ed. and Sergent, Et. (1906). Sur le second hôte de *l'Haemoproteus* (Halteridium) du pigeon. *Comptes Rendus de la Société de Biologie, Paris*, **61**, 494–6.

Sergent, Ed. and Sergent, Et. (1952). Recherches experimentales sur l'infection latente et la prémunition dans le paludisme. *Archives de l'Institut Pasteur d'Algerie*, **30**, 203–39.

Siccardi, F.J., Rutherford, H.O., and Derieux, W.T. (1974). Pathology and prevention of *Leucocytozoon smithi* infection of turkeys. *Avian Diseases*, **18**, 21–32.

Sherman, I.W. and Hull, R.W. (1960). Serum alterations in avian malaria. *Journal of Protozoology*, **7**, 171–6.

Skidmore, L.V. (1931). *Leucocytozoon smithi* infection in turkeys and its transmission by *Simulium occidentale* Townsend. *Journal of Parasitology*, **18**, 130.

Smith, D.C. and Sanford, L.B. (1985). Laveran's germ: the reception and use of a medical discovery. *American Journal of Tropical Medicine and Hygiene*, **34**, 2–20.

Stoskopf, M.K. and Beier, J. (1979). Avian malaria in African Black-footed penguins. *Journal of the American Veterinary Medicine Association*, **175**, 944–7.

Telford, S.R. Jr. and Forrester, D.J. (1975). *Plasmodium* (*Huffia*) *hermani* sp.n. from wild turkeys (*Meleagris gallopavo*) in Florida. *Journal of Protozoology*, **22**, 324–8.

van Riper III, C., van Riper, S.G., Goff, M.L., and Laird, M. (1986). The epizootiology and ecological significance of malaria in Hawaiian land birds. *Ecological Monograph*, **56**, 327–44.

Walker, D. and Garnham, P.C.C. (1972). Aberrant *Leucocytozoon* infection in parakeets. *Veterinary Record*, **91**, 70–2.

Warner, R.E. (1968). The role of introduced diseases in the extinction of the endemic Hawaiian avifauna. *Condor*, **70**, 101–20.

Webber, L.A. and Edman, J.D. (1972). Anti-mosquito behavior of ciconiiform birds. *Animal Behaviour*, **20**, 228–32.

Wehr, E.E. (1962). Studies on Leucocytozoonosis of turkeys, with notes on schizogony, transmission, and control of *Leucocytozoon smithi*. *Avian Diseases*, **6**, 195–210.

White, E.M., Greiner, E.C., Bennett, G.F. and Herman, C.M. (1978). Distribution of the hematozoa of Neotropical birds. *Revista de Biologia Tropical*, **26**, 3–102.

Williams, L.R. Jr. and Austin, D.H. (1988). *Studies of the wild turkey in Florida*. Florida Game and Fresh Water Fish Commission, Division of Wildlife, Technical Bulletin no. 10.

Yamada, K.A. and Sherman, I.W. (1979). *Plasmodium lophurae:* composition and properties of hemozoin, the malarial pigment. *Experimental Parasitology*, **48**, 61–74.

3 *The direct and indirect effects of the caecal nematode* Trichostrongylus tenuis *on red grouse*

PETER J. HUDSON and ANDREW P. DOBSON

Introduction

Until recently it has generally been accepted that a successful parasite maintains a harmless level of infection and lives in a symbiotic relationship with its host. Epidemics were considered the exception where density-independent factors perturbed the relationship between host and parasite and led to massive host mortality. An alternative view is now held by many evolutionary biologists, who realize that, since parasite fitness depends heavily on transmission, the individual parasite that successfully transmits its offspring to other hosts will be at a selective advantage, even if the parasite kills both itself and the host. Moreover, through the synthesis of epidemiology and ecology, empirical treatments suggest that parasites may act in a density-dependent manner and can regulate the size of the host population (Anderson and May 1978; May and Anderson 1978).

The overall impact of a macroparasite on the size of the host population depends on both parasite pathogenicity and the parasite's distribution within the host population (Anderson 1979; Fig. 3.1; see van

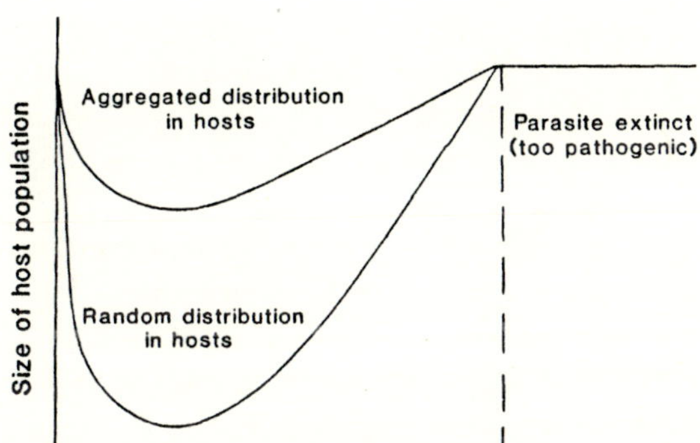

Fig. 3.1. The effect of macroparasite pathogenicity and pattern of distribution on the host population. Note that moderately pathogenic species tend to have the greatest effect on the host population size (after Anderson 1979).

Riper III, Chapter 8, this volume). Highly pathogenic parasites kill the host when parasite intensity is low, leading to low transmission rates and an insignificant impact on the host population. For moderately pathogenic species, the level of intensity increases so transmission rate is high and the parasite has a large impact on host population size. However, the effects on the host population will be even greater if the parasite also reduces host fecundity. Indeed, pathogens which reduce host fecundity without a large reduction in host survival will tend to have the greatest impact on host equilibrium size (see Atkinson and van Riper III, Chapter 2, this volume).

This chapter examines the population dynamics of the caecal nematode *Trichostrongylus tenuis,* a parasite of moderate pathogenicity, that is lethal at high worm intensities but has a significant sublethal impact on the breeding production of red grouse, *Lagopus lagopus scoticus* (Hudson 1986*a,b*). These direct effects are sufficient to generate the population cycles of red grouse, although indirect effects of parasitism may also influence aspects of the ecology and behaviour of the grouse.

*The history of research on the grouse–*T. tenuis *system*

There is anecdotal evidence of grouse numbers suffering high mortality in the eighteenth century and substantial evidence from the nineteenth century of grouse populations declining with many moribund and emaciated grouse, a condition known as 'grouse disease'. Cobbold (1873, quoted in Lovat 1911) noted that a parasitic nematòde was associated with this condition and felt the nematode (originally called *Strongylus perigracilis*) was the sole cause.

By the turn of the century outbreaks of the disease were common and a Committee of Inquiry on Grouse Disease was established under Lord Lovat in 1905. A number of scientists were appointed including Edward Wilson, who later died with Captain Scott on the ill-fated expedition to the South Pole. Wilson and Leiper (in Lovat 1911) analysed large numbers of samples and established the life cycle of *T. tenuis*; and they agreed with Cobbold that the nematodes were the principal cause of the disease.

A second enquiry was initiated by The Scottish Landowners Federation in 1956 to investigate the cause of the generally reduced, postwar grouse stocks. Detailed population data were collected from Glen Esk, Angus (Jenkins *et al.* 1963) and limited information collected on the intensity of infection. They concluded that parasitic diseases and the associated mortality were not fundamentally important in controlling the size of overwintering and breeding populations and high parasite intensity was a consequence of social status.

A third enquiry started in 1979, initiated by The Game Conservancy and landowners in northern England, investigated the influence of *T. tenuis* on the population dynamics of red grouse (Hudson 1986*a*). This chapter summarizes information from this research.

Population cycles in red grouse

Middleton (1934) collected long-term data on numbers of red grouse shot per annum from three estates, and noted that these showed a tendency to produce cyclic fluctuations. More records were collected by MacKenzie (1952) and, in an accompanying paper, Moran (1952) conducted some preliminary time series analysis and concluded that red grouse populations have a slight but heavily damped oscillatory tendency. Poole (1974) and later Williams (1985) conducted more detailed analysis but only used data up to 1939. In addition, they aggregated the data by regions, took no account of population trends, and analysed the data as average number of grouse shot per estate without a correction for area of moorland habitat.

A more detailed estate by estate analysis of English bag records was conducted by Potts *et al.* (1984) which demonstrated that 83 per cent of grouse moors produced bag records with a tendency to produce cycles. Potts *et al.* (1984) also proposed that population cycles in grouse are caused by the effects of nematode parasitism acting on the fecundity of the grouse in a delayed density-dependent manner.

Red grouse and the life cycle of T. tenuis

The red grouse host

Red grouse are the main host of *T. tenuis* although the parasite has been recorded in a number of gamebird species. Portal and Collinge (1932) reported an epidemic of *T. tenuis* in grey partridge, *Perdix perdix*. Workers in America have described and studied *T. tenuis* in bobwhite quail, *Colinus virginianus* (Moore *et al.* 1986), but both the prevalence and intensity of infection is low compared to that in red grouse. Within the habitat of red grouse, partridges and other gamebirds are essentially absent so grouse can effectively be considered the only host for *T. tenuis*.

Red grouse inhabit the heather moorlands of the British Isles, their distribution influenced by the presence of an important food plant, heather (*Calluna vulgaris*). Grouse density is increased through careful

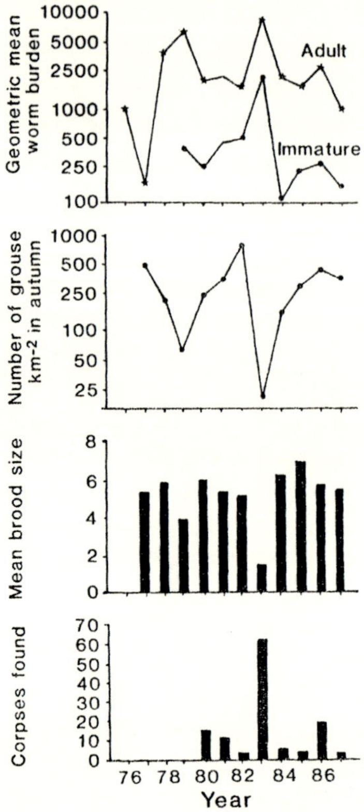

Fig. 3.2. Longitudinal study of grouse–*T. tenuis* system illustrating changes in (top to bottom): mean intensity of infection; autumn host density; grouse breeding production; number of corpses found within 1.5 km^2. Large reductions in grouse abundance and breeding production are associated with high levels of parasite intensity.

management of the heather habitat and the control of predatory foxes and crows. The birds are harvested in autumn with the letting of the shooting forming a significant income to many upland estates (Hudson and Renton 1988).

A longitudinal study of parasite intensity and grouse numbers (Fig. 3.2) on a 17 km^2 area of moorland in the central Pennines showed that high levels of infection are associated with poor breeding and survival and are followed by a fall in grouse density.

Basic life cycle

Within the moorland habitat *T. tenuis* has a simple direct life cycle with no secondary hosts or carriers of infective stages. The eggs pass from the grouse in the bird's caecal faeces and the first two larval stages stay within the caecal pat before the L3 ascends the heather plant to the growing tip and is ingested by the grouse. Following ingestion the L3 develops into an L4 and then adult stage, although development can be arrested after the L2 cuticle has been shed (Hudson 1986*a*; Shaw 1988).

The adult stage

The *T. tenuis* adult stage inhabits the particularly large caecum of the grouse. Prevalence of infection is high, 100 per cent of old grouse (> 15 months old) and 99 per cent of birds more than 3 months old were found to be infected (Hudson 1986*a*). A rapid increase in prevalence is associated with the development of the young bird's caecum and an increase in mean intensity of infection. Initially, the distribution of parasites within the population of young grouse conforms to a Poisson distribution but the variance rises faster than the mean and the distribution becomes aggregated (Hudson 1986*a*). This type of distribution is typical for macroparasites but the degree of aggregation is low probably because there is little resistance (H. Watson *et al.* 1988) to infection and mean parasite intensity is consequently high.

Parasite egg production

Egg production is density-dependent with a fall in production from infections greater than 4500 worms per bird. Egg development within the caecal pat is dependent on temperature and when temperature is greater than 15°C, more than 90 per cent of eggs hatch (Hudson 1986*a*).

The free-living stages

The L1 and L2 larvae remain in the caecal pat and feed on bacteria, accumulating reserves used by the L3 to reach and infect a host. Survival of free-living stages of trichostrongyle nematodes is generally dependent on humidity and the first two larval stages cannot survive 24 hours of desiccation.

The infective stage

The L3 infective stage retains the outer cuticle of the L2 and does not feed. Lipid reserves, accumulated during the first two larval stages, are used to actively swim and ascend the heather. Infectivity is dependent on the lipid reserves which are used up faster at high temperatures (H. Watson 1988).

There are two periods of infection, the first between July and September and the second in late winter. The abundance of infective larvae on the growing tips of the heather plant between May and August conforms well with the uptake of worms in young grouse (H. Watson and Hudson 1988). Under optimal conditions, larvae will develop from egg to infective stage within 7 days..This period of summer infection is dependent on both grouse density and the level of infection in the adult birds.

As with several Trichostrongylid nematodes, the larvae can become arrested and remain as L3 without the L2 cuticle. Shaw (1988) has shown that arrested larvae are present in November and December and then increase significantly up to March when they develop into adults, although she was unable to account for variations in worm intensity through the development of these larvae. Year-to-year variation in the period of winter infection is dependent on the density of grouse in late summer (Hudson and Dobson, in press).

The direct impact of T. Tenuis *on red grouse*

Pathophysiological impact

The adult nematodes burrow into the caecal mucosa causing localized flattening, disruption of the plicae, and internal bleeding (H. Watson *et al.* 1987). These activities probably cause a decrease in absorptive area, a decrease in brush-border enzyme activity, and a loss of plasma through the lesions. Treatment of grouse with an antihelmintic reverses these pathophysiological conditions.

The relative size of the caecum increases with the intensity of infection (Hudson 1986*a*). Within each caecum the nematodes tend to be aggregated although similar numbers are found in each caecum.

The effects of the parasite on grouse fecundity

The impact of parasites on weight gain and breeding production has been examined in a series of field experiments (Hudson 1986*a*,*b*). The basic technique was to catch grouse at random and treat half the birds with an anthelmintic to reduce worm intensity and the remaining birds with water as a control and to then monitor the parasite effects.

Weight gains of males over the winter months were greater for males with reduced worm intensities than for untreated males (Fig. 3.3). Furthermore, the weights of treated birds shot 5 months later were significantly greater than those of untreated birds (Table 3.1). Clutch size, hatching success, and chick survival were all greater from the hens with experimentally reduced worm intensities (Table 3.2: between-year comparisons). Replication of the experiment from a series of areas throughout England and Scotland confirms that parasites are reducing breeding production to a similar extent, demonstrating that parasitism is the principal factor influencing breeding production.

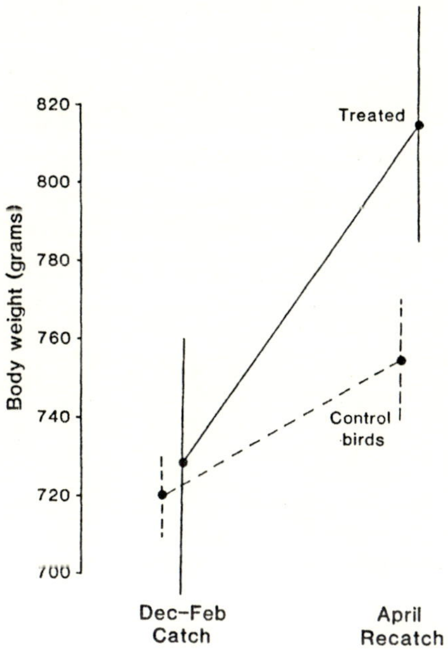

Fig. 3.3. Changes in mean weights of treated and untreated male grouse between autumn and spring; bars show standard errors (after Hudson 1986*a*).

The effects of parasites on grouse survival

Although it is accepted from laboratory studies that the impact increases with parasite intensity there are few estimates of parasite-induced host mortality. Anderson and May (1978) suggest that an indirect measure of parasite-induced mortality can be estimated by dividing the natural intrinsic growth rate of the population by the mean parasite intensity. In the grouse–*T. tenuis* system, worm intensity is high so we may expect the overall impact on survival to be low.

The most direct evidence for the survival of grouse being influenced by the parasites comes from the experimental manipulation of parasite intensities. Treated birds were significantly more likely to survive and be shot in the following autumn and less likely to be found dead than untreated birds (Table 3.3).

Table 3.1. Weights of treated and untreated hen grouse 5 months after treatment. Significance determined with one-tailed *t* test and sample sizes are shown in parenthesis

Sex	Treated	Untreated	Significance
Male	784.1 (72)	680.0 (124)	$p < 0.05$
Female	638.5 (100)	618.0 (57)	$p < 0.01$

Table 3.2. Breeding parameters of red grouse in relation to experimentally reduced worm intensities (after Hudson 1986*b*). Figures show means; significance was determined by the one-tailed *t* test

Year	Parameter	Treatment	Control	Significance
1982	Clutch	8.25	7.93	NS
	Hatching success	97%	77%	NS
	Brood @ 10 days	5.5	3.64	<0.05
1983	Clutch	8.0	5.28	<0.01
	Hatching success	75%	38%	<0.02
	Brood @ 10 days	3.23	0.45	<0.01
1984	Clutch	8.8	7.59	<0.05
	Hatching success	96%	92%	NS
	Brood @ 10 days	7.44	4.90	<0.01

NS, not significant

Indirect Impact of T. tenuis *on red grouse*

Since *T. tenuis* reduces the condition of the host it is reasonable to expect this reduced condition to influence other aspects of the bird's life history. Reduced competitive ability for resources such as food and territories, increased vulnerability to predation, and reduced opportunities for matings may all be caused by parasitism and have a direct influence on host equilibrium size.

Territorial success and intensity of parasite infection

In the studies by Jenkins *et al.* (1963) and A. Watson (1985) grouse were classified in autumn as either with or without territories. Significantly fewer of the territorial birds were subseqently found dead or killed by predators and in most cases died a non-violent death, probably from a

Table 3.3. Survival of adult grouse as measured by the number surviving to the following autumn and subsequently shot. No survival difference was found between territorial and non-territorial individuals ($p=0.45$) while there was a significant difference between treated and untreated individuals ($p<0.05$). Significance determined by chi-squared method

	Treated		Not treated	
	Not shot	Shot	Not shot	Shot
Territorial	64	30	18	3
Non-territorial	49	20	31	5

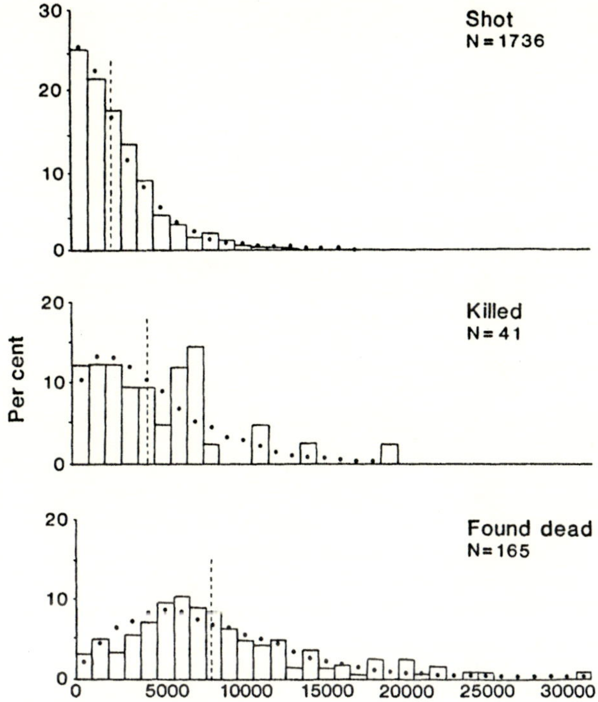

Fig. 3.4. The frequency distribution of worm intensities of grouse found dead and probably killed by parasites in comparison with grouse killed by predators and shot.

high intensity of infection. Indeed most birds found dead during winter months by Hudson (1986*a*) had died with a high intensity of infection (Fig. 3.4) and even grouse killed by predators carried large parasite intensities.

However, these studies do not show whether it was social status or parasite intensity that were the real cause of mortality. Parasite manipulation experiments and details of infection are needed. As shown earlier, Hudson (1986*a*) experimentally reduced worm intensities and thereby increased survival chances of grouse irrespective of social status. A more rigorous approach is to compare survival chances of territorial and non-territorial individuals in relation to parasite intensity. This experiment is currently being conducted and preliminary analysis has shown that parasite intensity and not social position influences survival chances in northern England (Table 3.3).

Studies on social status have shown that territorial individuals are more likely to survive than non-territorial ones. However, during years of heavy loss, A. Watson *et al.* (1984*b*) recorded that territorial birds lost were not replaced. This would imply that social status may not determine survival during periods of decline; other factors such as parasitism may become important at this time. This scenario would conform with the general model presented by Davies (1978; see Fig. 3.5)

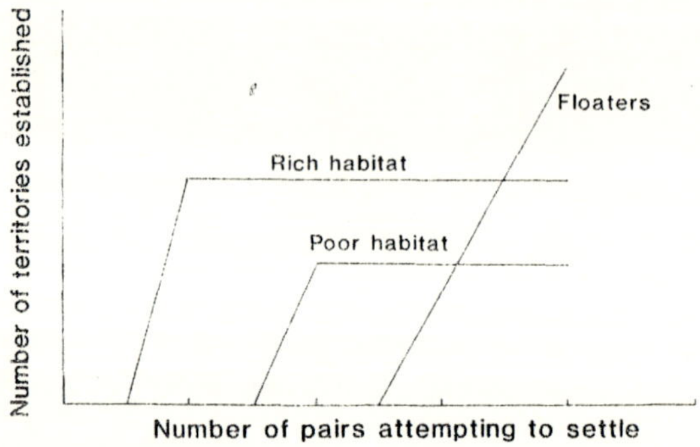

Fig. 3.5. The density limiting model of Davies (1978). The model assumes fixed minimum territory size; the birds occupy the rich habitat first until it is full, subsequent birds are forced into the poor habitat, and, when this is occupied, remaining birds are non-breeding 'floaters'.

that territorial behaviour limits density but other factors, such as parasitism, would cause population density to fall.

The two seasonal periods of infection occur when grouse are not divided into territorial and non-territorial groups so it seems unlikely that variations in infection rate are the result of social status. The first period of infection between July and September is before territorial behaviour commences and infections could act to influence territorial behaviour. The second period occurs between November and March when snow is present on the ground and both territorial and non-territorial grouse are in flocks.

Vulnerability and intensity of parasite infection

The reduced condition produced by parasites may also make the host slower, weaker, and thus more vulnerable to predation. Studies in England (Hudson 1986*a*; Hudson, Dobson, and Newborn, in preparation) found evidence that grouse killed by predators tended to be heavily infected. One mechanism by which parasites could make hosts more vulnerable was explored during our studies in England. Grouse exhibit a range of predator-avoidance adaptations including reduced scent emission during incubation. The scent probably arises from the bird's caecum which ceases producing caecal faeces prior to incubation. Hens with reduced levels of infection were less likely to be found during incubation by trained dogs (find nests by scent: 15 per cent) than by us (random visual search: 45 per cent) suggesting that infected hens emit more scent and are consequently more vulnerable to predation.

Models for the dynamics of T. tenuis *in grouse*

A clearer understanding of the effects of *T. tenuis* on the grouse population can be obtained by examining the properties of a mathematical model of the nematode system. The framework of the model follows the format developed by Anderson and May (see the appendix to this chapter) and considers the dynamics of three different populations: (1) the grouse; (2) adult parasites; and (3) free-living (including arrested larvae) parasite stages. Parameter estimates for the various birth and death rates for each population have been estimated from field and laboratory studies.

The relative influence of the different processes on the size of each population can be found by considering the expressions for each of the populations. This analysis suggests that grouse density varies inversely with the parasite's ability to influence host survival and fecundity. The magnitude of the population of parasite free-living stages varies indirectly with the pathogenicity of the parasite and directly with the life expectancy of the free-living larvae.

Local stability analysis may be used to explore the dynamical properties of the model. This exercise suggests that the behaviour of the host–parasite system is influenced by four parameters: (1) the degree of parasite aggregation within the grouse population; (2) the life expectancy of the free-living stages; (3) the reduction in host fecundity caused by the parasite; and (4) the mortality rate of grouse caused by the parasite.

When the life expectancy of the free-living stages is long and the effects on grouse fecundity are greater than host mortality, the system tends to oscillate. In the absence of other constraints on host population growth, these oscillations are essentially unstable and may be sufficiently large that either the host or parasite population are driven to extinction. If we now incorporate the effects of territorial behaviour using the model of Davies (1978; see Fig. 3.5) then stable limit cycles are produced. The period of these cycles is determined by the life expectancy of the free-living stages and the intrinsic growth rate of the grouse population (Fig. 3.6). Parameter estimates obtained from field studies suggest that the model populations would oscillate with a frequency of around 4–5 years, a figure that is very similar to the 4.8 years recorded in the north of England (Potts *et al.* 1984).

When the life expectancy of the parasite is low, worm intensities do not increase and the population fails to oscillate. On the east side of England where the rainfall is less than 1000 mm per annum, the survival of free-living stages is probably low, worm intensities are significantly lower, and the bag records from the grouse populations fail to demonstrate significant cycles of abundance (Hudson *et al.* 1985).

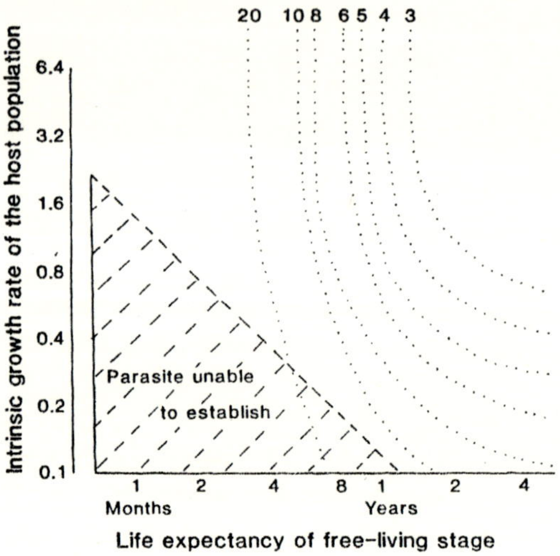

Fig. 3.6. Cycle length of model fluctuations in relation to the intrinsic growth rate of the population and life expectancy of free-living stages. Within the hatched area *T. tenuis* is unable to become established.

As it stands, the model provides a reasonable explanation of mechanisms causing cycles in the abundance of red grouse in the north of England and suggests an explanation for both the geographical variation in cycle length and why some populations do not cycle.

Population cycles and *T. tenuis*

The hypothesis proposed by Potts *et al.* (1984) that population cycles in red grouse are caused by the delayed density-dependent effects of parasites on the breeding production of red grouse has not been refuted. Parasites do reduce breeding production and infection acts in a delayed density-dependent manner, which, according to the Anderson and May model, is sufficient to generate population cycles. Incorporating a density-dependent mortality that mimicked the effects of territorial behaviour stabilized the cycles but was not the cause of year-to-year changes in density as proposed by Moss and Watson (1985).

Not all grouse populations produce cyclic bag records; Hudson *et al.* (1985) show that infections on dry grouse moors are significantly lower than on the cyclic wet moors. Hence the hypothesis now becomes 'Population cycles in red grouse are caused by the delayed density-dependent effects of parasites on the fecundity of grouse where environmental effects allow sufficient infection rates.' Unlike other hypotheses on grouse population fluctuations, this hypothesis can account for regional differences in the tendency of populations to cycle. Furthermore, it agrees with early observations (Lovat 1911; MacKenzie 1952) that not all population declines are caused by parasites but that these provide a sufficient explanation.

Alternative hypothesis for population cycles in grouse

A range of hypotheses have been, and could be, generated to explain the population cycles in red grouse. Several of those proposed for red grouse and other grouse species are now discussed in relation to their possible interaction with parasitism.

Predation

Scandinavian studies of black grouse (*Tetrao tetrix*) have provided sufficient information on temporal and spatial variations in predation pressure to explain population cycles (Angelstam *et al.* 1985). In Britain the major predators of red grouse, the fox (*Vulpes vulpes*) and carrion crow (*Corvus corone*), are controlled by keepers so it seems unlikely that predation is sufficient to cause cycles.

Evidence has been presented indicating that predators selectively catch heavily infected birds. The interaction of predation and parasitism on the dynamics of grouse populations is currently being investigated (Dobson and Hudson, in preparation) and shows that low levels of predation effectively increase the pathogenicity of the parasite and allow the grouse population to escape the regulatory role of the parasite (see Fig. 3.1) and subsequently increase to a greater equilibrium population. Higher levels of predation increase grouse mortality and lead to a fall in the grouse population to a low equilibrium. In the absence of predator control it seems quite likely that parasites would not be a significant factor in generating population cycles.

Food

Grouse feed principally on the growing tips of heather although Savory (1978) has estimated that the quantity eaten is usually less than 3 per cent. Variations in food quality are correlated with breeding production in north-east Scotland on freely drained heather moorland (Moss *et al.* 1975) although, on wetter moors, flowers of cotton grass (*Eriophorum vaginatum*) are selectively eaten by hen grouse prior to laying and heather quality is probably of less importance. Hudson *et al.* (1985) show that cycles are more likely to occur on wetter moors.

Food quality could cause population cycles if changes acted in a delayed density-dependent manner sufficient to cause cycles. The role of food quality during a population decline was studied by A. Watson *et al.*

(1984*a*) who fertilized areas of moorland and increased nitrogen and phosphorus content of heather. However, fertilization did not prevent numbers of grouse on the treated area from falling to extinction.

A. Watson and Moss (1972) proposed the maternal nutrition hypothesis that poor food quality reduced chick production and chicks from small broods became aggressive and took larger territories. While there was evidence that food quality influenced breeding production, aggression was not related inversely to brood size so Moss *et al.* (1984) refuted this hypothesis.

Parasites may interact synergistically with the nutritional status of the host, competing for resources and reducing the hosts' immunological response. If competition for food between hosts is density-dependent so that pathogenicity increases with density, some interesting dynamical conditions can arise (Anderson 1981).

Spacing behaviour and intrinsic mechanisms

Moss and Watson (1985) consider cycles to be adaptive irruptions caused by changes in spacing behaviour that have evolved to avoid outbreaks of trichostrongylosis. To understand the role of spacing behaviour in causing population cycles two fundamental questions must be answered.

1. Why do some populations of grouse fail to produce population cycles?
2. What is the mechanism generating fluctuations in spacing behaviour?

The second question has been addressed and studied during intensive research in north-east Scotland since 1955. Two main hypotheses have been developed and refuted: (1) the maternal nutrition hypothesis, described above and refuted by Moss *et al.* (1984); (2) the polymorphic behavioural hypothesis developed by Chitty (1967).

Chitty's polymorphic behavioural hypothesis proposed that the genetic composition of a population changes during a cycle. At high density, selection favours aggressive individuals with low reproductive rates and, at low density, selection favours less aggressive individuals with high reproductive rates. Moss *et al.* (1984) found that differences in dominance were inherent but changes in 'relative dominance index' were inconsistent with the hypothesis. They concluded that selection for dominance occurred as a consequence of, rather than as a cause of population change. Parasite intensity may also be negatively correlated with dominance (see this volume: Zuk, Chapter 16; Burley *et al.*, Chapter 19; Johnson and Boyce, Chapter 20; Spurrier *et al.*, Chapter 21) and, since captive birds used during dominance tests were known to be infected with parasites (Wilson 1979), it is possible that changes in dominance ranking were influenced by infection intensity.

The demographic study of two population fluctuations by A. Watson *et al.* (1984*b*) recorded changes in density, breeding production, and number of carcasses found on a study area of 1.77 km^2. Breeding production and the number of dead birds found could not account for changes in density so the authors assigned all other losses as being net emigrants. No estimate was made of the number of birds caught by predators and removed from the site and no information on parasitism was published. Changes in the extent of emigration are believed to be the result of changes in grouse spacing behaviour although this was only measured as the rate of emigration.

To demonstrate that grouse density is limited by spacing behaviour A. Watson and Moss (1970) considered three conditions that must be demonstrated.

1. *The presence of a non-breeding proportion of the population.* Territorial ownership is known to confer increased survival chances (A. Watson 1985) although most of these non-territorial birds are dead or have disappeared before breeding commences and a proportion of the birds that disappear may have simply bred in poor habitat (as per Davies model; Fig. 3.5). While A. Watson (1985) demonstrated the presence of non-territorial birds on Glen Esk and Kerloch up to 1967, territorial birds apparently disappeared during the population decline between 1968 and 1977 and were not replaced. Hence socially induced mortality was not the sole factor limiting the breeding population.

2. *Non-breeders are physiologically capable of breeding.* This has been demonstrated in studies of territorial limitation in a number of bird species (e.g. Stewart and Aldrich 1951; Krebs 1971) through removal experiments coupled with observations on replacement individuals. If spacing behaviour causes fluctuations in cyclic species it is necessary not only to demonstrate replacements during a population decline but such experiments should be conducted on the limiting sex. In red grouse the sex ratio nearly always favours males so experiments must be conducted on hens. The only removal experiments on hens were conducted at Kerloch between 1963 and 1965 (A. Watson and Jenkins 1968) when population density was increasing; this provides no information on the role of spacing behaviour during a population decline.

3. *No other resource, such as food, nest sites, space, or females, is limiting breeding density.* To eliminate all possible alternatives is

difficult and, while both nest sites and food are unlikely to be limiting, it is quite possible that hens or space could be.

There appears to be insufficient evidence to refute the null hypothesis that: changes in spacing behaviour of individuals are not necessary to cause cyclic fluctuations in population density (Chitty 1967). This does not suppose population changes were caused by parasites since this was not examined although there is no evidence to refute competing hypotheses.

Parasites are likely to have an effect on the territorial behaviour of the grouse directly and indirectly. Work on mammals has found that parasites influence aggressiveness (Freeland 1981). In red grouse Marquiss (1977) found links between nutrition and dominance which would imply that parasites may also influence dominance. Indeed, in some preliminary observations, Wilson (1979) found dominant grouse carried lower levels of worm intensity and dominance position in paired birds could be reversed by altering worm intensity for a period of 3 weeks.

The effects of parasitism on breeding success may also have an indirect effect on territorial behaviour. Levels of aggression can be expected to vary as a function of the coefficient of relatedness between neighbours which in turn is dependent on breeding success and dispersal (Charnov and Finnerty 1980). There is evidence that young cocks establish territories alongside related birds (Hudson, unpublished) so that changes in levels of aggression may simply relate to changes in relatedness between neighbours.

Conclusions

Evidence has been presented which supports the hypothesis that population cycles in red grouse are caused by reduced fecundity and survival through the delayed density-dependent effects of parasites and that these effects are sufficient to cause population cycles in grouse. Populations that did not show cycles carried lower levels of infection. While such effects do not preclude the action of other factors causing population cycles the evidence for other extrinsic factors or intrinsic changes in spacing behaviour is considered weak.

Appendix: structure of the basic model

The dynamics of the host–parasite interaction may be described by the set of three coupled differential equations

$$dH/dt = (a-b)H - (\alpha + \zeta)P, \tag{4.1}$$

$$dP/dt=\beta WH-(\mu+b+\alpha)P-\frac{\alpha P^2}{H}\frac{(k+1)}{k}, \tag{4.2}$$

$$dW/dt=\lambda P-\gamma W-\beta WH \tag{4.3}$$

where H is the host population, P the adult parasite population, W the population of the free-living stages, a the grouse fecundity, b the grouse mortality, α the parasite pathogenicity, ζ the parasite-induced reduction in host fecundity, β the infection rate, μ the adult worm mortality, k the aggregation of parasites in host population, λ the parasite birth rate, and γ the mortality of free-living stages.

Within the model the adult parasites are distributed in an aggregated manner and the degree of aggregation is given by k of the negative binomial.

As with other epidemiological studies it is useful to obtain expressions for R_0 the basic reproductive rate of the parasite and for H_t, the threshold number of hosts required to sustain the infection. When the parasite is first introduced into the population, pathogenicity and other density-dependent constraints on its rate of increase are effectively trivial. Collapsing eqns (4.1)–(4.3) down under these conditions gives an expression for this initial growth rate

$$R_0=\frac{H}{(\gamma+\beta H)(\mu+b+\alpha)}=T_1M_1M_2. \tag{4.4}$$

Here $T_1=H$, the transmission rate of eggs from mature adult female to establishment in another grouse, $M_1=1/(\gamma+\beta H)$, the life expectancy of a free-living larva, and $M_2=1/(\mu+b+\alpha)$, the life expectancy of an adult worm. Estimates of some of these parameters are available from either our own work, or the published literature, on *T. tenuis*. As is usually the case with epidemiological studies, few data are available with which to estimate rates of transmission. Crude estimates of these parameters may be obtained by determining the threshold host density at which we would expect the parasite to just maintain itself in the host population.

The threshold number of grouse required for the parasite to establish may be obtained by setting R_0 to unity and rearranging to give

$$H_t=\frac{\gamma/\beta}{\lambda M_2-1}. \tag{4.5}$$

Data from our main study area suggests that at this site *T. tenuis* would be effectively absent when grouse density dropped below 20 grouse/km^2. Thus a crude estimate of H_t suggests a figure for β_t of the order of 10^{-6} ($2\cdot10^{-6}>\beta_t<4\cdot10^{-5}$). Substitution of this estimate into (4.4) suggests a value for R_0 in the range 5 to 10 for a typical grouse moor in the north of England.

References

Anderson, R.M. (1979). The influence of parasitic infection on the dynamics of host population growth. In *Population dynamics* (ed. R.M. Anderson, B.D. Turner, and L.R. Taylor), pp. 245–81. Blackwell Scientific Publications, Oxford.

Anderson, R.M. (1981). Population ecology of infectious diseases. In *Theoretical ecology* (ed. R.M. May), pp. 318–55. Blackwell Scientific Publications, Oxford.

Anderson, R.M. and May, R.M. (1978). Regulation and stability of host–parasite interactions: I Regulatory processes. *Journal of Animal Ecology*, **47**, 219–49.

Angelstam, P., Lindstrom, E., and Widen, P. (1985). Synchronous short-term population fluctuations of some birds and mammals in Fennoscandia—occurrence and distribution. *Holarctic* Ecology, **8**, 285–98.

Charnov, E.L. and Finnerty, J.P. (1980). Vole population cycles: a case for kin selection. *Oecologia*, **45**, 1–2.

Chitty, D. (1967). The natural selection of self-regulatory behaviour in animal populations. *Proceedings of the Ecological Society of Australia*, **2**, 51–78.

Davies, N.B. (1978). Ecological questions about territorial behaviour. In *Behavioural ecology. An evolutionary approach* (ed. J.R. Krebs and N.B. Davies), pp. 317–50. Blackwell Scientific Publications, Oxford.

Freeland, W.J. (1981). Parasitism and behavioural dominance among male mice. *Science*, **213**, 461–2.

Hudson, P.J. (1986*a*). *Red grouse: the biology and management of a wild gamebird.* The Game Conservancy Trust, Fordingbridge.

Hudson, P.J. (1986*b*). The effect of a parasitic nematode on the breeding production of red grouse. *Journal of Animal Ecology*, **55**, 85–92.

Hudson, P.J. (in press). Red grouse population cycles and the population dynamics of the caecal nematode, *Trichostrongylus tenuis*. In *Proceedings of the grouse workshop, Imperial College* (ed. A. Lance). Blackwell Scientific Publications, Oxford.

Hudson, P.J. and Renton, J. (1988). What influences changes in grouse numbers? *Game Conservancy Annual Review*, **19**, 41–9.

Hudson, P.J. Dobson, A.P., and Newborn, D. (1985). Cyclic and non-cyclic populations of red grouse: a role for parasitism? In *Ecology and genetics of host-parasite interactions* (ed. D. Rollinson and R.M. Anderson), pp. 79–86. Academic Press, New York.

Jenkins, D., Watson, A., and Miller, G.R. (1963). Population studies on red grouse *Lagopus lagopus scoticus* (Lath.) in north-east Scotland. *Journal of Animal Ecology*, **32**, 317–76.

Krebs, J.R. (1971). Territory and breeding density in the Great Tit *Parus major* L. *Ecology*, **52**, 2–27.

Lovat, Lord (1911). *The grouse in health and disease.* Smith, Elder and Co, London.

MacKenzie, J.M.D. (1952). Fluctuations in the numbers of British tetraonids. *Journal of Animal Ecology*, **21**, 128–53.

Marquiss, M. (1977). The effects of post-hatching nutrition on the behaviour of red grouse in captivity. Unpublished D.Phil. thesis, University of Aberdeen.

May, R.M. and Anderson, R.M. (1978). Regulation and stability of host–parasite population interactions. II Destabilizing processes. *Journal of Animal Ecology*, **47**, 249–67.

Middleton, A.D. (1934). Periodic fluctuations in British game populations. *Journal of Animal Ecology*, **3**, 231–49.

Moore, J., Freehling, M., and Simberloff, D. (1986). Gastrointestinal helminths of northern Bobwhite in Florida: 1968 and 1983. *Journal of Wildlife Diseases,* **22,** 497–501.

Moran, P.A.P. (1952). The statistical analysis of gamebird records. *Journal of Animal Ecology,* **18,** 222–52.

Moss, R. and Watson, A. (1985). Adaptive value of spacing behaviour in population cycles of red grouse and other animals. In *Behavioural ecology. Ecological consequences of adaptive behaviour* (ed. R.M. Sibley and R.H. Smith), pp. 275–94. Blackwell Scientific Publications, Oxford.

Moss, R., Watson, A., and Parr, R. (1975). Maternal nutrition and breeding success in red grouse in relation to the age of the plant. *Journal of Applied Ecology,* **9,** 771–81.

Moss, R., Watson, A., and Rothery, P. (1984). Inherent changes in the body size, viability and behaviour of a fluctuating red grouse (*Lagopus lagopus scoticus*) population. *Journal of Animal Ecology,* **53,** 171–89.

Poole, R.W. (1974). *An introduction to quantitative ecology.* McGraw-Hill, Tokyo.

Portal, M. and Collinge, W.E. (1932). *Partridge disease and its causes.* Country Life, London.

Potts, G.R., Tapper, S.C., and Hudson, P.J. (1984). Population fluctuations in red grouse: analysis of bag records and a simulation model. *Journal of Animal Ecology,* **53,** 21–36.

Savory, C.J. (1978). Food consumption of red grouse in relation to the age and productivity of heather. *Journal of Animal Ecology,* **47,** 269–82.

Shaw, J.L. (1988). Arrested development of *Trichostrongylus tenuis* as third stage larvae in red grouse. *Research in Veterinary Science,* **45,** 256–8.

Stewart, R.E. and Aldrich, J.W. (1951). Removal and population of breeding birds in a spruce-fir forestry community. *Auk,* **68,** 471–82.

Watson, A. (1985). Social class, socially induced loss, recruitment and breeding of red grouse. *Oecologia,* **67,** 493–8.

Watson, A. and Jenkins, D. (1968). Experiments on population control by territorial behaviour in red grouse. *Journal of Animal Ecology,* **37,** 595–614.

Watson, A. and Moss, R. (1970). Dominance, spacing behaviour and aggression in relation to population limitation in vertebrates. In *Animal population in relation to their food resources,* Symposia of the British Ecological Society, no. 10. (ed. A. Watson), pp. 167–218. Blackwell Scientific Publications, Oxford.

Watson, A. and Moss, R. (1972). A current model of population dynamics in red grouse. *Proceedings of the International Ornithological Congress,* **15,** 134–49.

Watson, A., Moss, R., and Parr, R. (1984*a*). Effects of food enrichment on numbers and spacing behaviour of red grouse. *Journal of Animal Ecology,* **53,** 663–78.

Watson, A., Moss, R., Rothery, P., and Parr, R. (1984*b*). Demographic causes and predictive models of population fluctuations in red grouse. *Journal of Animal Ecology,* **53,** 639–62.

Watson, H. (1988). The ecology and pathology of *Trichostrongylus tenuis* (Nematoda), a parasite of red grouse (*Lagopus lagopus scoticus*). Unpublished Ph.D. thesis, University of Leeds.

Watson, H. and Hudson, P.J. (1988). The uptake of parasitic worms by red grouse. *Game Conservancy Annual Review,* **18,** 158–61.

Watson, H., Lee, D.L., and Hudson, P.J. (1987). The effect of *Trichostrongylus tenuis* on the caecal mucosa of young, old and antihelmintic-treated wild red grouse, *Lagopus lagopus scoticus. Parasitology,* **94,** 405–11.

Watson, H., Lee, D.L., and Hudson, P.J. (1988). Primary and secondary infections of the domestic chickens with *Trichostrongylus tenuis* (Nematoda), a parasite of red grouse, with observations on the effect on the caecal mucosa. *Parasitology*, **97**, 89–99.

Williams, J. (1985). The statistical analysis of fluctuations in red grouse bag data. *Oecologia*, **65**, 269–72.

Wilson, G.R. (1979). Effects of caecal threadworm *Trichostrongylus tenuis* on red grouse. Unpublished Ph.D. University of Aberdeen.

4 *The effects of ectoparasites on cliff swallow growth and survival*

BRIAN R. CHAPMAN and JOHN E. GEORGE

Introduction

Although there are many selective advantages associated with membership in an avian nesting colony, there are numerous disadvantages as well (Wittenberger and Hunt 1985). Among the 'costs' of coloniality is the inevitable increase of ectoparasite populations at a regularly reused colony site (Alexander 1974). Colonial aggregations of birds provide dense, reliable concentrations of hosts for ectoparasites (Rothschild and Clay 1952), which can affect development rates and mortality levels of nestling birds (Foster 1968; Moss and Camin 1970; Hoogland and Sherman 1976; Brown and Brown 1987). High ectoparasite populations have also resulted in nest abandonment and total colony desertion in a variety of species (Moss and Camin 1970; Chapman 1973; Converse *et al.* 1975; Feare 1976; King *et al.* 1977*a,b*; Samuel *et al.* 1982; Duffy 1983; Brown and Brown 1986; Emlen 1986).

Ticks have caused mortality in several avian species by affecting eyesight (Thomas 1941; Worth 1942) and general physical condition (Putzig 1939; Chapman 1973). Certain ectoparasites of domestic fowl produce an anaemic state in the host (Olson 1935, 1965). Jellison and Kohls (1938) suggested that ectoparasite-induced anaemia is not only an experimental disease, but occurs with some frequency in nature. It may be the immediate cause of death in some animals. However, in addition to the direct damage caused by the loss of blood, destruction of tissue, and allergic responses, ectoparasites may transmit epizootic diseases to their hosts (Philip *et al.* 1935; Locke *et al.* 1962; Moss 1972; Feare 1976; Clifford *et al.* 1980).

This study was undertaken to ascertain the effects that ectoparasites may have on cliff swallow (*Hirundo pyrrhonota*) productivity. Cliff swallows are small, migratory passerines that nest in colonies throughout much of western North America. Throughout their range, cliff swallows construct gourd-shaped nests of mud in dense aggregations on cliffs, under eaves, and beneath bridges and culverts (Emlen 1952). Its breeding biology has been well-documented (Emlen 1941, 1952, 1954; Mayhew 1958; Samuel 1971; Grant and Quay 1977; Withers 1977). The birds, which feed on insects caught in flight, are highly social in all of their activities including feeding, mud-gathering, loafing, and preening (Emlen 1952; Brown 1986).

Nesting activities within a colony are often synchronized (Chapman 1973; Brown and Brown 1986; Loye and Carroll, Chapter 12, this volume). Clutch size varies from three to five eggs and the eggs are incubated 12 to 16 days by both parents. After hatching, the altricial young are brooded at night for 5 to 10 days. The young fledge 22 to 25 days after hatching. The species is normally single-brooded, but second broods have been reported in some areas (McCann 1936; Gross 1942; Samuel 1971; Chapman 1973). Much of the observed nestling mortality results from haematophagous ectoparasites (Foster 1968; Chapman 1973; Brown and Brown 1986).

The nest surfaces and whatever cracks and crevices exist in the substrate provide ample habitat for the development of sizeable endemic ectoparasite populations. These haematophagous arthropods survive prolonged cliff swallow absences from colony sites by physiological or behavioural adjustments (Cook 1973) or by utilizing alternative hosts. Bats and other species of birds frequently occupy cliff swallows nests when the swallows are away on migration approximately 8 months each year and, therefore, may serve as alternate hosts which contribute to the maintenance of substantial ectoparasite densities from season to season (Cook 1972; Loye 1985*a*). Ectoparasite populations may survive in deserted swallow colonies for as long as 4 years (Loye and Hopla 1983).

Many species of haematophagous arthropods are commonly associated with cliff swallow colonies, including fleas (Eads 1956; Wheeler *et al.* 1970; Haas and Wilson 1979; Hopla and Loye 1983), cimicids (Myers 1928; Foster and Olkowski 1968; Chapman 1973; Smith and Eads 1978; Loye and Hopla 1983; Loye 1985*b*; Brown and Brown 1986), other insects (Hicks 1959), and acarines (Baerg 1944; Kohls and Ryckman 1962; Chapman 1973; Howell and Chapman 1976; Hopla and Loye 1983). In north-western Texas, four ectoparasite species commonly reach high densities in cliff swallow colonies: the swallow bug, *Oeciacus vicarius* (Hemiptera: Cimicidae); the swallow tick, *Argas cooleyi* (Acari: Argasidae); another tick, *Ornithodoros concanensis* (Acari: Argasidae); and the cliff swallow tick, *Ixodes baergi* (Acari: Ixodidae). Of these, only *Ixodes baergi* was not common in our study colonies.

The life cycles of these ectoparasites are closely coordinated with the 3- to 5-month cycle of swallow colony occupancy. The presence of cliff swallows at a colony site in some manner 'activates' the endemic ectoparasite populations (see Howell 1975; Webb *et al.* 1977; George 1987). After obtaining a blood meal, adult ticks and swallow bugs lay eggs on the nest surfaces, the substrate, or in substrate crevices. After hatching, swallow bugs develop through five nymphal stages before moulting to the adult stage (Myers 1928). Blood meals are obtained during each stage and the cycle may be completed during a swallow nesting season (Myers 1928; Loye 1985*b*). The adult females of both tick species present in our study colonies lay eggs after obtaining a blood meal. Larval and nymphal ticks overwinter with adults in substrate crevices and are available to parasitize swallows upon their arrival at a colony site (Hopla and Loye 1983; George 1987).

To ascertain the role that ectoparasites play in development and mortality among swallows, we compared the growth rates, blood cytology, and mortality rates of nestling cliff swallows in colonies treated with a pesticide to colonies in which the parasite populations were allowed to develop normally. We also speculated on the role that ectoparasites play in colony site abandonment. In our study area, cliff swallows frequently abandon colony locations after 2 or 3 years of consecutive usage. These abandonments may be related to the development of sizeable parasite populations (Grinnell *et al.* 1930; Grinnell 1937; Sikes and Arnold 1984; Brown 1985; Emlen 1986; Loye and Carroll, Chapter 12, this volume).

Study area and procedures

The six cliff swallow colonies studied from 15th March 1971 to 10 August 1973 were located in the canyons or under highway culverts associated with the Double Mountain Fork of the Brazos River or its tributaries near Justiceburg, Garza County, Texas. Three colonies 0.3 km south of the river on the sandstone cliffs of Rocky Creek were designated RC-1, RC-2, and RC-3, respectively, approaching the river. Two sites, one hereafter called 'the Point' and the other 'the Cliff', were located on the cliffs along the river itself. A sixth study colony was located under a double-bent culvert on a highway (Texas FM 2458) approximately 0.6 km E of Justiceburg. Two Rocky Creek colonies, RC-1 and RC-2, were studied in 1971; all of the other colonies were examined in 1972. All of the study colonies were accessible with ladders and were of approximately the same population size (300–500 nests). No other species of birds nested in these colonies during the study.

One cliff site along the Brazos River and one bent of the culvert were selected as treatment sites. These were sprayed twice weekly with Dibrom^{R}, a short-life acaricide that is not toxic to avian hosts (Moss and Camin 1970). The acaricide was diluted to 1 part per 170 parts water and was applied as a fine mist on to the nests and nest substrate with an air-pressure type garden sprayer from a distance of about 1 m. The nests and

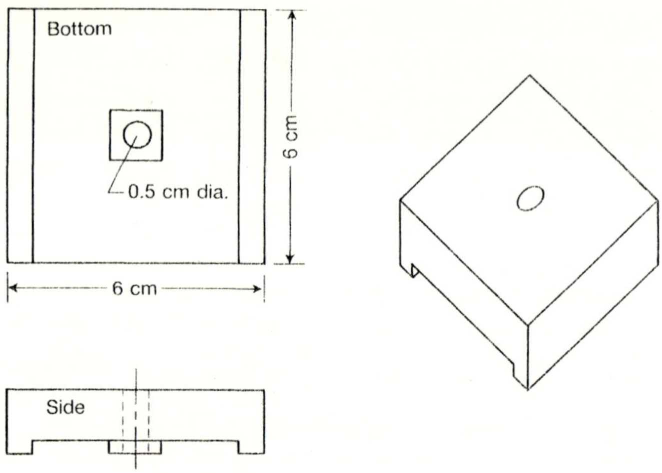

Fig. 4.1. Artificial ectoparasite harbourage. Attached to cliff with bolt.

surrounding substrate were lightly moistened externally with the acaricide. This treatment began 1 week after approximately 90 per cent of the occupied nests in a colony contained young and was continued until all young had fledged. 'Stickem' (polymerized butene, isobutene, and butane; Michel and Pelton Co, Emeryville, California) was smeared around the circumference of the treatment colonies to hinder immigration of ectoparasites.

No method for accurately estimating total ectoparasite populations at a given site was ever devised, but artificial ectoparasite harbourages used in 1972 provided a comparative basis to index tick and swallow bug populations. Ten plaster harbourages (Fig. 4.1) were attached with bolts to the cliff face or with tape to the culvert surface 10 cm below each study colony. The recessed area in the harbourage provided a dark crevice into which ectoparasites moved after obtaining a blood meal and leaving the nests. The harbourages were removed daily and counts made of each species per trap. After the count, the ectoparasites were released near the base of the cliff. Harbourages were placed on the cliff face immediately following cliff swallow occupancy of a site and remained there for 2 months after the site was abandoned. Harbourages were placed beneath the nests in the culvert before the swallows arrived.

In each cliff colony, 80 nests were individually marked with spray enamel in a double colour code. The marked nests were adjacent to each other in a strip transect three to five nests deep from the periphery to the centre of the colony. All of the nests in each bent of the culvert were identified by numbers written on the culvert wall beneath each nest. The nests were inspected daily and each egg was removed and numbered as it appeared. A spoon with a long, bent handle was used to remove the eggs and nestlings of each nest after their position in the nest had been observed using a dental mirror and a flashlight. No alteration of the nest entrance was necessary to observe or remove the inhabitants. Coloured cloth-tape, tarsal bands were used to individually identify the young. Prior to fledging, serially numbered metal US Fish and Wildlife Service bands were substituted for the cloth bands. On alternate days, each nestling was removed from a designated set of 50 marked nests and weighed to an accuracy of 0.1 g, selected feathers were measured to the nearest mm, and notes were taken on the development of feather tracts and other plumage and developmental characteristics.

Blood samples were taken from nestlings in the 30 marked nests not used in the weight-growth trials every fourth day and analysed in several ways. Haemoglobin concentration was determined with a Spencer Serial 1010D Hemoglobinometer and measured in g haemoglobin blood per 100 ml. Three trials were made per drop of blood and the mean of the trials recorded. A second drop of blood was utilized to make a thin smear on a glass slide, dried, and later stained with Wright's staining solution. A third portion of blood was collected in a heparinized micro-haematocrit tube, sealed with Plasticene, and stored for a short time in a container surrounded with ice. The tubes were spun for 5 min at 11 500 r.p.m. in an International Micro-capillary Centrifuge before measuring the haematocrit.

Intercolony associations of cliff swallows were also located and examined for free ectoparasites in an attempt to assess intercolony parasite transfer. Ten birds each, from three study colonies, were removed from their nests at night. The rectrices of each bird were sprayed lightly with a high visibility enamel; the tail colour for birds of each colony was different. The movements of these birds were followed to determine if birds from different colonies occupy the same foraging areas, mud-gathering sites, or loafing and dusting points. The mud-gathering and dusting sites were examined visually by one of us (BRC) on hands and knees using a × 10 hand lens immediately after cliff swallow occupancy. Ectoparasites were identified and counted at each site.

Results

Colony site selection

The first cliff swallows to appear in the Justiceburg area typically arrived in mid-March in groups of about three or four birds. These small groups apparently were the precursors of larger distinct flocks. The movements

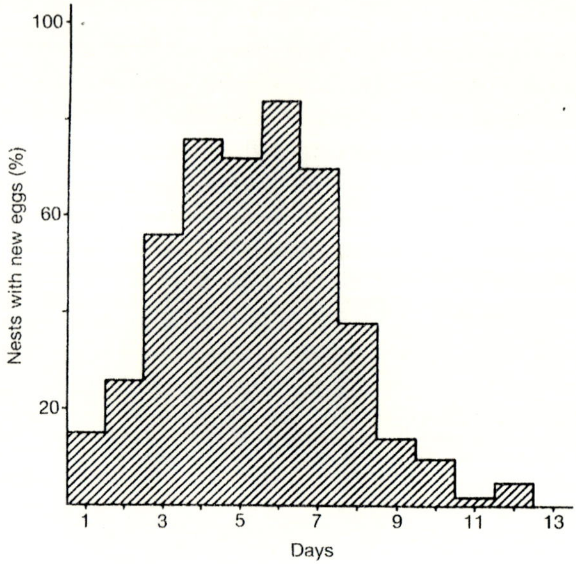

Fig. 4.2. Daily percentage of nests with new eggs. Data from all study colonies were pooled.

of each flock for a period of about a week became concentrated around one or more potential nesting sites. Just prior to site selection, a flock flew for several hours a day in a tight vortex about a prospective site and periodically landed on extant nests and nest remnants. Some birds entered the nests. This behaviour was most pronounced in the evening and the birds roosted in the immediate area at night. Some may have roosted in the nests. Copulation took place during these initial stages of nest occupancy and territory formation. Nest construction occurred for approximately a week.

Colony selection behaviour and subsequent nesting activities also occurred in mid-June. These colonies were initiated after most of the swallow colonies that were established in March had completed their nesting cycle. Whereas most, but not all, of the March-started colonies were situated on cliffs, all of the June colonies were located under bridges and culverts. These late-season colonies were active until young swallows fledged in mid- to late-August.

Intracolony synchrony

Egg laying data for the six Justiceburg study sites were pooled (Fig. 4.2) to determine the degree of laying synchrony. Regardless of the date, the day that the first egg was laid in a colony was recorded as day 1 of the laying period. The number of marked nests having new eggs each day after laying started was noted. Most of the active nests (84 to 92 per cent) in each colony contained their first egg within five days of each other. The majority of eggs (74.6 per cent) were laid within a 5-day period (days 3–7 of the laying period) in each colony. These data indicate a high degree of intracolony synchrony. Apparently, swallows that arrived in the area after a colony had been established at a particular site joined a colony formation flock at a nearby site.

In the vicinity of our study (10 km radius) we counted nine cliff swallow colony sites located under bridges and culverts. Numerous other sites were available on the sandstone cliffs along the river and its tributaries. For example, 22 sites were located along the 6 km stretch of river between Justiceburg and our study site. Nine sites were situated within 1 km of the 'Point' colony.

Annual pattern of colony occupation

Most of the mud nests in a cliff swallow colony remain intact at a colony site for several years even if swallows do not use the nests or repair them. The colony sites designated RC-1 and RC-2 were occupied by swallows only in 1971 and were not used in 1970, 1972, or 1973. The 'Cliff' and RC-3 were active in 1972 and were not used in 1970, 1971, or 1973. The 'Point' was inhabited by a small colony in 1971 (52 of 479 nests were occupied) and was fully occupied in 1972 and 1973. Both culvert bends were utilized in all 3 years of the study.

Ectoparasite populations

Only a few active ectoparasites were observed on the nests or the nest substrate prior to the arrival of the swallows. Early in the nest construction phase, paired birds began to occupy the nests at night. Egg laying and incubation were usually initiated just prior to nest completion. As soon as the swallows began to occupy the nests, the number of ectoparasites active about the nests increased dramatically.

Ectoparasite populations, measured using the artificial harbourages, increased with cliff swallow tenure at the culvert (Fig. 4.3). A similar pattern was noted at the cliff sites, but data from the early stages of nesting were not available. The accumulation of ectoparasites about the nests increased to a peak that corresponded temporally to the fledging period. The initial increases in ectoparasite density were probably a function of increased activity of endemic ectoparasites responding to host availability followed by further increases from reproduction. Large numbers of larval and nymphal ectoparasites were noted later in the swallow nesting season.

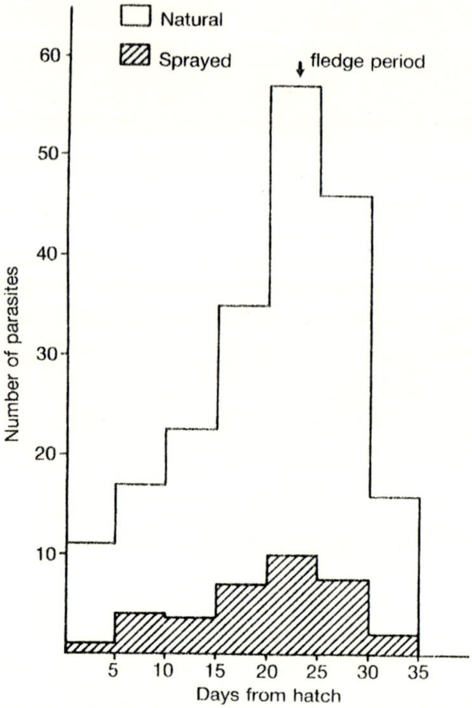

Fig. 4.3. Average number of ectoparasites per harbourage in each 5-day increment of the brooding season. Data from the two culvert bends are utilized for the comparison.

Table 4.1. Ectoparasite populations at the cliff swallow study colonies (1972)

Colony	Trap-nights	Total ectoparasites	Ectoparasites per trap-night
Natural	108	1171	10.84*
Point	38	849	22.34
Rocky Creek 3	24	118	4.92
Culvert 1	46	204	4.43
Sprayed	76	53	0.70*
Cliff	32	14	0.44
Culvert 2	44	39	0.88

* $\chi^2 = 10.14$ ($p < 0.01$) 1 *df*.

A comparison of ectoparasite populations in the natural and chemically treated sites is shown in Table 4.1. The number of ectoparasites per trap per night was calculated to determine an ectoparasite index. These data were subjected to a chi-square test for independence which indicated that there was a significant ($p < 0.01$) difference in the ectoparasite aggregations.

The swallow bug was the most abundant of the three dominant ectoparasite species at the 1972 study colonies (Table 4.2), but our visual observations indicate that the bugs may have been underrepresented in the artificial harbourages. Of the ticks *A. cooleyi* occurred in the highest numbers at the 'Point' colony. This tick appeared to be particularly sensitive to the Dibrom treatment. After spraying, only a few swallow bugs and *O. concanensis* were observed, but *A. cooleyi* were absent. Aggregations of *O. vicarius* were significantly reduced by the spray, but their numbers tended to increase during the periods between pesticide applications.

A total of 29 ectoparasites were found when nine mud-gathering sites were examined. Of these 24 were swallow bugs, three were *A. cooleyi*, and two were *O. concanensis*. Marked birds from all three colonies were observed at two of the sites. No ectoparasites were found at three dusting and four loafing sites.

Clutch size and hatching success

A single egg was laid on each consecutive day until clutch completion. Clutch size varied from three ($n=6$) to five ($n=4$) with most active marked nests ($n=236$) having four eggs. Intermittent incubation began after the first egg was laid. Both the male and female spent the night in the nest and may have alternated during the day to incubate the eggs. The incubation period, measured from the day the last egg was laid until all eggs hatched, varied from 15 to 16 days, 15 days being the mode (97.3 per cent).

The sequence of hatching was determined by oviposition sequence. The first egg laid hatched up to 72 hours before the last egg. The first and second egg laid hatched within the first 24-hour period; the third and fourth egg laid hatched within a second 48-hour period.

Table 4.2. Comparative population indexes for the three dominant ectoparasite species (1972)

Colony	Index of abundance		
	Argas cooleyi	*Ornithodors concanensis*	*Oeciacus vicarius*
Natural	4.53	0.15	6.17
Point	12.26	0.37	9.71
Rocky Creek-3	0.79	0.04	4.08
Culvert 1	0.09	0.02	4.33
Sprayed	0.00	0.04	0.66
Cliff	0.00	0.06	0.38
Culvert 2	0.00	0.02	0.86

Nestling growth and development

Fledging period and success

The period from hatching until leaving the nest averaged 25.3 ± 4.2 days for 82 cliff swallow broods in colonies treated with the acaricide. The fledging period was shorter (23.2 ± 3.4 days) for 106 broods in parasitized colonies.

Fledging success, the percentage of hatchlings to successfully leave the nest, was calculated for each of the Justiceburg colonies (Table 4.3). The lowest fledging success (14.8 per cent) was experienced by a Rocky Creek colony. The Rocky Creek colonies may have experienced much higher ectoparasite densities, especially of *Oeciacus vicarius*, than any other colonies studied, but no data were obtained to substantiate this observation. A chi-square test of independence compared the percentage of nestlings successfully fledging from unsprayed versus treatment colonies; a significantly ($p < 0.001$) higher level of success occurred in the sprayed colonies ($\chi^2 = 108.83$).

Table 4.3. Per cent fledging success of cliff swallows in the study colonies

Colony	Number of nestlings	Number that fledged	Fledging success (per cent)
Natural	338	182	53.85*
Point	88	59	67.04
RC-1	61	15	24.59
RC-2	54	8	14.82
RC-3	76	53	69.73
Culvert 1	59	47	79.66
Sprayed	252	236	89.47*
Cliff	167	160	89.55
Culvert 2	85	76	89.41

* $\chi^2 = 108.83$ ($p < 0.001$) 1 *df.*

Weight

At hatching, or within 6 hours thereafter, the weights of young cliff swallows varied from 1.8 to 2.1 g ($\bar{Y} = 1.0$, SD $= 0.71$, $n = 186$). Nestlings showed an early, rapid growth phase which reached a maximum on about the twelfth day after hatching (Fig. 4.4). After a peak was achieved, a gradual loss of weight occurred until fledging in both parasitized and treated colonies.

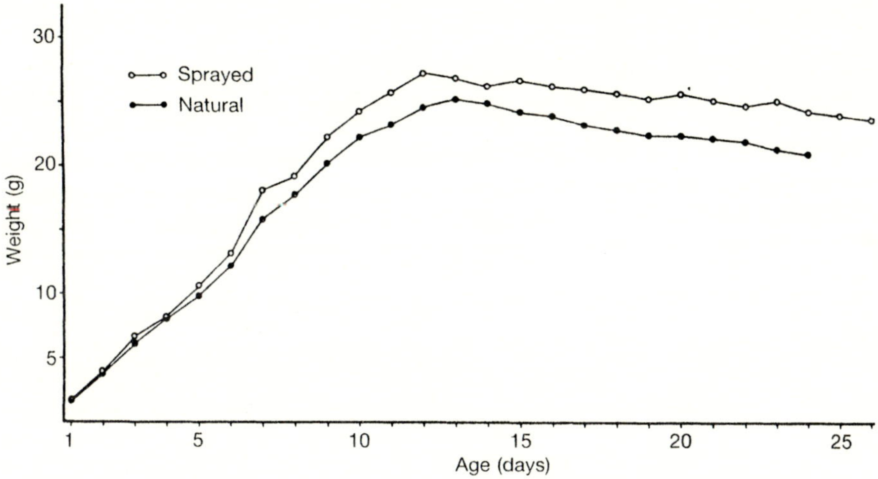

Fig. 4.4. Weight growth curve based on means for nestling cliff swallows in sprayed and parasitized colonies.

Nestlings exposed to ectoparasites did not achieve as high a maximum weight as birds from treated colonies. Disparity in weight continued throughout the brooding period. Ectoparasitized young fledged lighter than parasite-free nestlings. Because the sample sizes from day 24 to fledging were small, day 23 of nestling age was chosen for a statistical comparison of nestling body weight near the fledging period. The weight at day 23 of nestlings in sprayed ($\bar{Y} = 25.0$, SD $= 2.13$, $n = 41$) and parasitized ($\bar{Y} = 21.3$, SD $= 2.78$, $n = 58$) colonies was compared using a single classification analysis of variance; the difference was highly significant ($F = 14.6$; $p < 0.001$).

Hatch sequence effects

Weight curves were plotted for nestlings on the basis of individual position in the hatch sequence of each nest (Fig. 4.5). The first young to hatch maintained a developmental advantage over its cohort throughout the brooding period. Correspondingly, the second nestling hatched weighed more than the third and fourth; the third was heavier than the fourth.

Hatching sequence influenced the rate of weight gain in nestling cliff swallows. A higher mortality rate was noted for late hatching individuals in both treated and parasitized colonies (Table 4.4). Chi-square tests of independence were employed to determine if there was a significant difference in mortality before fledging between nestlings of corresponding hatching sequence. The differences in mortality between early-hatching

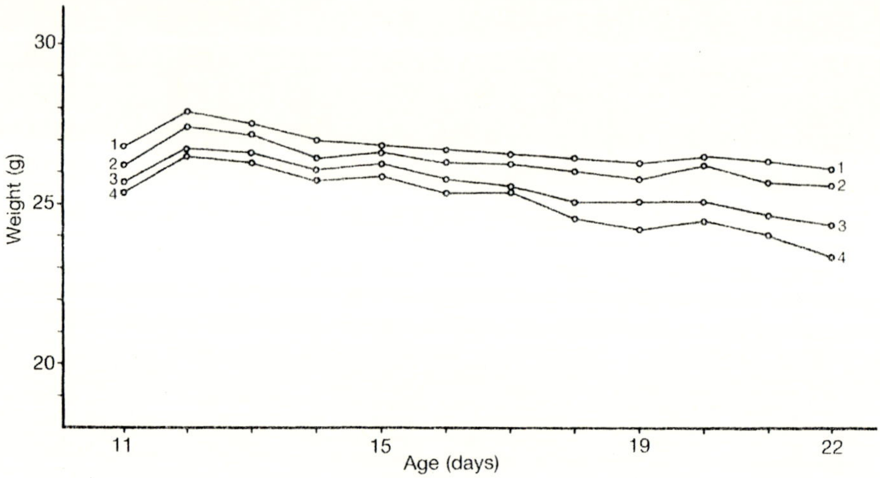

Fig. 4.5. Weight growth curve for nestlings based on hatching sequence (line 1 = first hatched bird in clutch; line 2, second hatched bird; etc.). Data were pooled from the sprayed colonies. Time of hatching is plotted as day 1 to facilitate weight comparisons at common ages. Weights at days 11–23 are shown; prior to day 11 overlap of individual weights prevented graphic clarity.

Table 4.4 Cliff swallow mortality according to hatching sequence in parasitized and sprayed colonies

Hatching sequence	Mortality (per cent)		χ^2
	Natural	Sprayed	
1	5.2	4.1	0.23 (NS)
2	8.8	5.4	1.29 (NS)
3	21.1	8.2	7.84*
4	45.6	11.3	25.76†

NS not significant.
* $p<0.01$, 1 *df*.
† $p<0.005$, 1 *df*.

young were not statistically significant. However, significant differences ($p<0.01$) were obtained between mortality rates in late-hatching cohort members in both sprayed and natural colonies.

Mortality tended to increase among late-hatching young in both colonies. To test the trends, RXC tests of independence using the *G*-statistic were employed. In the ectoparasitized colonies a *G*-value of 11.977 was obtained. The *G*-value was significant ($p<0.01$) indicating that hatching sequence did influence nestling mortality. In the treated colonies a *G*-value of 0.751 was obtained which was not significant.

Primaries 1 and 9

The primary feathers emerged as minute rounded points on the fourth day. The innermost and outermost primaries, 1 and 9 respectively, had almost the same growth rate, approximately 2.8 mm per day until day 13. Thereafter, the growth rate of the outermost primary increased to about 3.7 mm per day; the innermost primary growth rate decreased to 1.2 mm per day.

When mean daily growth increments for primary 1 were compared, consistent and increasing differences were noted between birds from natural and treated colonies. The length of the inner primary at day 23 was subjected to a single classification analysis of variance (Table 4.5).

Table 4.5. A comparison of selected feather lengths in parasitized and colonies of cliff swallows at day 23 of nestling age. All data from the Point and sprayed cliff colony sites

Feather	Sprayed			Natural			Percentage difference
	Y	*SD*	*n*	Y	*SD*	*n*	
Primary 1	49.3	1.91	41	46.1	2.13	58	6.5
Primary 9	74.5	1.20	41	73.2	1.85	58	1.8
Rectrice 5	42.2	0.96	37	39.8	1.26	43	5.7

There was a significant difference ($F = 2.58$, $p < 0.001$) in fledging length of primary 1 between the treated and natural colonies. A similar comparison of primary 9 also revealed consistent differences in their mean lengths ($F = 2.01$, $p < 0.01$).

Most (84.3 per cent) of the nestlings in the treated colonies remained in the nest longer before they fledged. Therefore, their primary feathers had 1 or 2 days of extra growth before they began to fly. If the additional time for feather growth was considered, the inner primary was an average of 7.0 mm longer and the outer primary 4.1 mm longer at fledging in treated than in natural colonies.

Rectrices

Both the inner and outer rectrices were measured. However, due to the similarity of growth rates between the two feathers, only the outermost rectrix, rectrix 5, was utilized for statistical analysis. Rectrix 5 appeared as a small dark knob on the fourth day of nestling development. Growth of the feather was slow until day 7 when it typically attained a length of 1 mm. After day 7, the growth rate was 2.5 mm per day in the treated colonies and 2.4 mm per day in natural colonies. A 3.6-mm difference in mean length of the retrix at fledging was noted between natural and treated colonies. A single classification analysis of variance revealed a significant difference ($F = 2.37$, $p < 0.001$) between the lengths at day 23 in the two types of colonies (Table 4.5).

Tarsometatarsus and ulna

The average tarsometatarsus length was 3.1 mm ($n = 98$) on the day of hatching. The growth rate showed great variation within each colony. On the fifth day of nestling growth, the mean tarsometatarsus length in treatment colonies was 7.8 mm (range 7–9 mm); the length in parasitized colonies was 7.6 mm (range 7–10 mm). The maximum tarsometatarsus length of 11 mm was attained on day 11 or 12 and did not increase beyond that point. Until day 12 mean tarsometatarsus lengths were consistently lower in ectoparasitized colonies. However, the degree of variation obscured any possible statistically significant differences and tarsometatarsus length at fledging was equal in both types of colonies.

For 26 nestlings measured within 8 hours of hatching, the average ulna length was 5.0 mm. The growth rate of the ulna also varied. A maximum length of 26.1 mm was reached on day 20. No difference occurred between the growth rate of the ulna in treated and parasitized colonies.

Blood studies

Because of the haematophagous habits of cliff swallow ectoparasites, the blood of nestlings from natural colonies was compared to that of sprayed colonies to determine if quantitative differences occurred in the concentration of haemoglobin, haematocrit, and cellular elements.

The concentration of haemoglobin in the blood of 21-day-old nestling cliff swallows, expressed as g haemoglobin per 100 ml blood, averaged 10.8 for 40 nestlings in colonies exposed to ectoparasites. The mean value for 46 young in sprayed colonies was 14.6, a significant difference ($F = 1.98$. $p < 0.005$) from haemoglobin concentrations in parasitized nestlings. The mean per cent haematocrit was 51.5 for 77 nestlings in sprayed colonies versus 47.3 for 67 young in parasitized colonies, a significant difference in haematocrit value ($F = 1.67, p < 0.005$). A gradual increase in haematocrit values with age was noted during the fledging period. The increase was most obvious toward the end of the brooding period. A two-way analysis of variance with replication was utilized to compare the influence of ectoparasites on haematocrit in 16-day-old and 21-day-old nestlings; the difference in haematocrit due to age was significant ($F = 2.69$, $p < 0.001$). Significant differences in haematocrit were also associated with ectoparasite levels in the colonies ($F = 3.62$, $p < 0.025$). The age, or stage of development, interacted ($F = 4.94$,

$p < 0.25$) with ectoparasite density. The haematocrit of nestling cliff swallows increased with age, but was suppressed by ectoparasite density within the colony.

A total of 156 dried blood smears was read and the number of each cell type mm^2 in sprayed and parasitized colonies was calculated (Table 4.6).

Table 4.6. A comparison of the cellular elements of nestling cliff swallow blood per mm^2 in parasitized and sprayed colonies

Cell type	Cells per mm^2		
	Sprayed	Natural	χ^2
Erythrocytes	11 697	10 602	26.88†
Leucocytes (total)	314	493	21.96†
Lymphocytes	105	213	18.29†
Monocytes	104	135	1.91 (NS)
Thrombocytes	46	62	1.17 (NS)
Polymorphonuclear leucocytes (total)	68	106	4.15*
Heterophil	33	50	1.76 (NS)
Eosinophil	7	28	5.83*
Basophil	19	5	4.03*
Sample size	67	91	

NS, not significant.
* $p < 0.05$.
† $p < 0.005$.

Nestlings from the sprayed colonies characteristically possessed more erythrocytes and fewer of each leucocyte type than their counterparts from ectoparasitized colonies. Significant differences based on chi-square tests of goodness of fit ($p < 0.005$) were obtained in the number of erythrocytes, total leucocytes, and lymphocytes in the two types of colonies. The peripheral blood of the nestlings also exhibited significant differences ($p < 0.05$) in the total number of eosinophils and basophils.

Mortality

Mortality was calculated for each colony. The data were grouped according to the treatment the colonies received and are summarized in Table 4.7. A highly significant difference ($p < 0.005$) based on chi-square tests of goodness of fit indicated that increased nestling mortality was induced by the ectoparasites in the untreated sites. Therefore, productivity was adversely affected at those sites.

A 2×2 chi-square test of independence was also employed to determine if brood production during the second half of the summer

Table 4.7. Total mortality in the parasitized and sprayed colonies of cliff swallow

Colony	Number of eggs	Number fledged	Percentage mortality
Natural	755	182	75.9*
Point	150	59	60.7
RC-1	147	15	89.9
RC-2	109	8	92.7
RC-3	167	53	68.3
Culvert 1	58	47	18.9
Sprayed	260	236	9.2
Cliff	182	160	12.1
Culvert 2	80	76	5.0

* Chi-square sprayed versus parasitized, $\chi^2 = 76.69$ ($p < 0.005$) 1 *df*.

compensated for low productivity in the first half. A significant difference ($p < 0.005$) in mortality was obtained between production during the first and second portions of the summer in parasitized sites. The difference in mortality between the Point, which had the lowest mortality among parasitized colonies (60.7 per cent), and culvert 1 (18.9 per cent), suggested that the trend among the Justiceburg cliff swallows to use man-made structures for nesting during the second part of the summer aided in compensating for high early summer mortality.

Discussion

Throughout the swallow nesting period, the abundance of haematophagous ectoparasites increased dramatically. These increases in ectoparasite density resulted primarily from insect and tick reproductive efforts and emergence of the endemic population from substrate harbourages. The occurrence of ticks and swallow bugs at intercolony mud-gathering sites suggests that limited exchanges of ectoparasites between colonies may occur (Foster and Olkowski 1968). However, there is probably little net increase of ectoparasites at any colony site due to this type of exchange. Ectoparasites transferred in this manner may provide a newly established swallow colony site with a pioneer population of haematophagous arthropods.

The impact of ectoparasites on the development of nestling cliff swallows was clearly demonstrated. The irritation caused by the ticks and swallow bugs undoubtedly contributed to the shortened fledging period seen in untreated colonies (see Loye and Carroll, Chapter 12, this volume). Similar early-fledging, accompanied by large populations of the mite, *Ornithonyssus sylviarum*, caused significant mortality in phoebes

(Klaas, in Moss and Camin 1970). The survival potential of prematurely fledged young is lower than in those birds remaining in the nest until maximum development is achieved (Royama 1966).

The weight growth curve for nestling cliff swallows was similar to that reported by Stoner (1945). Weight recession prior to fledging is characteristic of swallows, swifts, and oceanic species; it is related to reduction in the proportion of body water in preparation for flight (Ricklefs 1968). Differences in daily weights of nestlings between the treated and untreated colonies were good indicators of ectoparasite influence on nestling weights. A positive relationship between maximum nestling weight and survival after fledging as shown for other species (Perrins 1965; Moss and Camin 1970; Brown and Brown 1986) seems likely in cliff swallows, although we lack direct evidence to support this contention.

Daily measurements of the inner and outer primaries and rectrices also reflected the influence of ectoparasites. The lengths of the flight feathers were consistently lower in the nestlings from unsprayed colonies. This was regarded as evidence that blood loss due to ectoparasites produced energy and nutrient drains on the young swallows. Nutrients that would normally contribute to feather growth were presumably diverted into blood maintenance and infection inhibition. If so, nestlings from parasitized colonies may have been less aerodynamically prepared for flight when they fledged and for this reason had a lowered survival potential. The disparity in hatching sequence among brood cohorts resulted in a behavioural and developmental lag among the late-hatching nestlings. Early-hatched young tended to dominate the nest entrance (B. R. Chapman, unpublished manuscript) and probably obtained more food than their younger nest mates. This further increased the weight difference within a brood. Developmental difference among a nest cohort resulted in a slight trend toward higher mortality among late-hatched nestlings in sprayed colonies, but the trend was not statistically significant. However, the mortality increments were statistically significant in ectoparasitized colonies. These differences were attributed to the ticks and swallow bugs. Apparently, the smaller nestlings of a brood were generally weaker and affected to a greater degree by blood loss than were the early-hatched, heavier young in a nest. This effect was accentuated because late-hatched nestlings were in the nest longer, and thus were exposed to higher ectoparasite concentrations during the latter stages of brooding than were early-hatched young.

That heavy infections of haematophagous arthropods may seriously weaken or kill avian hosts has been noted by several authors (Herman 1937; Putzig 1939; Boyd 1951; King *et al.* 1977*a*; Brown 1985). Such ectoparasites may induce weight loss, oedema, hyperaemia, or introduce bacterial infections (Jellison and Kohls 1938; Smith and Cheatum 1944; Feare 1976), although Baerg (1944) was unable to recognize any of the above conditions in nestling cliff swallows. However, anaemia was consistently prevalent among the ectoparasitized nestlings at Justiceburg. The significant differences in haemoglobin concentration, haematocrit, and erythrocyte counts were the most direct measurements of the influence of the haematophagous ectoparasites. Leucocytosis, which may indicate bacterial infections, was common among the nestlings in parasitized colonies. Significant increases of lymphocytes and polymorphonuclear leucocytes in the blood of parasitized nestlings indicate tissue inflammation and infection (Lucas and Jamroz 1961). Heterophils may increase with parasitic infestations and are found in the tissues in certain allergic states (Olson 1965). Thus, infections and severe anaemia may have resulted in nestling mortality in the parasitized cliff swallow colonies at Justiceburg.

All other factors being approximately equal, total mortality was greatest among nestlings in ectoparasitized colonies (75.9 versus 9.2 per cent). Thus, seemingly normal concentrations of ticks and swallow bugs have an appreciable effect on cliff swallow productivity.

Complex social interactions are involved in the selection of nesting sites and breeding among colonial birds (Darling 1938; Alexander 1974; Wittenberger and Hunt 1985). Emlen and Demong (1975) and Brown and Brown (1986) have suggested that members of a breeding colony may derive beneficial information regarding the location of food resources and Hoogland and Sherman (1976) demonstrated that colony members benefited from increased predator detection. However important these advantages accruing from group membership are, the colony members must also contend with at least one major disadvantage—ectoparasite-induced brood mortality, nest disruption, and reduced fitness.

The ectoparasites associated with cliff swallow colonies are so specific in their habits, it is likely that the association between the ectoparasites and the hosts has a long evolutionary history (Myers 1928; Eads 1956; Loye and Hopla 1983; George 1987). Cliff swallows have undoubtedly developed strategies to counteract the effectiveness of the ectoparasites. In north-western Texas the cliff swallows employ alternate colony site usage patterns, intracolony nest synchrony, and second brood rearing as mechanisms to overcome the effects of ectoparasites. Alternate colony site usage may have evolved as a result of population increases among endemic ectoparasites. Loye and Carroll (Chapter 12, this volume) have suggested that cliff swallows evaluate the density of ectoparasites clustered around nest entrances at potential colony sites during the initial phases of colony site selection. Such potential interactions between other species of swallows and ectoparasites have also been noted by Rothschild and Clay (1952), Hoogland and Sherman (1976), and others. Annual utilization of the same site would be advantageous from the standpoint of nest construction, but continual inhabitation would allow the ectoparasite population to increase to the point where the parasites might cause total or excessive brood mortality. Moss and Camin (1970) demonstrated that annual usage of purple martin colonies increased ectoparasite-induced mortality. Populations of cliff swallows, which have developed alternate site selection patterns, may have a selective advantage over swallow populations that do not alternate sites due to the higher survival potential of the offspring of alternate-nesters (Brown and Brown 1986; Loye and Carroll, Chapter 12, this volume).

Similarly, the evolution of second-brood rearing in cliff swallows may have been influenced by ectoparasite-reduced productivity in initial broods. Individuals suffering heavy first-brood mortality and then successfully renesting within the same season would have a selective advantage; individuals not renesting probably dwindled away. Driven below a critical minimum population size, social facilitation in cliff swallows seems to cease and reproduction may fail (Emlen 1952).

There is some dispute concerning renesting and second-brood rearing in cliff swallows. It seems likely that two broods were produced at Justiceburg because there were two distinctly sequential periods of brood rearing by cliff swallows. Fledging and colony abandonment of the early summer brood was followed within a week by nest site selection and subsequent brood rearing. The temporal coordination between first-brood fledging and the initiation of a second reproductive cycle could hardly be due to chance and second broods were probably reared by some of the same adults that produced the first. No difference was noted between average clutch sizes in early and late summer broods.

A distinct subpattern of swallow colony site usage was superimposed upon the two brood sequence. Not only were the same cliff sites not used from year to year, but there was also a shift from predominantly cliff usage during first-brood production to entirely bridge and culvert usage for the second brood. Although it would be nice to be able to attribute this shift in site usage to the effects of ectoparasites, we cannot do so. Environmental factors such as temperature or humidity cycles that vary through the nesting season may exert an unknown influence. However, the abundance of ectoparasites at the unsprayed culvert colony was much lower than that observed in the unsprayed cliff colonies and the fledging success was higher.

All of the swallow colonies in our study exhibited a high degree of intracolony nest synchrony. We propose that reproductive synchrony within a colony reduces the loss of young to ectoparasites. Because ectoparasite concentrations continually increase during the swallow nesting cycle, late broods are exposed to much higher parasite densities throughout their development and have higher mortality rates (Brown and Brown 1986). Swallow aggregations exhibiting synchrony have a selective advantage in terms of brood production, that is, more young survive in synchronous colonies (Emlen and Demong 1975; Brown and Brown 1986; Loye and Carroll, Chapter 12, this volume). In this instance the influence of ectoparasites seems to outweigh other considerations such as benefits derived from group foraging (Emlen and Demong 1975; Brown and Brown 1986) because birds from many nearby colonies foraged together and utilized the same habitats over an extended nesting period.

Endemic populations of ectoparasites can result in severe limits on the reproductive success of their hosts and, thus, can affect host fitness. Parasite hosts must constantly evolve new strategies to adapt to the limitations imposed by the parasites. For some reason, these interactions have been largely overlooked by biologists who have examined factors regulating host populations. Our study demonstrates that haematophagous arthropods can be as important a factor in affecting host populations as the more commonly studied aspects such as food supply, space requirements or predation.

Future work

Although there are many studies that document the fact that parasites reduce host nest success or cause nest abandonment, few studies have provided data from controlled experiments. Olson (1935, 1965) conducted a series of experiments using domestic fowl (chickens), but these are among the few laboratory approaches to the study of the relationships between haematophagous arthropods and their avian hosts. Given the various responses of hosts to ectoparasites outlined by the chapters in this volume, it is apparent that some controlled experiments are in order.

Our study suggests that high densities of haematophagous arthropods create changes in the blood composition of young swallows. However, it is difficult to assess what these changes mean in relation to the physiological condition of the birds. Although we know that the cellular composition of the blood of a chick changes with age (Olson 1935. 1965; Lucas and Jamroz 1961), the extent of such changes (and the cellular composition of the blood of most adults) is not known for most species. Obviously, to more fully assess the impact of haematophagous arthropods on the development and potential survivability of avian chicks, much more information must be made available on the blood–cellular systems of the birds.

A major difficulty associated with field studies, aside from the near impossibility of quantifying ectoparasite abundance, is the interpretation of results when several different ectoparasite species are involved. Different parasite species probably cause different effects on the host. When the composition of the parasite guild varies from colony to colony, and perhaps from region to region within a colony, it is difficult to make comparisons with any degree of confidence. To what extent are the differences in nestling weight gain in our study and that by Loye and Carroll (Chapter 12, this volume) dependent upon differences in ectoparasite guild composition and species abundance?

To obtain a more complete understanding of the effects of ectoparasites on the growth and survival of young birds, evidence from laboratory experiments under controlled conditions must be compared to field studies. Loye and Carroll (Chapter 12, this volume) provide an excellent synopsis of questions that must also be answered with additional field studies.

Acknowledgements

This research was supported in part by Public Health Service Research Grant No. AI-09591-04 from the National Institutes of Allergy and Infectious Diseases. Funds were also provided by the Frank M. Chapman Memorial Fund, American Museum of Natural History; a Grant-in-Aid from The Society of the Sigma Xi; and The Graduate Student Development Fund, Texas Tech University.

Special acknowledgement is extended to Drs Bart Cook and Fred G. Howell for their encouragement and collaboration in the field. We are extremely grateful to Mr Walter Boren and his two sons, James and John, of Justiceburg, Texas, for the unselfish provision of access to their property and the interest they expressed. Dr Eric G. Bolen critically read the manuscript and provided helpful suggestions.

References

Alexander, R.D. (1974). The evolution of social behavior. *Annual Review of Ecology and Systematics*, **5**, 323–83.

Baerg, W.J. (1944). Ticks and other parasites attacking Northern cliff swallows. *Auk*, **61**, 413–14.

Boyd, E.M. (1951). The external parasites of birds: a review. *Wilson Bulletin*, **63**, 363–9.

Brown, C.R. (1985). The costs and benefits of coloniality in the cliff swallow. Unpublished D.Phil. thesis. Princeton University, Princeton, New Jersey.

Brown, C.R. (1986). Cliff swallow colonies as information centers. *Science*, **234**, 83–5.

Brown, C.R. and Brown, M.B. (1986). Ectoparasitism as a cost of coloniality in cliff swallows (*Hirundo pyrrhonota*). *Ecology*, **67**, 1206–18.

Brown, C.R. and Brown, M.B. (1987). Group living as an advantage in avoiding predators. *Behavioural ecology and sociobiology*, **21**, 97–107.

Chapman, B.R. (1973). The effects of nest ectoparasites on cliff swallow populations. Unpublished Ph.D. thesis. Texas Tech University, Lubbock, Texas.

Clifford, C.M., Hoogstraal, H., Radovsky, F.J., Stiller, D., and Keirans, J.E. (1980). *Ornithodoros* (*Alectorobius*) *amblus* (Acarina: Ixoidea: Argasidae): Identity, marine bird and human hosts, virus infections and distribution in Peru. *Journal of Parasitology*, **66**, 312–23.

Converse, J.D., Hoogstraal, H., Moussa, M.I., Feare, C.J., and Kaiser, M.N. (1975). Solado virus (Hughes group) from *Ornithodoros* (*Alectorobius*) *capensis* (Ixoidea: Argasidae) infesting sooty tern colonies in the Seychelles, Indian Ocean. *American Journal of Tropical Medicine and Hygiene*, **24**, 1010–18.

Cook, B. (1972). Hosts of *Argas cooleyi* and *Ornithodoros concanensis* (Acarina: Argasidae) in a cliff-face habitat. *Journal of Medical Entomology*, **9**, 315–17.

Cook, B. (1973). The effects of repeated desiccation and rehydration on lipid and haemoglobin concentrations in fasting *Ornithodoros concanensis* (Acarina: Argasidae). *Comparative Biochemistry and Physiology*, **44**, 1141–8.

Darling, F.F. (1938). *Birds flocks and the breeding cycle*. Cambridge University Press, Cambridge.

Duffy, D.C. (1983). The ecology of tick parasitism on densely nesting Peruvian seabirds. *Ecology*, **64**, 110–19.

Eads, R.B. (1956). Ectoparasites from swallow nests with the description of a new ceratophyllid flea. *Journal of Parasitology*, **42**, 73.

Emlen, J.T. Jr (1941). Cliff swallow colonies of the central Sacramento valley in 1941. *Condor*, **43**, 248.

Emlen, J.T. Jr (1952). Social behavior in nesting cliff swallows. *Condor*, **54**, 177–99.

Emlen, J.T. Jr (1954). Territory, nest building and pair formation in the cliff swallow. *Auk*, **71**, 16–35.

Emlen, J.T. (1986). Responses of breeding cliff swallows to nidicolous parasite infestations. *Condor*, **88**, 110–11.

Emlen, S.T. and Demong, N.J. (1975). Adaptive significance of synchronized breeding in a colonial bird: a new hypothesis. *Science*, **188**, 1029–31.

Feare, C.J. (1976). Desertion and abnormal development in a colony of sooty terns *Sterna fuscata* infested by virus-infected ticks. *Ibis*, **118**, 110–15.

Foster, W.A. (1968). Total brooding mortality in late-nesting cliff swallows. *Condor*, **70**, 275.

Foster, W.A. and Olkowski, W. (1968). The natural invasion of artificial cliff swallow nests by *Oeciacus vicarius* (Hemiptera, Cimicidae) and *Ceratophyllus petrochelidoni* (Siphonaptera, Ceratophyllidae). *Journal of Medical Entomology*, **5**, 488–91.

George, J.E. (1987). Field observations on the life cycle of *Ixodes baergi* and some seasonal and daily activity cycles of *Oeciacus vicarius* (Hemiptera: Cimicidae), *Argas cooleyi* (Acari: Argasidae), and *Ixodes baergi* (Acari: Ixodidae). *Journal of Medical Entomology*, **24**, 683–8.

Grant, G.S. and Quay, T.L. (1977). Breeding biology of cliff swallows in Virginia. *Wilson Bulletin*, **89**, 286–90.

Grinnell, J. (1937). The swallows at the life sciences building. *Condor*, **39**, 206–10.

Grinnell, J., Dixon, J.S., and Linsdale, J.M. (1930). Vertebrate natural history of a section of northern California through the Lassen Peak region. *University of California Publications in Zoology*, **35**, 1–594.

Gross, A.O. (1942). Northern cliff swallow. In *Life histories of North American flycatchers, larks, swallows and their allies*, US National Bulletin, no. 1979 (ed. A.C. Bent), pp. 463–89.

Haas, G.E. and Wilson, N. (1979). Fleas (*Siphonaptera*) from nests of the cliff swallow (*Petrochelidon pyrrhonota*) in Alaska. *Wasmann Journal of Biology*, **37**, 59–63.

Herman, C.M. (1937). Notes on hippoboscid flies. *Bird-Banding*, **8**, 161–6.

Hicks, E.A. (1959). *Checklist and bibliography on the occurrence of insects of birds' nests*. Iowa State University Press, Ames. (1962 Supplement.)

Hoogland, J.L. and Sherman, P.W. (1976). Advantages and disadvantages of bank swallow (*Riparia riparia*) coloniality. *Ecological Monographs*, **46**, 33–58.

Hopla, C. and Loye, J.E. (1983). The ectoparasites and microorganisms associated with cliff swallows in west-central Oklahoma. I. Ticks and fleas. *Bulletin of the Society of Vector Ecology*, **8**, 111–21.

Howell, F.G. (1975). The roles of host-related stimuli in the behavior of *Argas cooleyi* (Acarina: Argasidae). *Journal of Medical Entomology*, **11**, 715–23.

Howell, F.G. and Chapman, B.R. (1976). Acarines associated with cliff swallow communities in northwestern Texas. *Southwestern Naturalist*, **21**, 275–80.

Jellison, W.L. and Kohls, G.M. (1938). Tick–host anemia: A secondary anemia induced by *Dermacentor andersoni* Stiles. *Journal of Parasitology*, **24**, 143–54.

King, K.A., Blankenship, D.R., Paul, R.T., and Rice, R.C.A. (1977*a*). Ticks as a factor in the 1975 nesting failure of Texas Brown Pelicans. *Wilson Bulletin*, **89**, 157–158.

King, K.A., Keith, J.O., Mitchell, C.A. and Keirans, J.E. (1977*b*). Ticks as a factor in nest desertion of California Brown Pelicans. *Condor*, **79**, 507–9.

Kohls, G.M. and Ryckman, R.E. (1962). New distributional records of ticks associated with cliff swallows, *Petrochelidon* spp., in the United States. *Journal of Parasitology*, **48**, 507–8.

Locke, L.N., Scanlon, J.E., Byrne, R.J., and Knisley, J.O. Jr (1962). Occurrence of eastern encephalitis virus in House Sparrow. *Wilson Bulletin*, **74**, 263–6.

Loye, J.E. (1985*a*). Host-effects on feeding and survival of the polyphagous cliff swallow bug, *Oeciacus vicarius* (Hemiptera Cimicidae). *Bulletin of the Society of Vector Ecology*, **10**, 7–13.

Loye, J.E. (1985*b*). The life history and ecology of the cliff swallow bug, *Oeciacus vicarius* (Hemiptera: Cimicidae). *Cahiers Office de la Recherche Scientifique et Technique Outre Mer, Serie Entomologie Médicale et Parasitologie*, **23**, 133–9.

Loye, J.E. and Hopla, C.E. (1983). Ectoparasites and microorganisms associated with the cliff swallow in west-central Oklahoma. II. Life history patterns. *Bulletin of the Society of Vector Ecology*, **8**, 79–84.

Lucas, A.M. and Jamroz, C. (1961). Atlas of avian hematology. *Agricultural Monographs*, **25**, 10271.

Mayhew, W.W. (1958). The biology of the cliff swallow in California. *Condor*, **60**, 7–37.

McCann, H.D. (1936). A colony of cliff swallows in Chester County, Pennsylvania. *Auk*, **53**, 84–5.

Moss, W.W. (1972). Some ecological relationships of purple martins and their acarine ectoparasites. *Journal of Medical Entomology*, **9**, 599.

Moss, W.W. and Camin, J.H. (1970). Nest parasitism, productivity, and clutch size in purple martins. *Science*, **168**, 1000–3.

Myers, L.E. (1928). The American swallow bug, *Oeciacus vicarius* Horvath (Hemiptera: Cimicidae). *Parasitology*, **20**, 159–72.

Olson, C. (1935). The effect of certain ectoparasites on the cellular elements and hemoglobin of the blood of the domestic chicken. *Journal of the American Veterinary Medicine Association*, **87**, 559–561.

Olson, C. (1965). Avian hematology. In *Diseases of Poultry*, (5th edn) (ed. H.E. Biester and L.H. Schwarte), pp. 100–19. Iowa State University Press, Ames.

Perrins, C.M. (1965). Population fluctuations and clutch size in the great tit, *Parus major* L.Journal of Animal Ecology, **34**, 601–47.

Philip, C.B., Jellison, W.L., and Wilkins, H.F. (1935). Epizootic tick-borne tularemia in sheep in Montana. *Journal of the American Veterinary Medicine Association*, **86**, 726–30.

Putzig, P. (1939). Rauchschwalben-Sterben durch Zecken (Ixodidae) verursacht. *Vogelzug*, **10**, 25–7.

Ricklefs, R.E. (1968). Weight recession in nestling birds. *Auk*, **85**, 30–5.

Rothschild, M. and Clay, T. (1952). *Fleas, flukes and cuckoos: a study of bird parasites*. Collins, London.

Royama, T. (1966). Factors governing feeding rate, food requirements and brood size of nestling great tits, *Parus major*. *Ibis*, **108**, 313–47.

Samuel, D.E. (1971). The breeding biology of barn and cliff swallows in West Virginia. *Wilson Bulletin*, **83**, 284–301.

Samuel, W.M., Williams, E.S., and Rippin, A.B. (1982). Infestations of *Piagetiella peralis* (Mallophaga: Menoponidae) on juvenile white pelicans. *Canadian Journal of Zoology*, **60**, 951–3.

Sikes, P.J. and Arnold, K.A. (1984). Movement and mortality estimates of cliff swallows in Texas. *Wilson Bulletin*, **96**, 419–25.

Smith, G.C. and Eads, R.B. (1978). Field observations on the cliff swallow, *Petrochelidon pyrrhonota* (Vieillot), and the swallow bug, *Oeciacus vicarius* Horvath. *Journal of the Washington Academy of Sciences*, **68**, 23–6.

Smith, R.H. and Cheatum, E.L. (1944). Role of ticks in the decline of an insular cottontail population. *Journal of Wildlife Management*, **8**, 311–17.

Stoner, D. (1945). Temperature and growth studies of the northern cliff swallow. *Auk*, **62**, 207–16.

Thomas, R.H. (1941). Ticks affecting birds' eyesight. *Auk*, **58**, 590–1.

Webb, J.P. Jr, George, J.E., and Cook, B. (1977). Sound as a host-detection cue for the soft tick *Ornithodoros concanensis*. *Nature*, **265**, 443–4.

Wheeler, J.R., Forcum, D.L., and Rael, C.D. (1970). Two records of bird fleas for New Mexico. *Southwestern Naturalist*, **14**, 351–75.

Withers, P.C. (1977). Energetic aspects of reproduction by the cliff swallow. *Auk*, **94**, 718–25.

Wittenberger, J.F. and Hunt, G.L. Jr (1985). The adaptive significance of coloniality in birds. In *Avian biology*, Vol. 8 (ed. D.S. Farner, J.R. King, and K.C. Parkes), pp. 1–78. Academic Press, New York.

Worth, C.B. (1942). Ticks affecting birds' eyesight. *Auk*, **59**, 576–7.

5 *Philornis parasitism and nestling survival of the Puerto Rican sharp-shinned hawk*

CARLOS A. DELANNOY and ALEXANDER CRUZ

Introduction

Avian myiasis is the infestation of nestling and adult birds by dipterous fly larvae of the families Calliphoridae, Muscidae, and Neottiophilidae (Hicks 1962; Zumpt 1965; Rogers *et al.*, Chapter 7, this volume). Dipteran parasitism has been shown to contribute to the mortality of some nestling birds (Plath 1919; Zumpt 1965; Winterstein and Raitt 1983; Arendt 1985*a,b*). In the New World, many *Philornis* species (Muscid botflies) are parasitic on nestling and, to a lesser extent, adult birds in subtropical and tropical regions (e.g. Nielsen 1911; Dodge 1955, 1963; Aitken *et al.* 1958; Dodge and Aitken 1968; Smith 1968, 1980; Pont 1972; Hector 1982; Oniki 1983; Winterstein and Raitt 1983; Arendt 1985*a,b*; LaRue 1987; Snyder *et al.* 1987).

Philornis larvae penetrate the host integument, undergo a period of development and growth, and establish as third instar larvae between the dermis and the body musculature (Fig. 5.1(b),(c)). Movement into this location plus growth and development of the fly larvae, appears to be linked to the ingestion of host tissues including red blood cells, mononuclear cells which infiltrate from focal accumulations adjacent to the lesions, and necrotic cellular debris which accumulate in the lesion (Nielsen 1911; Uhazy and Arendt 1986). The larvae develop just under the skin of the host until they are about 2 cm long and about 4–6 days old. When development is completed, the larvae exit from the skin and drop into the host's nest to pupate (Fig. 5.1(d)). Adult flies emerged from pupae about 2 weeks after pupation (Fig. 5.1(a)). Larvae develop only in living chicks, and when a chick dies they leave its body within a few minutes (Snyder *et al.* 1987).

Whether female flies deposit eggs or already-hatched larvae is unknown but, if eggs are laid, the incubation period must be very short as Snyder *et al.* (1987) have found larvae on nestling pearly-eyed thrashers (*Margarops fuscatus*) only 2 days old and Arendt (1983) found that thrasher nestlings only a few hours old were often parasitized. Uhazy and

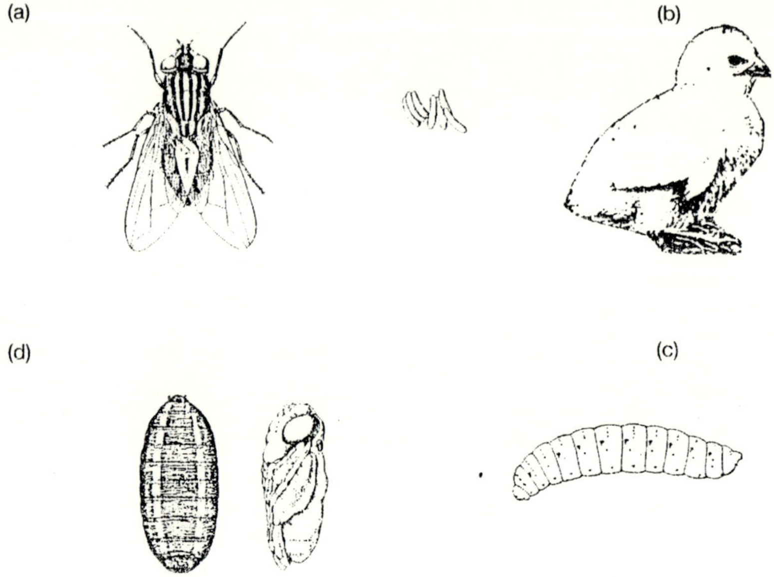

Fig. 5.1. The life cycle of *Philornis*. It is thought that adult flies (a) deposit eggs on nestling sharp-shinned hawks (b). The larvae penetrate the host integument, undergo a period of development and growth, and establish as third instar larvae (c), between the dermis and the body musculature. When development is completed, the larvae exit from the skin and drop into the host's nest to pupate (d).

Arendt (1986) observed fly eggs on nestlings dissected for histological purposes, however, they could not be specifically attributed to *Philornis*.

Three *Philornis* species (*Philornis pici*, *Philornis obscura*, and *Philornis deceptivus*) have been reported from Puerto Rico (Wolcott 1948; Arendt 1985*a*,*b*). The larvae of these species feed subcutaneously on nestlings of at least 30 bird species (Wolcott 1948; Pont 1972; Perez-Rivera and Collazo 1976; Arendt 1985*a*,*b*; Uhazy and Arendt 1986; Snyder *et al.* 1987; personal observation). The endangered Puerto Rican parrot (*Amazona vittata*) is parasitized by *P. pici,* which have caused the death of parrot nestlings in the Luquillo Experimental Forest, eastern Puerto Rico (Snyder *et al.* 1987). In the same forest, Arendt (1983, 1985*b*) found that *P. deceptivus* parasitism was responsible for the deaths of 97 per cent of 224 nestling pearly-eyed thrashers.

In this chapter we report on the prevalence and effect of *Philornis* parasitism on the survival of sharp-shinned hawk (*Accipiter striatus venator*) nestlings in the Maricao Commonwealth Forest, western Puerto Rico, and discuss the ecological implications of this host–parasite interaction. The Puerto Rican sharp-shinned hawk occurs locally in montane areas, and the island population is estimated (conservatively) to be 230–50 individuals (Cruz and Delannoy 1986). Specific objectives were to: (1) determine the prevalence (percentage of host individuals infested; terminology of Margolis *et al.* 1982), intensity or parasite load (mean number of larvae/infested host), and impact of *Philornis* parasitism on sharp-shinned hawk nestlings; (2) evaluate the effects of philornid parasitism on the selection of nesting habitats; (3) determine if host nestling survival was related to the number of *Philornis* larvae present; and (4) assess the potential of *Philornis* as a regulator of sharp-shinned hawk populations.

Methods

This study took place within the 4150-ha Maricao Commonwealth Forest located at the western end of the Cordillera Central (18°09′N, 66°59′W) of Puerto Rico. Elevations in the study area ranged from 550 m in the subtropical moist forest to 900 m in the subtropical lower montane wet forest life zones. Precipitation ranged from 57 mm in January to 374 mm during October with a mean annual precipitation of 2345.4 mm. Mean monthly temperatures varied from 20.1°C in January to 23.2°C in August with a mean annual temperature of 21.8°C (National Oceanic and Atmospheric Administration 1971–82). The life zones and vegetation of this forest have been described in detail by Ewel and Whitmore (1973) and Cruz and Delannoy (1984).

From 1979 to 1983 and in 1985 we determined the prevalence, parasite load, and impact of botfly parasitism on sharp-shinned hawk nestling survival. Prevalence was expressed as percentage of nests and nestlings parasitized and parasite load as mean number of botfly larvae per nestling. Nesting sites were found by systematic searches of woods in the Maricao Forest. Nestlings ($n = 75$) were checked for parasites by climbing trees with active nests every day. We recorded the number of old and newly implanted larvae on their bodies until the nestlings died or fledged. At each visit we recorded the condition of the nestlings, i.e. emaciation, weak, and the number of days that parasitized nestlings survived from the onset of parasitism were recorded. Several larvae were removed from nestlings and allowed to pupate in sand-filled jars covered with a cloth. The adult botflies that emerged from puparia were collected and later identified to genus at the Agricultural Experimental Station, University of Puerto Rico.

We used the Mann–Whitney *U* test to determine significance between means at the 0.05 level. The chi-square statistic was used to determine statistical independence of parasite prevalence and host survival in relation to nesting habitat, and survival in relation to parasitized and unparasitized nestlings. All statistical procedures followed Sokal and Rohlf (1981).

Results

Prevalence, parasite load, and impact of parasitism

The prevalence, parasite load, and impact of *Philornis* parasitism on sharp-shinned hawk nestlings varied considerably during the study period (Table 5.1). Philornid parasitism of nestlings was high in 1980,

Table 5.1. Prevalence, parasite load, and impact of *Philornis* parasitism on sharp-shinned hawk nestlings

Year	No. active nests	No. nests with parasitized nestlings (%)	No. of nestlings	No. nestlings parasitized (%)	No. of dead parasitized nestlings* (%)	No. of larvae (parasite load)
1979	4	2(50)	10	4(40)	1(25)	21(5.25)
1980	3	3(100)	8	7(87.5)	6(85.7)	No data
1981	8	7(87.5)	18	17(94.4)	16(94.1)	275(16.2)
1982	5	1(20)	15	1(6.7)	1(100)	6(6)
1983	4	4(100)	12	8(66)	0(0)	13(1.63)
1985	6	3(50)	12	4(33)	1(25)	25(6.25)
Total	30	20(66.7)	75	41(54.7)	25(61)	340(10)

* Mortality caused by *Philorinus* parasitism.

1981, and 1983, moderate in 1979 and 1985, and low in 1982. Twenty of 30 nests (66.7 per cent) and 41 of 75 nestlings (54.7 per cent) were parasitized during the study period. Parasite-induced mortality was high from 1980 through 1982, low in 1979 and 1985, and none of the nestlings found to harbour *Philornis* larvae died in 1983. Twenty-five of the 41 parasitized nestlings (61 per cent) died during the 6-year study. Parasite loads were highest in 1981, lower in 1979, 1982, and 1985, and lowest in 1983. A total of 34 nestlings harboured 340 botfly larvae (nestlings in 5 of 6 years sampled) for an average load of 10 per nestling.

Habitat differences

Sharp-shinned hawks used two nesting habitats in the Maricao Forest—plantations of Maria tree (Guttiferae) and natural forest. The prevalence of *Philornis* parasitism on sharp-shinned hawk nestlings was 56.3 and 53.5 per cent in plantation and natural forest, respectively (Table 5.2). Prevalence was statistically independent ($\chi^2=0.02$, $df=1$, $p>0.05$) of the nesting habitat selected. *Philornis*-induced mortality on sharp-

Table 5.2. Prevalence of *Philornis* parasitism on sharp-shinned hawk nestlings in two nesting habitats, Maricao Forest, Puerto Rico

Nesting habitat	Number parasitized (%)	Number unparasitized (%)	Total nestlings
Plantations	18 (56.3)	14 (43.7)	32
Natural forest	23 (53.5)	20 (46.5)	43

shinned hawk nestlings was approximately 61 per cent in both habitats (Table 5.3). Nestling mortality related to *Philornis* parasitism was also statistically independent ($\chi^2 = 0.94$, $df = 1$, $p > 0.05$) of nesting habitat used.

Nestling mortality

Mortality to the fledging stage was significantly higher ($\chi^2 = 11.4$, $df = 1$, $p < 0.05$) in parasitized than unparasitized nestlings (Table 5.4). Proportionally, there was almost a fourfold difference in mortality between parasitized and unparasitized nestlings.

We examined host parasite load and compared it to survival among parasitized nestlings (Table 5.5). In general, parasitized nestlings which survived sustained significantly lower parasite loads (Mann–Whitney $U = 279$, $p = 0.001$) than nestlings that perished. Parasite loads were more uniform on nestlings that survived than those that died. None of the parasitized nestlings that sustained parasite loads of more than six larvae survived. Note, however, that as few as two larvae were enough to kill a sharp-shinned hawk nestling.

Nestling size

It is reasonable to assume that the effect of the parasite load could be inversely proportional to the nestling size. Larger nestlings could have better chances of surviving or could survive longer than smaller nestlings. We found that, although larger nestlings survived longer, the difference

Table 5.3. Impact of *Philornis* ectoparasitism on sharp-shinned hawk nestling survival in two nesting habitats, Puerto Rico

Nesting habitat	Total nestlings parasitized	Number died (%)	Number survived (%)
Plantations	18	11 (61.1)	7 (38.9)
Natural forest	23	14 (60.9)	9 (39.1)

Table 5.4. Mortality among parasitized and unparasitized sharp-shinned hawk nestlings

Nestlings	Total nestlings	Number died (%)	Number survived* %
Parasitized	41	25 (61.0)	16 (39.0)
Unparasitized	34	6 (17.6)	28 (82.4)

* Survival to fledging.

was not statistically significant (Mann–Whitney $U = 20.5$, $p > 0.05$) (Table 5.6).

Discussion

In general, sharp-shinned hawk nestlings did not tolerate *Philornis* parasitism, and nestling mortality was high following infestation. We believe there are at least two possible reasons why mortality was high following infestation. First, even though sharp-shinned hawk populations and philornid flies may have interacted over relatively long periods of time, the former may have not evolved adaptations to attenuate the effects of the latter. Alternatively, due to habitat fragmentation this host–parasite interaction may be relatively recent and, given sufficient time, the host may evolve adaptations against parasitism. These alternatives pose different implications.

In the first case the host may lack the genetic potential to evolve defence mechanisms against recurrent parasitism. In the second case the host may respond to the selective pressure imposed by the parasite, given enough time. However, the long generation time of the sharp-shinned hawk in relation to *Philornis* flies may delay the evolutionary response of the former. As suggested by Levin *et al.* (1982, p. 237), ' . . . the generation time and the initial frequency of the alleles that confers resistance will affect the rate of selection in host–parasite interactions.'

Table 5.5. Botfly parasite load in sharp-shinned hawk nestlings that survived versus those that died

	No. of nestlings	Total number of parasites	Parasite load	SD	Range
Survived	15	36	2.86	± 0.66	1–6
Died	19	304	16.0	± 13.03	2–60

Table 5.6. Average number of days that parasitized nestlings survived from the onset of parasitism as a function of nestling size

	No. of nestlings	Mean no. days survived*	SD	Range (days)
Nestlings <25 g	12	4.9	± 1.68	3–8
Nestlings >25 g	7	6.0	$+ 1.55$	5–9

* Survival to fledging.

Possible adaptations against parasitism

Sharp-shinned hawks could adopt different 'strategies' to minimize the effects of *Philornis* parasitism in the Maricao Forest. For instance, hawks could select nesting habitats where parasitism is not as intense. This could be a viable 'option' if nesting habitats differed as to the occurrence and effects of botfly parasitism on nestlings. However, prevalence and nestling mortality in the habitats currently used by sharp-shinned hawks were not significantly different. Since sharp-shinned hawks occupied the same nesting areas each year (Delannoy 1984; Delannoy and Cruz 1988) this nest site tenacity may preclude the evolution of a spatial mechanism to minimize parasitism. This nest site fidelity may also contribute to recurrent infestations as in the cliff swallow (*Hirundo pyrrhonota*) and various dependent ectoparasites (Loye and Hopla 1983; Loye 1985).

Alternatively, the hawks could breed when the *Philornis* fly population is at a seasonal low. For this 'strategy' to be successful the parasite should follow a predictable seasonal pattern of population growth and the host must have the 'flexibility' to accommodate the initiation of the nesting activities accordingly. The first requirement of this 'strategy' cannot be evaluated because the population dynamics of *Philornis* flies are not known. There was a marked seasonality in rainfall and avian prey resources in the Maricao Forest (Delannoy 1984; Delannoy and Cruz 1988) which may account for the restricted reproductive pattern of sharp-shinned hawks. The seasonality of these factors (e.g. food availability) may set limits to the possibility of the hawks shifting the timing of the nesting activities in response to a possible seasonal variation in the population density of botflies.

Although Arendt (1985*b*) established that between-year and seasonal changes in the pearly-eyed thrasher nestling mortality and botfly prevalence were related to changes in the rainfall pattern, it is unclear exactly how variations in the rainfall pattern affected the population dynamics of *P. deceptivus*. To understand the population dynamics of *Philornis* in the Maricao Forest, it might be necessary to examine this host–parasite system at the 'suprapopulation' level (*sensu* Esch *et al.* 1975)—the population in all stages of development within all hosts, or at least in more abundant and heavily parasitized hosts.

Impact of *Philornis* ectoparasitism

One major viewpoint on the impact of parasites on their host populations is that any mortality produced is almost entirely compensatory, i.e. replaces other mortality; hence their impact is minimal (Holmes 1982). The results of this study, however, indicate that mortality induced by *Philornis* is additive, i.e. in addition to other mortality factors. The fact that mortality was significantly higher in parasitized than unparasitized nestlings suggests that *Philornis* parasitism had an additive effect on the overall mortality. Philornid parasitism maintains sharp-shinned hawk nestling populations below parasite-free levels, and thus has the potential to regulate sharp-shinned hawk populations in Maricao Forest.

The impact of parasitism is believed to be regulated by the ability of the host species to disperse (Price 1980). On islands, even an abundant host species may be restricted and concentrated because of limited suitable habitats, allowing for an unnaturally high instance and impact of parasitism. Botfly parasitism of sharp-shinned hawks in the Maricao Forest may result from the presence of the sharp-shinned hawk in this limited habitat and from the hawk's inability to disperse adequately because of the absence of suitable habitats outside the forest (Cruz and Delannoy 1986). In addition, on oceanic islands (e.g. Puerto Rico) vertebrate predators are often absent and nest predation may be greatly reduced (Arendt 1983). Under such circumstances, it is not unreasonable to assume that avian parasitism may serve as a regulating factor or as an additional reproductive stress on island birds.

Since the parasite-induced mortality is directly related to parasite load, the parasite and not other factors appears to be the cause of nestling mortality. However, the large variability associated with the parasite load in nestlings that die suggests that the relationship between the host's chances of survival and the parasite load is complex. The differences could be due partly to nestling size at the onset of parasitism, i.e. the effect of parasite numbers is inversely proportional to the host's size. Although sample sizes were small in this study, larger nestlings did not survive significantly longer than smaller nestlings (see this volume: Chapman and George, Chapter 4; Scott and Edman, Chapter 10). This contrasts with findings that showed that larger pearly-eyed thrasher siblings that died from botfly parasitism survived longer and withstood greater loads than smaller nestlings (Arendt 1985*b*). However, Arendt also found that only a few larvae, if implanted in sensitive areas of the body, would kill nestlings at any age. We also found that larvae implanted near the eyes or the throat of sharp-shinned hawk nestlings would kill them in a short period of time regardless of the parasite load. In addition, the nutritional condition of the nestlings and the predisposition to other pathogens probably contributed to the variability in the capacity to withstand the pathogenic effect of botfly larvae and thus acted synergistically in causing nestling mortality.

Acknowledgements

We thank A. Graves, S. Graves, and J.A. Rivera for their help with the field work. H. Diaz-Soltero, R. Schmidt, and L. Sanchez, Department of Natural Resources, Commonwealth of Puerto Rico, generously granted permission to work in the Maricao Commonwealth Forest and provided living facilities. S. Medina-Gaud, Research Entomologist, Agricultural Experimental Station, University of Puerto Rico, identified botfly specimens to genus. We gratefully acknowledge the support by the United States Fish and Wildlife Service Contract 14-16-0004-82-047, the University of Colorado, National Academy of Sciences, Alexander Wetmore Award of the American Ornithologists' Union, and the Frank M. Chapman Fund of the American Museum of Natural History. W.A. Arendt, C.E. Bock, Y.B. Linhart, J.A. Moreno, and C.P. Ortega reviewed the manuscript and made valuable suggestions.

References

Aitken, T.H.G., Downs, W.G., and Anderson, C.R. (1958). Parasitic *Philornis* flies as possible sources of arbor virus infections (Diptera, Anthomyidae). *Proceedings of the Society of Experimental Biology and Medicine*, **99**, 635–7.

Arendt, W.J. (1983). The effects of dipteran ectoparasitism on the growth and development of nestlings of the Pearly-eyed Thrasher (*Margarops fuscatus*) in the Luquillo mountains, Puerto Rico. Unpublished MS thesis, University of Missouri, Columbia.

Arendt, W.J. (1985*a*). *Philornis* ectoparasitism of pearly-eyed thrashers. I. Impact on growth and development of nestlings. *Auk*, **102**, 270–80.

Arendt, W.J. (1985*b*). *Philornis* ectoparasitism of pearly-eyed thrashers. II. Effects on adults and reproduction. *Auk*, **102**, 281–92.

Cruz, A. and Delannoy, C.A. (1984). Ecology of the Elfin Woods Warbler (*Dendroica angelae*). I. Distribution, habitat usage, and population densities. *Caribbean Journal of Science*, **20**, 89–96.

Cruz, A. and Delannoy, C.A. (1986). *Status, breeding biology, and conservation needs of the Puerto Rican sharp-shinned hawk, Accipiter striatus venator*. United States Fish and Wildlife Service Report, Washington, DC.

Delannoy, C.A. (1984). The Puerto Rican sharp-shinned hawk (*Accipiter striatus venator*): the ecology and breeding biology of a neotropical island bird of prey. Unpublished Ph.D. thesis, University of Colorado, Boulder.

Delannoy, C.A. and Cruz, A. (1988). Breeding biology of the Puerto Rican sharp-shinned hawk (*Accipiter striatus venator*). *Auk*, **105**, 649–62.

Dodge, H.R. (1955). New muscid flies from Florida and the West Indies. *Florida Entomologist*, **38**, 147–51.

Dodge, H.R. (1963). A new *Philornis* with coprophagous larva, and some related species (Diptera; Muscidae). *Journal of the Kansas Entomological Society*, **36**, 239–45.

Dodge, H.R. and Aitken, T.H.G. (1968). *Philornis* flies from Trinidad. *Journal of the Kansas Entomological Society*, **41**, 134–54.

Esch, G.W., Whitfield–Gibbons, J., and Borque, J.E. (1975). An analysis of the relationship between stress and parasitism. *American Midland Naturalist*, **93**, 339–53.

Ewel, J.J. and Whitmore, J.L. (1973). *The ecological life zones of Puerto Rico and US Virgin Islands*, Forest Service Research Paper ITF-18 Rio Piedras, Puerto Rico.

Hector, D.P. (1982). Botfly (Diptera; Muscidae) parasitism of nestling aplomado falcons. *Condor*, **84**, 443–4.

Hicks, E.A. (1962). Check-list and bibliography on the occurrence of insects in birds' nests. *Iowa State Journal of Science*, **46**, 123–338.

Holmes, J.C. (1982). Impact of infectious disease agents on the population growth and geographical distribution of animals. *Population biology of infectious diseases* (ed. R.M. Anderson and R.M. May), pp. 37–51. Springer-Verlag, New York.

LaRue, E.L. (1987). Premethrin control of *Philornis pici* (Diptera: Muscidae) on pearly-eyed thrasher nestlings in Luquillo Forest, Puerto Rico. Unpublished MS thesis, Ohio State University.

Levin, B.R., Allison, A.C., Bremermann, H.J., Covalli-Sforza, L.L., Clarke, B.C., Frentzel-Beyme, R., Hamilton, W.D., Levin, S.A., May, R.M., and Thieme, H.R. (1982). Evolution of parasites and hosts, group report. In *Population biology of infectious diseases* (ed. R.M. Anderson and R.M. May), pp. 213–43. Springer-Verlag, New York.

Loye, J.E. (1985). The life history and ecology of the cliff swallow bug (*Oeciacus vicarius*) (Hemiptera: Cimicidae). *Cahiers Office de la Recherche, Scientifique et Technique Outre Mer, Serie Entomologie Médicale et Parasitologie*, **23**, 133–9.

Loye, J.E. and Hopla, C.E. (1983). Ectoparasites and microorganisms associated with the cliff swallow in west-central Oklahoma. II. Life history patterns. *Bulletin of the Society of Vector Ecology*, **8**, 79–84.

Margolis, L., Esch, G.W., Holmes, J.C., Kuris, A.M., and Schad, G.A. (1982). The use of ecological terms in parasitology (report of an ad hoc committee of the American Society of Parasitologists). *Journal of Parasitology*, **68**, 131–3.

National Atmospheric and Oceanic Administration, N.O.A.A. (1971–82). Climatological data–Puerto Rico and the US Virgin Islands. National Climatic Center, US Department of Commerce, Asheville, North Carolina.

Nielsen, J.C. (1911). *Mydaea anomala* Jaenn., a parasite of South American birds. *Videnskabelige Meddelelser Dansk Naturhistorisk Forening Kjobenhavn*, **63**, 195–208.

Oniki, Y. (1983). Notes on fly (Muscidae) parasitism of nestlings of South American birds. *Le Gerfaut*, **73**, 281–6.

Perez-Rivera, R. and Collazo, J. (1976). Aves en Puerto Rico parasitadas por la larva de la mosca del genero *Neomusca*. *Science–Ciencia*, **4**, 50–1.

Plath, O.E. (1919). Parasitism on nestling birds by fly larvae. *Condor*, **21**, 30–8.

Pont, A.C. (1972). *A catalogue of the diptera of the Americas south of the United States*. Museo Zoologia Universidade de Sao Paulo, Brazil.

Price, P.W. (1980). *Evolutionary biology of parasites*, Monographs in Population Biology, no. 15. Princeton University Press, Princeton, New Jersey.

Smith, N.G. (1968). The advantage of being parasitized. *Nature*, **219**, 690–4.

Smith, N.G. (1980). Some evolutionary, ecological, and behavioural correlates of communal nesting by birds with wasps or bees. *Acta 17th Congress of International Ornithology*, **2**, 1195–205.

Snyder, N.F.R., Wiley, J.W., and Kepler, C.B. (1987). *The parrots of Luquillo: natural history and conservation of the Puerto Rican parrot*. Western Foundation of Vertebrate Zoology, California.

Sokal, R.R. and Rohlf, J. (1981). *Biometry: the principles and practice of statistics in biological research*. Freeman, San Francisco.

Uhazy, L.S. and Arendt, W.J. (1986). Pathogenesis associated with philornid myiasis (Diptera: Muscidae) on nestling pearly-eyed thrashers (Aves: Mimidae) in the Luquillo rain forest, Puerto Rico. *Journal of Wildlife Diseases*, **22**, 224–37.

Winterstein, S.R. and Raitt, R.J. (1983). Nestling growth and development and the breeding biology of the Beechey Jay. *Wilson Bulletin*, **95**, 256–68.

Wolcott, G.N. (1948). The insects of Puerto Rico. *Journal of Agriculture of the University of Puerto Rico*, **22**, 1–974.

Zumpt, F. (1965). *Myiasis in man and animals in the Old World*. Butterworth, London.

6 Occurrence and demography of mites of tree swallow, house wren, and eastern bluebird nests

EDWARD H. BURTT, JR, W. CHOW, and G. A. BABBITT

Introduction

Debate over the evolutionary determinants of avian clutch size has ignored the potential of nest-dwelling, parasitic arthropods to significantly reduce the reproductive success of birds (McKilligan 1987; this volume: Delannoy and Cruz, Chapter 5; Rogers *et al.*, Chapter 7). Parasitic mites (Arthropoda, Arachnida, Acari) occur in bird nests (Hicks 1953, 1959, 1971, 1975; Wilson 1965; Philips and Dindal 1977, 1979; Welch 1977), especially nests sheltered from weather and nests built in cavities (Woodroffe 1953; Moss 1978; Chow *et al.* 1983). Such mites are known to feed on blood, which causes weight loss and mortality of nestlings (Moss and Camin 1970) and may cause desertion of whole broods (Moss 1966).

Little is known about the host–parasite relationship, which can be viewed as a reproductive race between the avian host and its parasitic mite: the bird to hatch and rear the maximum number of young to the optimum weight for postfledging survival and the parasitic mite to efficiently use a temporarily abundant food supply, provided by the nestlings' blood, to raise the maximum number of offspring before the resource fledges. The distribution of nidicolous mites among species of birds is imperfectly known from studies that indicate presence (e.g. Moss 1978), but provide little or no indication of abundance or interaction with other species of nest-dwelling arthropods. The demography of such mites is known only from laboratory studies (e.g. Hastings and Wollkind 1982) and a single study (Phillis 1972) of the population dynamics of *Dermanyssus* spp. in the nests of house sparrows (*Passer domesticus*). Such information is critical to assessing the potential importance of parasitic mites or insects as selection agents in the evolution of clutch size and parental (in this volume: Clark, Chapter 11; Loye and Carroll, Chapter 12; Møller, Chapter 17) and social (Emlen 1986; Shields and Crook 1987) behaviour.

The purpose of this study was to document the distribution of nidicolous mites in three common, cavity-nesting birds: the eastern bluebird (*Sialia sialis*); tree swallow (*Tachycineta bicolor*); and house wren (*Troglodytes aedon*). Further, we sought to evaluate the environmental factors affecting the occurrence and abundance of mites and influencing their demographics. Finally, we looked at possible effects on the reproductive success of the birds.

Methods

Occurrence and distribution

In 1977 16 wooden bluebird houses were placed in Delaware State Park, Delaware County, Ohio. The number of houses was increased subsequently, with 93 available in 1980. Nest boxes were checked every third day during nest construction and incubation and every other day from hatching through departure of the young from the nest. The reproductive status of the occupants was recorded at each check. Eastern bluebirds, tree swallows, and house wrens were the most common occupants (Tuttle 1987), with occasional occupancy by Carolina chickadees (*Parus carolinensis*) and tufted titmice (*P. bicolor*).

In the summer of 1980, we collected invertebrates from 13 eastern bluebird nests, 18 tree swallow nests, and 61 house wren nests taken from the next boxes. The nest was removed intact from its box within 24 h after the young departed. It was tagged, sealed in a plastic bag, and transported to the laboratory where it was placed in a Berlese funnel or, if all funnels were in use, placed in a refrigerator until a funnel was available. All nests were placed in funnels not more than 48 h after removal from the nest box.

The Berlese funnels were of galvanized steel and had a diameter (23.2 cm) and depth (27.9 cm) that accommodated the largest nests without distorting their shape. Before a nest was placed in the funnel, a jar containing a 5 per cent formaldehyde solution was placed beneath the funnel so that the spout of the funnel projected 2–3 cm into the jar without touching any part of the jar or the formaldehyde solution. The funnel was lined with a single layer of cheesecloth to prevent debris from falling into the jar. The nest was removed from the plastic bag, placed on the cheesecloth, and any loose material remaining in the bag was shaken onto the nest. With the nest in place the top of the funnel was closed. The top contained a 60-watt light bulb that heated and dried the nest thereby driving invertebrates down through the nest and into the funnel where they slid into the jar of formaldehyde. Once a day the nest in its cheesecloth was lifted from the funnel and anything clinging to the cloth was gently brushed into the metal funnel. After 72 h the jar was sealed and the nest was put in a paper bag and frozen. Nests and jars received a code number, but were not labelled with information on the species of bird or its reproductive success. Thus we hoped to avoid biasing our identification and counting of invertebrates.

The volume of the nest was measured by placing the nest in an empty aquarium and pressing it gently into a corner. The height, width, and depth of the nest were measured against the glass sides of the aquarium. Construction materials and the amount of residual faecal material were described briefly.

A sample of mites from each jar was mounted on glass slides and identified as to family under a compound microscope using keys available in Krantz (1978) and McDaniel (1979). Representative specimens were sent to D. E. Johnston at the Acarology Laboratory, Ohio State University for species identification. Following species identification, the contents of each jar were sorted and counted under a dissecting microscope. Insects and spiders were identified as to family using Borror *et al.* (1976).

Demography

In 1983 20 wooden bluebird houses were placed in Alum Creek State Park, Delaware County, Ohio. By 1987 the number of bluebird houses in the park was 46. In that summer we studied the population dynamics of mites inhabiting tree swallow nests in Alum Creek State Park. Tree swallows were the most numerous species nesting in bird houses in the park (Tuttle 1987). To study the growth dynamics of mite populations in swallow nests, we divided our sample around six collection times within the swallow's reproductive cycle. Nests were collected at 10 days after the last egg was laid, at hatching, 5, 10, or 15 days after hatching, or within 24 h of the nestlings' departure from the nest. Nest boxes were checked every third day until laying began and every other day during laying to determine the day on which the last egg was laid. When the clutch was complete, the nest was assigned a collection time within the reproductive cycle (only the original nest of each pair of swallows was collected). Assignments were random, except that no time was assigned a second nest until all had received one nest, no time was assigned a third until all had two, and so on.

At the time of collection the side of the box was removed, the nest lifted out, the eggs or nestlings placed in a handmade replacement nest, the replacement put in the box, and the box closed. The real nest was sealed in a plastic bag. A few feathers were included in the replacement nest, but as we withdrew from the area, we threw several white feathers into the air. The swallows immediately caught these and, usually within five minutes, carried the feathers into the nest box. The use of feathers combined with our practice of delaying the first exchange until late in incubation (Burtt and Tuttle 1983) resulted in all swallows accepting the nest exchange.

Arthropods were extracted from the nests as explained above. Few arthropods other than mites were found in these nests and these were not identified or counted.

Statistics

The Spearman rank correlation coefficient (Sokal and Rohlf 1981) was used to evaluate the relationship of mite populations to each other, other arthropod populations, and avian brood size. Regression lines were compared using the small-sample *t*-test for parallelism (Kleinbaum and Kupper 1978). Because many correlations and comparisons were calculated, alpha was set at 0.01 to reduce the possibility of a type two error. The rate of population growth was estimated from the equation

$$\mathrm{d}N/\mathrm{d}t = rN. \qquad (6.1)$$

Average population sizes were compared using *t*-tests (Sokal and Rohlf 1981). Expected proportions of mite species in the nests of each avian species were calculated by assuming that the proportion of each mite species in the nests of each avian species was the same as the proportion of each species of mite in the total population of mites. Chi-square (Sokal and Rohlf 1981) was used to compare expected and observed proportions of mite species. For these few comparisons alpha was set at 0.05.

Results and discussion

Occurrence and distribution

Dermanyssus hirundinis (Fig. 6.1(a)) occurred in all nests of all three avian species (Table 6.1). It has an oval body, long legs, and emerges from the nest material to feed on the blood of nestlings (Moss 1978). *D. hirundinis* may injure nestlings by frequent piercing of the skin, by triggering allergic reactions in the nestlings, by transmitting diseases among nestlings and across generations, or by blood loss (Moss 1966; Krantz 1978). After each feeding the mite returns to the nest material where the female lays about 20 eggs. Most *D. hirundinis, D. americanus,* and *D. prognephilus* emigrate from the nest soon after the nestlings depart, although a few females and nymphs remain behind to overwinter in the nest material (Moss 1966; Phillis 1972).

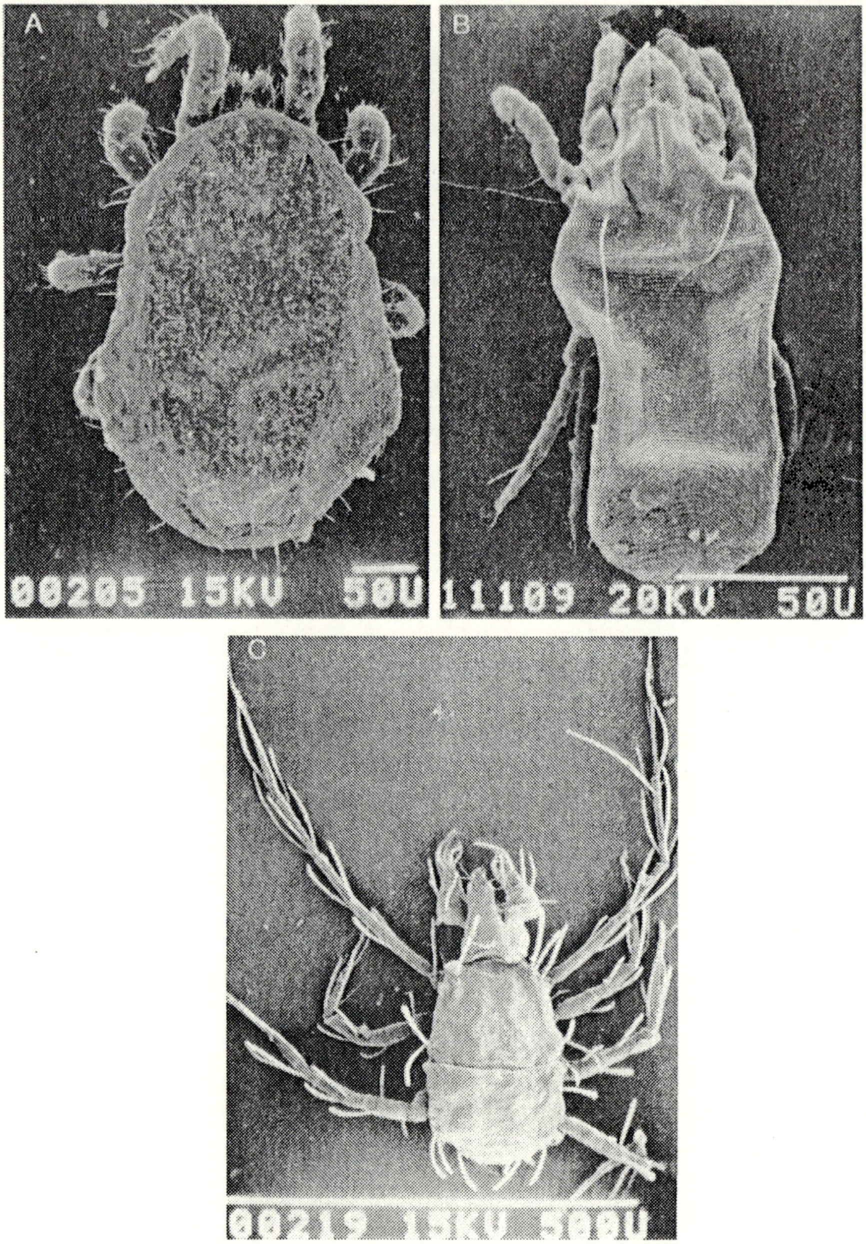

Fig. 6.1. (A) The parasitic mite *Dermanyssus hirundinis;* (B) the scavenging mite *Dermatophagoides evansi;* and (C) the predatory mite *Cheletomorpha lepidopterorum.*

Table 6.1. Mean number of mites/nest ± standard error (range in parentheses)

Avian host species	*n* (number of nests)	Mites		
		Dermanyssus	Dermatophagoides	Cheletomorpha (in nests where occurred)
Tree swallow	18	9745 ± 978 (1881–16579)	5908 ± 573 (1386–9332)	1206 ± 197 (396–1829)
House wren	61	12675 ± 676 (844–20819)	6714 ± 404 (393–12354)	2622 ± 212 (218–4240)
Eastern bluebird	13	5384 ± 568 (934–8271)	1763 ± 192 (311–3216)	0

Next to *D. gallinae,* which has been found in the nests of 30 species of birds, *D. hirundinis* is the most widespread species of the genus, occurring in the nests of 14 species (Moss *et al.* 1970; Moss 1978). The tree swallow and house wren are among its known hosts, but this is its first known association with the eastern bluebird as reported by Chow *et al.* (1983). *D. hirundinis* may have colonized bluebird nests by overwintering in a box formerly occupied by a swallow or wren and subsequently occupied by a bluebird. However, the number of *D. hirundinis* inhabiting bluebird nests (Table 6.1) suggests that bluebirds are an acceptable host species.

Dermatophagoides evansi (Fig. 6.1(b)) occurred in all nests of all three species of birds (Table 6.1). It was less numerous than *Dermanyssus hirundinis,* but like *Dermanyssus* was most common in nests of the house wren. *Dermatophagoides evansi* belongs to a genus of free-living mites that feed on organic debris and are a common component of house dust where they contribute to human dust allergies (Wharton 1976; Arlian *et al.* 1983). Members of the genus occur on skin, fur, and feathers of mammals and birds and in their nests (Krantz 1978).

Cheletomorpha lepidopterorum (Fig. 6.1(c)), the third species found in our study, belongs to the family Cheyletidae. It has been found in tree bark, organic debris, soil, grain storage, and in association with the *Proxenus* (Noctuidae) moth (Beer and Dailey 1956; Summers and Price 1970; Krantz 1978). *C. lepidopterorum* is a predator capable of capturing and consuming four grain mites per day at 20°C and 80 per cent relative humidity (Krantz 1978). It is known to feed on mites in the families Pyroglyphidae, which includes *Dermatophagoides evansi,* and Laelaptidae, which includes *Dermanyssus hirundinis* (Beer and Dailey 1956).

C. lepidopterorum was the least numerous of the three species of mites (Table 6.1), occurring in 0 of 13 bluebird nests, 6 of 18 (33 per cent) swallow nests, and 30 of 61 (49 per cent) wren nests. In nests colonized

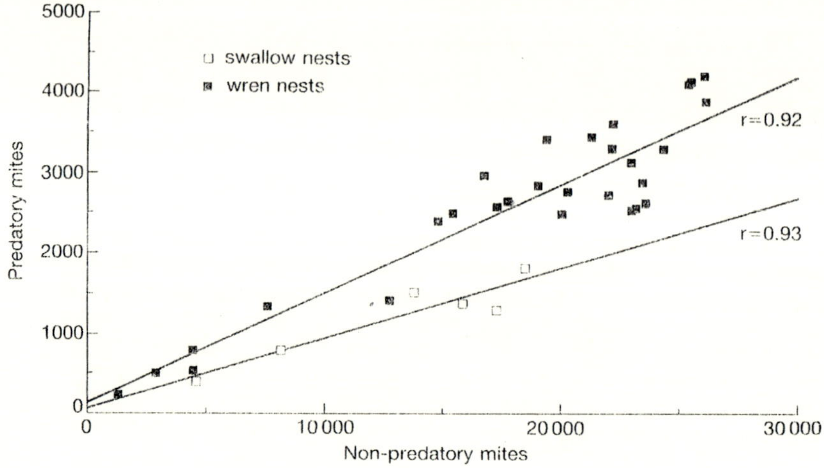

Fig. 6.2. Predatory mites (*Cheletomorpha lepidopterorum*) as a function of the population of non-predatory mites in nests colonized by *Cheletomorpha*. Both populations of mites were counted at fledging of the nestlings.

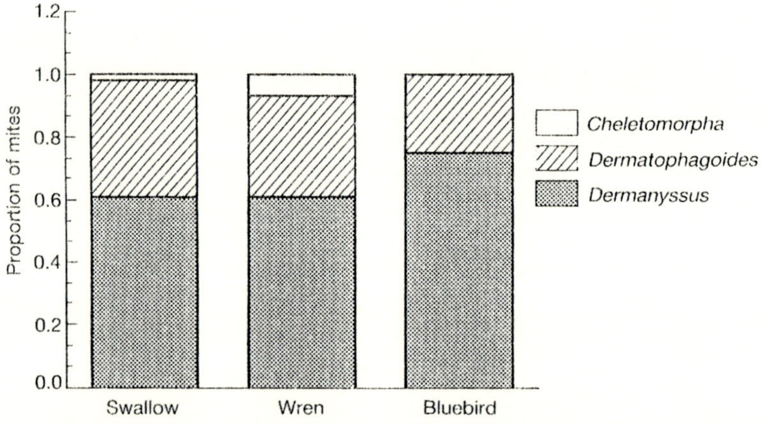

Fig. 6.3. The proportion of parasitic, scavenging, and predatory mites in nests of swallows, wrens, and bluebirds at the time the nestlings fledged.

by *Cheletomorpha*, its population increased in proportion to the number of non-predatory mites (Fig. 6.2), but always remained the smallest of the three populations of mites (Fig. 6.3). *Cheletomorpha* had little effect on populations of non-predatory mites in tree swallow nests (mean population without *Cheletomorpha*, 16 974; mean population with *Cheletomorpha*, 13 010; $t = 1.29$, $df = 16$, $p > 0.5$) and house wren nests (mean population without *Cheletomorpha*, 20 454; mean population

Table 6.2. Direction of deviation of the observed population of mites from the expected population

Avian host species	Mites		
	Dermanyssus	*Dermatophagoides*	*Cheletomorpha*
Tree swallow	−	+ +	−
House wren	−	−	+ +
Eastern bluebird	+ +	−	−

+ +, 10 000 or more mites than expected; −, 3500–9999 fewer mites than expected where 'expected' is the product of the total number of mites for that bird species and the total number of that species of mite in all nests divided by the total number of mites.

with *Cheletomorpha*, 18 287; $t=0.31$, $df=59$, $p>0.7$). Similar prey availability in nests with and without *Cheletomorpha* suggests that food was not a limiting factor in its distribution, except possibly in the case of bluebirds.

As suggested by the patchy distribution of *Cheletomorpha lepidopterorum*, the number of mites and the relative proportion of each species in the total population depends on the avian host species (Fig. 6.3; $\chi^2=17\,590$, $df=4$, $p<0.001$). Eastern bluebirds had the fewest mites (Table 6.1) and the highest proportion of *Dermanyssus hirundinis* (Table 6.2, Fig. 6.3), which may have contributed to the failure of *Cheletomorpha* to colonize bluebird nests. *Cheletomorpha* is a 'sit and wait' predator dependent on the movement of its prey, but *Dermanyssus* tends to aggregate low in the nest and remain quiescent between feedings (Davis and Camin 1972). Thus the prey population in bluebird nests is relatively small and inactive and may be less available than prey populations in swallow and wren nests. Tree swallow nests had intermediate numbers of mites (Table 6.1) with a disproportionate number of *Dermatophagoides evansi* (Table 6.2, Fig. 6.3). House wren nests had the most mites (Table 6.1) including a disproportionate number of predators (Table 6.2, Fig. 6.3), which supports the suggestion that prey availability may be the limiting factor in colonization of bluebird nests by *Cheletomorpha*. However, there is no obvious reason why 31 house wren nests with abundant prey and 12 tree swallow nest also with abundant prey contained no predatory mites.

Ecological factors

Distribution of the three species of mites varies with the avian host species. Some of the variation (e.g. the distribution of *C. lepidopterorum*) could derive from interactions among the species of mites, but factors that vary within and among the host species (e.g. distance

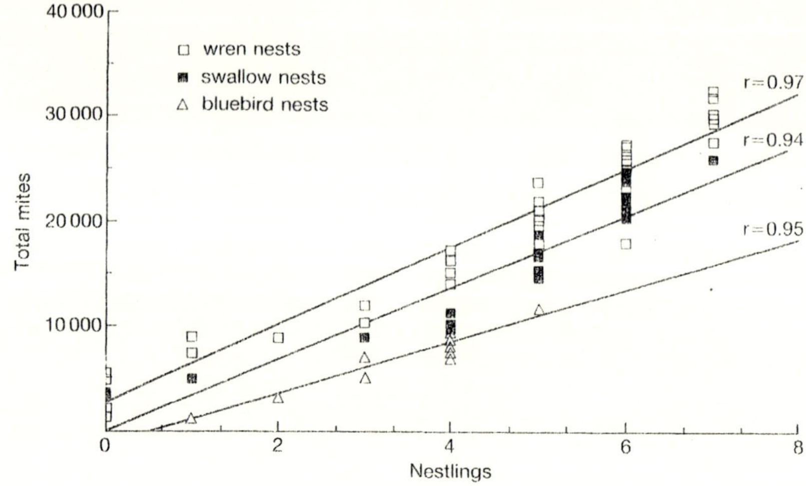

Fig. 6.4. Total population of mites at the time of fledging as a function of the number of nestlings in swallow, wren, and bluebird nests.

between conspecific nests, tendency of avian species to reoccupy nest cavity in subsequent years) may be important determinants of mite populations.

Nestlings per nest

The number of nestlings per nest varied from zero to five in bluebirds and up to seven in tree swallows and house wrens. The combined populations of mites are correlated significantly with the number of nestlings per nest in all three host species (Fig. 6.4). For eastern bluebirds the regression is given by the equation

$$M = 2440\ N - 1297 \tag{6.2}$$

where M is the total number of mites and N is the number of nestlings. The regression for tree swallows is

$$M = 3413\ N + 131 \tag{6.3}$$

and, for house wrens,

$$M = 3694\ N - 2814. \tag{6.4}$$

The slopes are significantly different from each other (small sample t-test for parallelism, wren to swallow: $t = 3375$, $df = 75$, $p < < 0.001$; wren to bluebird: $t = 6285$, $df = 70$, $p < < 0.001$, swallow to bluebird: $t = 4511$, $df = 27$, $p < < 0.001$). As brood size increases, the mite population increases least rapidly in bluebird nests, more rapidly in swallow nests, and most rapidly in wren nests. The increase in mites with increasing

Table 6.3. Correlation coefficients between the number of nestlings and the number of mites for each host species and each species of mite

Avian host species	Mite			
	Dermanyssus	*Dermatophagoides*	*Cheletomorpha*	Total
Tree swallows	0.92	0.92	0.05* 0.93†	0.94
House wrens	0.95	0.82	0.37* 0.92†	097
Eastern bluebird	0.93	0.96	—	0.95

* Based on all nests.
† Based only on nests with *Cheletomorpha*.

numbers of nestlings is rapid, but linear. Thus more nestlings mean more mites, but the number of mites per nestling remains roughly constant.

Taking the species of mites separately, populations of *Dermanyssus hirundinis,* the parasitic mite, are correlated significantly and positively with the number of nestlings per nest (Table 6.3) as are populations of *Dermatophagoides evansi,* the scavenging mite (Table 6.3). Populations of *Cheletomorpha lepidopterorum* are correlated poorly with the number of nestlings per nest (Table 6.3) unless populations of zero are removed from the sample. Thus populations of parasitic and scavenging mites correlate closely with the number of nestlings per nest (Table 6.3), whereas populations of predatory mites correlate closely with the number of nestlings per nest (Table 6.3) and populations of the other mites (Fig. 6.2) when colonization of the nest occurs. Colonization of nests is a limiting factor in the distribution of *Cheletomorpha lepidopterorum.*

Other arthropods

Other arthropods are potential prey, predators, and competitors. In addition to mites, species in one family of spiders and 11 families of insects occurred in the nests of swallows, wrens, and bluebirds. Jumping spiders (Arachnida, Araneae, Salticidae) are small (2–4 mm), predatory spiders that ambush small arthropodan prey. Jumping spiders occurred in most nests and averaged about four per nest in bluebird, swallow, and wren nests. Book lice (Psocoptera, Liposcelidae), which are scavengers, were patchily distributed, but averaged about two per nest. Gall midges (Diptera, Cedidomyiidae), march flies (Diptera, Bibionidae), and fungus gnats (Diptera, Mycetophilidae) all feed on decaying vegetation. All three families occurred in most nests with three to four of each family per nest. Blowflies (Diptera, Calliphoridae, *Protocalliphora* sp.) feed on living flesh and can be serious parasites of nestling birds (Roberts 1981). They occurred in a few nests with six to seven being found wherever they occurred. We may have underestimated their occurrence since we did not dissect the nests and most of those we found were pupae. Hymenoptera were represented by ants (Formicidae), which prey on smaller arthropods, ichneumon wasps (Ichneumonidae), which parasitize arthropods including blowflies, and chalcid wasps (Perlampidae), which parasitize arthropods including ichneumon wasps. Worker ants occurred in most nests (~11 per nest). Since the nest boxes were mounted on greased pipes, the ants were probably carried in on nest materials. Ichneumonids (~4 per nest) and chalcids (~3 per nest) were patchily distributed. In most cases they occurred in nests with blowflies, but not in all cases, which further suggests that we did not find all the blowflies and our estimate of blowfly occurrence and distribution is low. Flat bark beetles (Coleoptera, Silvanidae), which feed on vegetation, occurred in all wren nests (~7 per nest) and patchily in swallow and bluebird nests. House wrens fill the nest box with twigs, which may account for the larger numbers of flat bark beetles in their nests as compared to those of swallows and bluebirds, which use few sticks and more grass. All nests contained some bird lice (Mallophaga, Menoponidae, and Philopteridae, ~6 per nest).

Spiders and ants might have preyed on mites. Most of the insects present fed on vegetation and would have interacted minimally with the mites. The blowfly (*Protocalliphora* sp.) may compete with *Dermanyssus hirundinis* for nestling blood, but the blowfly population was so small and patchy that any effect in the present study could not be documented. The combined mite populations were not correlated with any other arthropod population (35 correlation coefficients, $p > 0.01$ in all cases; blowflies in tree swallow nests were too few to calculate a correlation coefficient) nor with the combined population of other arthropods (three correlation coefficients, $p > 0.01$ in all cases).

The nest

Mites live in the nest material. However, the combined mite populations were not correlated with the volume of the host species' nest. Nest volume, within species, is not a determinant of population size among mites inhabiting eastern bluebird, tree swallow, and house wren nests.

Nest structure and materials may be factors. With only three avian species for comparison the data are sketchy. Bluebird nests are flat and densely woven compared to swallow and wren nests. Thus *Dermanyssus* need move only short distances to feed, but engorged adults and nymphs should find ample shelter in the densely woven vegetation. The more active *Dermatophagoides* may be somewhat hampered in moving through the nest in search of organic debris (fewer *Dermatophagoides* than expected, Table 6.2) and the predatory *Cheletomorpha* (none in bluebird nests) may be severely restricted in its ability to detect and jump approaching prey in the dense mat of finely woven grasses. As the coarseness of the materials increases, the weave becomes looser and movement easier. Tree swallows have coarser materials and a looser weave than bluebirds and also have more mites. Tree swallow nests have fewer predatory mites than expected (Table 6.2), but some are present and populations of parasitic and scavenging mites are far above those of bluebird nests (Table 6.1). House wren nests have the coarsest materials, the loosest weave, and the most mites with a disproportionate number of predatory mites (Table 6.2). The comparison suggests that the structure of the nest, particularly the type of materials and the tightness of the weave, may be factors in the distribution and abundance of mites.

Bluebird nests had few or no faeces throughout the reproductive cycle and had relatively few scavenging mites (Table 6.1). House wren nests contained substantial deposits of nestling faeces and flakes from feather sheaths and large populations of scavenging mites (Table 6.1), although a smaller proportion than expected (Table 6.2). Tree swallow nests contained large deposits of faecal material and flakes of feather sheaths and disproportionately large populations of scavenging mites (Table 6.2). The population of *Dermatophagoides evansi* appears closely tied to the presence of organic wastes from the nestlings, suggesting that the correlation with nestling number is due to the increasing amount of waste material with increasing brood size.

Timing of avian reproduction

Numbers of mites should increase as the season advances. Not only will the mites in occupied nests reproduce, but mites from early nests will be able to colonize later nests and reproduce again. House wrens initiated their clutches (19 June ± 25.2 days) significantly later ($F = 20.78$, $df = 2, 133$, $p < 0.01$) than tree swallows (18 May ± 10.7 days) and eastern bluebirds (28 May ± 32.8 days) and had significantly larger populations of mites (Table 6.1), as expected. However, tree swallows had much larger populations of mites than eastern bluebirds (Table 6.1), but began laying eggs at about the same time. The data are equivocal. They do not provide strong support for the expectation of more mites in later nests, but they do not refute the expectation. The possibility that mites may influence the timing of reproduction by birds may repay systematic study.

Table 6.4. Population growth (mites/nestling) of *Dermanyssus hirundinis* and *Dermatophagoides evansi* in tree swallow nests during the swallows' reproductive cycle

Swallow's developmental stage	Number of nests	*Dermanyssus*		*Dermatophagoides*
		Adult	Nymph	
10 d incubation	7	1.7	3.3	72.2
Hatching	4	0.3	0.1	339.0
5 d nestlings	6	4.4	5.7	254.6
10 d nestlings	7	111.7	81.7	1354.9
15 d nestlings	5	61.8	60.0	2282.7
Postfledging	7	628.3	353.0	3613.9

Demography

During 1987 we studied the population dynamics of *Dermatophagoides evansi* and *Dermanyssus hirundinis* in tree swallow nests. Too few *Cheletomorpha lepidopterorum* were found for analysis. Because the number of mites per nest increases with an increase in the number of nestlings all discussion of demography is adjusted to mites per nestling. The population dynamics of both *Dermatophagoides* and *Dermanyssus* were related to the reproductive cycle of the tree swallow.

The initial population of *Dermatophagoides evansi* measured on the tenth day of incubation was 72.2 mites per nestling (Table 6.4). Between the tenth day of incubation and hatching the population of mites grew rapidly. Prior to hatching an r value (eqn 6.1) of 1.35 gives a close approximation to the growth rate of the mite population. The population appears to stabilize for the first 5 days the nestlings are in the nest and then enters a second growth phase during which r is initially 1.3, but declines to 0.9 as fledging approaches.

The pattern of population growth suggests two distinct growth phases. The first corresponds to an initial period of colonization and exploitation of the detritus to be found in the nest materials and the flakes of material shed by the skin and feathers of the female (only the female incubates). These resources are limited and the number of *Dermatophagoides* reaches carrying capacity about the time of or just prior to hatching of the nestlings. Initially the nestlings add little to the food resources available to *Dermatophagoides*. Newly hatched swallows are small and without feathers. They probably supply little, if any, dead skin or feather material to the nest habitat. Faeces from the nestlings are consumed by the parents throughout this period (Burtt, personal observation). By 10 days after hatching, the feathers of the nestlings are emerging from the sheaths in which they first penetrate the skin. Flakes of material

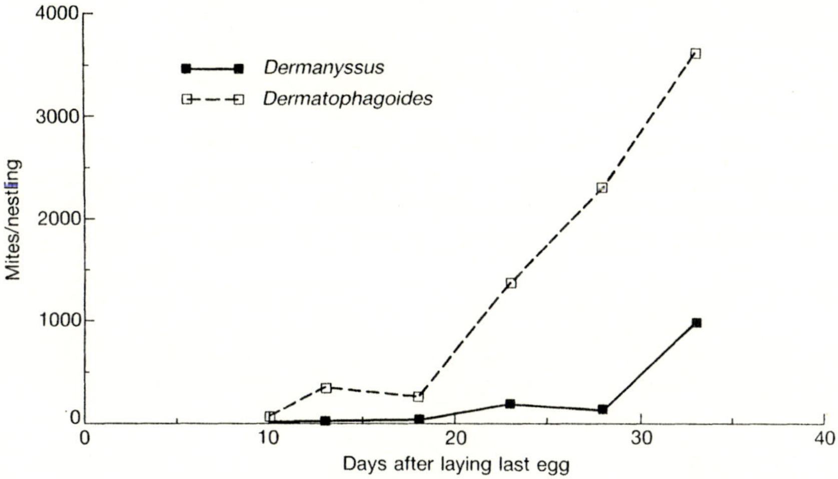

Fig. 6.5. Population growth of *Dermanyssus* and *Dermatophagoides* in relation to the reproductive cycle of the tree swallow.

from the disintegrating sheaths would increase food resources for *Dermatophagoides*. As little as 180 mg of shed skin can sustain large cultures of *D. pteronyssinus* for some months (van Bronswijk and Sinha 1971, cited in Krantz 1978). Deposition of material from growing feathers would increase after 10 days post-hatch as the feathers of nestlings increase their growth rates. Furthermore, after day 10 the parents gradually cease removing nestling faeces from the nest and this source of organic waste would become available to the scavenging *Dermatophagoides*. Despite increasing food resources the rate of growth in the population of *Dermatophagoides* declines from an r value of around 1.3 at 10 days after the nestlings hatch to an r value of about 0.9 at the time of fledging. The declining growth rate suggests that the population of *Dermatophagoides* is approaching its carrying capacity in the nests of tree swallows.

Few *Dermanyssus hirundinis* are present during incubation (Table 6.4). The population of *Dermanyssus* remains low through the first 5 days of the nestling period and then grows very rapidly for the next 5 days (Fig. 6.5). The population maintains a plateau until the nestlings are 15 days old, whereupon the *Dermanyssus* population grows rapidly until the nestlings fledge. From the fifth day after hatching through fledging of the nestlings the growth rate, r, of *Dermanyssus* is close to 1.75.

The population of *Dermanyssus* is substantially lower than that of *Dermatophagoides* during incubation and, unlike the population of *Dermatophagoides*, remains unchanged through the fifth day after hatching of the nestlings. *Dermanyssus* feeds on blood and throughout incubation only the female swallow is available as a food source. The newly hatched nestlings are not growing feathers and their blood supply may be too far below the skin for *Dermanyssus* to reach. Alternatively, ectothermic nestlings may not attract mites that feed on endothermic animals. The two growth surges evident in Fig. 6.5 may correspond to the development of separate generations of mites. *Dermanyssus prognephilus* has a generation time of 7.3 days at 24°C, 93 per cent relative humidity (Moss 1966). The generation time drops to 5 days at 32°C, 93 per cent relative humidity (Moss 1966). These generation times correspond roughly to the surges in population growth shown in Fig. 6.5. Unlike *Dermatophagoides*, the growth rate of the *Dermanyssus* population is accelerating at the time the nestlings fledge. Phillis (1972) found a similarly rapid population growth for *Dermanyssus* at the time of fledging in the house sparrow. Such rapidly increasing populations of ectoparasites could set a limit on the duration of the nestling period. Furthermore, population increases in the nests of the earliest breeders would provide large numbers of immigrants to the nests of those swallows breeding later in the season or trying to raise a second brood. Such large founding populations would have a much greater potential for population growth than the small founding populations occurring in the present study and might account for the rarity of second broods among tree swallows, the frequent desertion of broods by late-nesting barn swallows (*Hirundo rustica*) noted by Shields and Crook (1987), and the synchrony of breeding and desertion within colonies of cliff swallows (*H. pyrrhonota*) (Emlen 1986; in this volume Loye and Carroll, Chapter 12).

Effect of mites on avian reproductive success

The reproductive success of house wrens was not monitored. Eastern bluebirds laid 53 eggs in the 13 nests from which mites were collected. Forty-five eggs hatched and 45 nestlings fledged. All eight eggs that failed to hatch were infertile. Four additional nests with 12 bluebird eggs were usurped by house wrens, but mites were not collected from these nests. Furthermore, wrens took over the nests soon after the bluebirds began incubation when the mite population would have been very small.

Tree swallows laid 99 eggs in the 18 nests from which mites were collected. Ninety of these eggs hatched, eight were infertile, and one contained a partially developed embryo. Of the 90 hatchlings, 84 fledged. Six nestlings in one nest died from unknown causes, all on the same day. The mite population in the nest was 3420, well below average. Nests were checked every other day and the deaths may have occurred as much as 40 h before they were discovered and the nest collected. This may have given the mites time to disperse, but, even with dispersal, such a population seems to rule out mites as a cause of death. Five nests were deserted by tree swallows. Four of these, containing 14 eggs, were usurped by house wrens and the fifth, containing 2 eggs, by a house sparrow. Two nests were destroyed by vandals. Mites in these seven nests were not counted. As with the bluebirds, eviction by wrens and sparrows occurred early in the incubation period when mite populations would have been small.

Few nestling deaths occurred and none could be attributed to parasitism by *Dermanyssus hirundinis*. The number of mites increased as the number of nestlings increased, but the increase was arithmetic not geometric. Mites per nestling remained constant and, apparently, within tolerable limits. Moss and Camin (1970) found that purple martin (*Progne subis*) nestlings subject to parasitism by *Dermanyssus prognephilus* were about 7 per cent lighter at fledging than nestlings protected from such parasitism and suggest that such reduced weight may have important implications for survival after fledging. Based on the weight differential Moss and Camin (1970) estimate that parents in nests without mites could have raised an additional nestling. We did not protect nestlings from mites nor did we weigh our nestlings; however, the ubiquity of *Dermanyssus hirundinis* in nests of tree swallows, house wrens, and eastern bluebirds suggests the potential for serious effects on the reproductive success of these species. Replication of Moss and Camin's (1970) experiments are needed for eastern bluebirds, tree swallows, and house wrens particularly in warm, damp years when conditions favour rapid growth of mite populations.

Conclusions

Mites are the most frequent and most numerous of the arthropods inhabiting nests of the eastern bluebird, tree swallow, and house wren at our Ohio study site. In most nests mites also comprise the largest arthropod biomass. The population growth of *Dermanyssus hirundinis*, a parasitic mite, and *Dermatophagoides evansi*, a scavenging mite, is closely tied to the reproductive cycle of the tree swallow, the only species for which we have demographic data. *Dermanyssus* and *Dermatophagoides* probably overwinter in the nest or nest box. *Cheletomorpha lepidopterorum*, a predator of the other two species, occurred in 33 per cent of tree swallow nests, 49 per cent of house wren nests, and 0 per cent of eastern bluebird nests. It may overwinter in nest boxes with the other species or in bark (Summers and Price 1970; Krantz 1978) and enter nests on twigs used for nest construction by wrens, used occasionally by swallows, and not used by bluebirds. Such a method of colonization would account for the absence of *C. lepidopterorum* in bluebird nests, their occasional appearance in swallow nests, and their common occurrence in wren nests.

Dermanyssus has an extremely small founding population. Its population growth shows two separate growth phases: one at the time feathers are emerging, which is probably associated with the availability of blood in the feather papillae, and a second growth phase associated with the nestlings reaching their maximum weight. The growth rate of *Dermanyssus* is increasing at the time the nestlings leave the nest. The rapidly increasing population of the parasitic *Dermanyssus* may limit the time the nestlings can remain in the nest, although no mortality attributable to mites was found in this study.

Dermatophagoides has a larger founding population than *Dermanyssus* and its population has an initial growth phase before hatching. A second growth phase begins as the feathers emerge and continues through the departure of the young. This second growth phase begins with rapid growth ($r = 1.3$) that slows toward the time the nestling swallows leave the nest ($r = 0.9$). *Dermatophagoides* feeds on organic debris, which is available during incubation and abundant as the nestlings feather sheaths flake off and the parents stop removing nestling faeces. A growth plateau early in the nestling period suggests that *Dermatophagoides* may have reached a temporary carrying capacity prior to emergence of the nestlings feathers. The decreasing growth rate late in the nestling period suggests that the *Dermatophagoides* population is approaching a second, higher carrying capacity, which it does not reach prior to fledging of the young swallows.

The final populations of *Dermanyssus* and *Dermatophagoides* are larger in nests with more nestlings. The correlation is linear and significant. Nests with faeces and accumulated flakes from feather sheaths have more mites. However, the volume of the nest (i.e. living space for the mites) and the presence of other arthropods have no effect on the population size of *Dermanyssus* and *Dermatophagoides* at fledging of the nestlings.

The presence of *Cheletomorpha lepidopterorum* has little effect on the populations of *Dermanyssus* and *Dermatophagoides*. Bluebirds had relatively few mites and this may account for the failure of *Cheletomorpha* to colonize bluebird nests, but uncolonized wren and swallow nests had the same or more prey than colonized nests. Thus, abundance of prey was not the limiting factor and the occurrence of *Cheletomorpha* in some nests and not others may depend on the difficulty of colonization. Ideas from island biogeography may help explain the distribution of *Cheletomorpha lepidopterorum* among the nests of birds.

Acknowledgements

We are deeply grateful to Richard M. Tuttle who maintains the bluebird houses and monitors reproductive success of the birds. Dr Donald E. Johnston and members of the Acarology Laboratory at Ohio State University provided species identification, lent us reprints, and provided invaluable insight into the biology of mites. A. John Gatz, Dennis C. Radabaugh, and two anonymous referees offered important suggestions on earlier drafts. Through their organization of a symposium on avian ectoparasites Dale Clayton and Jenella E. Loye provided the motivation to analyse these data. We thank Jenella E. Loye especially for her organizational, diplomatic, and editorial skills which have made this chapter and volume possible.

References

Arlian, L.G., Woodford, P.J., Bernstein, I.L., Gallagher, J.S. (1983). Seasonal population structure of house dust mites, *Dermatophagoides* spp. (Acari: Pyroglyphidae). *Journal of Medical Entomology*, **20**, 99–102.

Beer, R.E. and Dailey, D.T. (1956). Biological and systematic studies of two species of cheyletid mites, with a description of a new species (Acarina, Cheyletidae). *University of Kansas Science Bulletin*, **38**, 393–437.

Borror, D.J., DeLong, D.M., and Triplehorn, C.A. (1976). *An introduction to the study of insects* (4th edn). Holt, Rinehart, and Winston, New York.

Burtt, E.H. Jr and Tuttle, R.M. (1983). Effect of banding on reproductive success of tree swallows. *Journal of Field Ornithology*, **54**, 319–23.

Chow, W., Burtt, E.H. Jr, and Johnston, D.E. (1983). Invertebrate fauna of eastern bluebird nests. *Sialia*, **5**, 53–7.

Davis, J.C. and Camin, J.H. (1977). Aggregation behavior in the martin mite, *Dermanyssus prognephilus* (Acari: Dermanyssidae). *Journal of Medical Entomology*, **14**, 373–8.

Emlen, J.T. (1986). Response of breeding cliff swallows to nidicolous parasite infestations. *Condor*, **88**, 110–11.

Hastings, A. and Wollkind, D. (1982). Age structure in predator-prey systems. I. A general model and a specific example. *Theoretical Population Biology*, **21**, 44–56.

Hicks, E.A. (1953). Observations on the insect fauna of birds nests. *Journal of the Kansas Entomological Society*, **26**, 111–18.

Hicks, E.A. (1959). *Checklist and bibliography on the occurrence of insects in birds nests*. Iowa State College Press, Ames.

Hicks, E.A. (1971). Supplement I. *Iowa State College Journal of Science*, **36**, 233–348.

Hicks, E.A. (1975). Supplement II. *Iowa State College Journal of Science*, **46**, 123–338.

Kleinbaum, D.G. and Kupper, L.L. (1978). *Applied regression analysis and other multivariate methods*. Wadsworth Publishing Co, Belmont, California.

Krantz, G.W. (1978). *A manual of acarology*. Oregon State University Press, Corvallis.

McDaniel, B. (1979). *How to know the mites and ticks*. Wm. C. Brown, Dubuque, Iowa.

McKilligan, N. (1987). Causes of nesting losses in the cattle egret *Ardeola ibis* in eastern Australia with special reference to the pathogenicity of the tick *Argas* (Persicargas) *robertsi* to nestlings. *Australian Journal of Ecology*, **12**, 9–16.

Moss, W.W. (1966). The biological and systematic relationships of the martin mite, *Dermanyssus prognephilus* Ewing. Unpublished Ph.D. thesis. University of Kansas, Lawrence, Kansas.

Moss, W.W. (1978). The mite genus, *Dermanyssus*: a survey, with description of *Dermanyssus trochilinis*, n. sp., and a revised key to the species (Acari: Mesostigmata: Dermanyssidae). *Journal of Medical Entomology*, **14**, 627–40.

Moss, W.W. and Camin, J.H. (1970). Nest parasitism, productivity, and clutch size in purple martins. *Science*, **168**, 1000–3.

Moss, W.W., Mitchell, C.J., and Johnston, D.E. (1970). New North American host and distribution records for the mite genus *Dermanyssus* (Acari: Mesostigmata: Dermanyssidae). *Journal of Medical Entomology*, **7**, 589–93.

Philips, J.R. and Dindal, D.L. (1977). Raptor nests as a habitat for invertebrates: a review. *Raptor Research*, **11**, 86–96.

Philips, J.R. and Dindal, D.L. (1979). The acarine community of nests of birds of prey. *Recent Advances in Acarology*, **1**, 559–62.

Phillis, W. (1972). Seasonal abundance of *Dermanyssus hirundinis* and *D. americanus* (Mesostigmata: Dermanyssidae) in nests of the house sparrow. *Journal of Medical Entomology*, **9**, 111–12.

Roberts, T.H. (1981). Parasites of the eastern bluebird. *Sialia*, **3**, 92–4.

Shields, W.M. and Crook, J.R. (1987). Barn swallow: a net cost for group breeding in the Adirondacks? *Ecology*, **68**, 1373–86.

Sokal, R.R. and Rohlf, F.J. (1981). *Biometry*. W.H. Freeman, San Francisco.

Summers, F.M. and Price, D. (1970). *Review of the mite family Cheyletidae*, University of California Publications in Entomology, no. 61.

Tuttle, R.M. (1987). A six-year study of nesting tree swallows in Delaware State Park, Delaware, Ohio 1979–1984. *Sialia*, **9**, 3–7.

Welch, R.C. (1977). An ecological study of the insects and mites in the nests of certain birds in Britain. *Bulletin of Entomological Research*, **44**, 739–72.

Wharton, G.W. (1976). House dust mites. *Journal of Medical Entomology*, **12**, 577–621.

Wilson, N. (1965). Occurrence of *Dermanyssus quintus* Vitzthum in North America (Acarina: Mesostigmata: Dermanyssidae). Proceedings of the Entomological Society Washington, **67**, 6.

Woodroffe, G.E. (1953). An ecological study of the insects and mites in the nests of certain birds in Britain. *Bulletin of Entomological Research*, **44**, 739–72.

7 *Patterns and effects of parasitism by* Protocalliphora sialia *on tree swallow nestlings*

CATHERINE A. ROGERS, RALEIGH J. ROBERTSON, and BRIDGET J. STUTCHBURY

Introduction

Altricial nestlings frequently host a great variety and number of ectoparasites (Herman 1936; Rothschild and Clay 1952). Energy loss through such parasitism can have pronounced deleterious effects on the nestlings (Rothschild and Clay 1952; Moss and Camin 1970; Wilson 1979; Chapman and George, Chapter 4, this volume) since they must mature rapidly to avoid depredation within the nest (Ricklefs 1968). We investigate the little known patterns and effects of blowfly (*Protocalliphora sialia*) parasitism on nestlings of the tree swallow (*Tachycineta bicolor*).

With one exception, larvae of the genus *Protocalliphora* are nest-dwelling, intermittent ectoparasites that feed solely on nestling blood (Gold and Dahlsten 1983). The genus is widely distributed in North America, as almost every avian species that rears its young in a nest is susceptible to infestation by one or more Protocalliphoran species (Bennett 1957).

Quantitative information on the regularity and severity of infestations is quite limited. Bennett (1957) suggests that species-specific distributions of Protocalliphoran larvae reflect a combination of host and habitat preferences. To date, the larvae of *P. sialia* have been found in a variety of nests including those of the screech owl (*Otus asio*), downy woodpecker (*Picoides pubescens*), great crested flycatcher (*Myiarchus crinitus*), house wren (*Troglodytes aedon*), European starling (*Sturnus vulgaris*), eastern bluebird (*Sialia sialis*), and several species of swallow (Hirundinidae; Bennett 1957; Hicks 1962). Tree swallows appear to experience particularly high rates of Protocalliphoran infestation (Mason 1944; Bennett 1957; Pinkowski 1977). Mason (1944) found that 82 per cent of tree swallow nests he examined were infested and in 1 year the average number of parasites per brood was 220.

Opinions regarding the effect of ectoparasites on their hosts are widely varied. Many authors infer that Protocalliphoran larvae severely jeopardize the development and survival of host nestlings (Plath 1919; Johnson 1930; Herman 1936; Mason 1944; Neff 1945; Hill and Work 1947; George and Mitchell 1948; Hill 1948; Boyd 1951; Esehuis-van der Voet and Kluyver 1971; Capreol 1983; Garrison *et al.* 1986; Shields and Crook 1987). Conversely, other studies show that many species of Protocalliphoran ectoparasites do not significantly harm their hosts under normal conditions (McAtee 1929; Wetherbee 1932; Bennett 1957; Whitworth 1976; Pinkowski 1977; Gold and Dahlsten 1983; Bortolotti 1985).

In this study, we investigate: (1) how tree swallow nesting density, habitat, and nesting chronology affect levels of infestation by *P. sialia;* (2) the effect of parasitic exploitation on the development and survival of nestling tree swallows; and (3) how experimentally manipulated parasite loads affect the rate of feeding nestlings by parent tree swallows.

Life histories of the study species

The parasite: *Protocalliphora sialia*

The life history of *P. sialia* (Diptera: Calliphoridae) has been described in detail by Bennett (1957). The life cycle begins when free-living female flies oviposit a small number of eggs (10–20) in the nest of a potential avian host. Since gravid flies may be guided to nests by an odour peculiar to young birds, oviposition can occur whenever the nest contains nestlings (Gold and Dahlsten 1983). However, oviposition usually occurs when the nestlings are 1/4 to 1/3 grown, which allows the parasitic larvae to complete development before their hosts fledge. The accustomed hosts of *P. sialia* spend 14 to 21 days in the nest, while the parasites develop in 11 to 13 days (Bennett 1957).

Within 24 hours of oviposition, minute slug-like larvae hatch and immediately burrow into the nesting material. When hungry, they approach the nestlings and attach themselves to accessible portions of the nestlings' bodies. Unless disturbed, the larvae feed on host blood for 15 min to 1 h before returning to the nesting material. Although some feeding occurs during the day, it is possible that most feeding occurs at night (Gold and Dahlsten 1983). This intermittent feeding continues for nearly 2 weeks, over which time each larva attains a weight of approximately 0.1 g. Roughly 2 days after entering a non-feeding stage, the larvae burrow into the nest material to pupate. Adult flies emerge 2 to 3 weeks later and begin to mate. Newly formed adults may produce offspring immediately if they oviposit in late season nests, or they may seek sheltered crevices for hibernation and delay reproduction until the following spring (Bennett 1957).

The host: *Tachycineta bicolor*

Tree swallows (Passeriformes: Hirundinidae) are small (22 g), aerial insectivores that breed throughout the northern United States and continental Canada (Kuerzi 1941). They are facultatively colonial, but defend a minimum area around their nests (Robertson and Gibbs 1982; Muldal *et al.* 1985). They build feather-lined nests of grass in natural and man-made cavities, either in open country or over water (Austin and Low 1932; Kuerzi 1941). Natural cavities are often reused for many years. Tree swallows are usually monogamous and, although renesting often occurs if nests are disturbed or destroyed early in the breeding season, they are usually single-brooded (Kuerzi 1941).

Females usually lay four to seven eggs (Stutchbury and Robertson 1988) and begin to incubate on the day the penultimate egg is laid (DeSteven 1978). When the nestlings hatch, approximately 14 days later (Kuerzi 1941), they weigh about 1.5 g each (Low 1933; personal observation). During the next 3 weeks, parent tree swallows capture flying insects and deliver them to the nest. The delivery rate is influenced by parental age and foraging experience (DeSteven 1978), environmental factors (e.g. climate, time of day or season), and demands on the nestlings (e.g. number and age of the nestlings, amount of vocalization: Knapton 1984; Quinney *et al.* 1986). Tree swallow nestlings gain weight continuously until they are about 12 days old and weigh roughly 24 g. Thereafter, the nestlings enter a brief period of weight recession, during which they approach the adult weight of about 22 g (Paynter 1954; Zach and Mayoh 1982; Quinney *et al.* 1986). Tree swallow nestlings fledge at about 21 days of age and become fully independent within 3 or 4 days of leaving the nest (DeSteven 1978). They usually do not return to the nest after fledging (Kuerzi 1941).

Juvenile and adult tree swallows moult in late July or early August. Thereafter, they flock increasingly until migration, which begins in late August or September (Kuerzi 1941). Thirty-five to 50 per cent of the adult tree swallows and 3–5 per cent of the juveniles will return to breed the following year (DeSteven 1978; Robertson and Stutchbury 1988).

Methods

Patterns and effects of parasitism

We conducted this study during the summers of 1983 through 1987 at the Queen's University Biological Station near Chaffey's Lock, Ontario.

Breeding populations of tree swallows have been established there, and naturally infested by *P. sialia* for approximately 10 years. The swallows nest in similar plywood nest boxes which are mounted on aluminium posts, at a height of 1.5 m. Most of the nest boxes are arranged in close proximity to one another (20–40 m between adjacent boxes) in shallow wetlands (colonial wetland site (CW): the North-east Sanctuary grid) or in open hayfields (small colonial field site (SCF): the Hughson's grid; large colonial field site (LCF): the New Barn, Sand Pit, and Bridget's grid). The remaining boxes are dispersed as singles, pairs, or triples in isolated clearings (dispersed field site: DF, rural boxes).

In early May, we captured the tree swallows in mist nets and marked them individually with coloured acrylic paint. In addition, we banded and sexed (Stutchbury and Robertson 1987) those not previously caught. From the onset of nest-building, we checked each nest every 2 to 4 days, noting the height of nest material, first and final egg date, clutch size, predation, and renesting attempts. We predicted hatch dates for each nest by counting 14 days foward from the final egg date. To estimate actual hatch dates, we weighed the young within a few days of hatch and compared their weights with those of nestlings whose age was known. Approximately 12 days after hatch, when nestlings are expected to attain maximum weight (Paynter 1954; Zach and Mayoh 1982; Quinney *et al.* 1986), we counted live nestlings in each nest and weighed them to within 0.5 g using a Pesola balance. Approximately 25 days after hatch, we returned to each nest to determine how many nestlings fledged. Average brood size at a nest was calculated by averaging the brood size at hatch, day 12, and at fledging.

After the nestlings fledged, usually within 2 weeks, we thoroughly cleaned out each nest box and transferred the contents into a plastic bag. After freezing the nest box contents, we searched them for Protocalliphoran larvae and puparia by thoroughly teasing apart the lining feathers, grass, lichen, twigs, and other nesting materials. Since small (immature) larvae have parasitized their hosts to a lesser extent than large or metamorphosing larvae, brood loading in each nest was evaluated as the total number of puparia + the total number of large (>0.5 cm) larvae + 1/2 the number of small (<0.5 cm) larvae.

Experimental studies

In an attempt to create a wider range of parasite loading among nests, we did an experimental study in 1986, using 11 nests in the small colonial field site. In each of three synchronous pairs of nests, one nest was randomly assigned experimentally decreased parasite loading (nest D), while the other in the pair was assigned experimentally increased parasite loading (nest I). When nestlings of each synchronous pair reached 8 days of age, parasites were transferred from D nests to I nests as follows. After removing the nestlings, we took nest D from its nest box and placed it in a 'shaker' (a nest box with a floor made of coarse mesh). We then gently agitated the shaker over a large tray in order to collect nest 'dust' and as many parasites as possible. After returning nest D and its nestlings to their nest box, we placed the collected 'dust' and parasites beneath the nesting material of the paired nest, nest I. The five remaining nests were maintained as unmanipulated nests.

To evaluate the effect of parasitic exploitation with respect to the nestlings' food demands on the parents, we subsequently monitored nestling feeding rates in the manipulated and unmanipulated nests. Weather permitting, we recorded the number of parental visits to each nest for 1 hour between 0.700 and 13.00 on alternate days, throughout the nesting period. We observed parental visits to each nest through a spotting scope, situated approximately 30 m from the nest box. We assumed that a feeding occurred each time a parent entered the nestbox or perched at its cavity entrance.

Data were analysed using SAS (Statistical Analysis System) and personal computer programs.

Results

Of 335 monitored nests, 242 (72 per cent) were infested by *P. sialia*. Brood loading (number of parasites per brood) ranged from 0 to 167, while nestling loading (brood loading divided by average brood size) ranged from 0 to 54.7.

Spatial and temporal patterns of brood loading

Brood loading was correlated with both nesting habitat and time of season, being higher in fields than wetland habitats, and higher as the season progressed. A non-parametric multiway ANOVA was performed on rank of brood loading to establish which nest parameters are significantly correlated with brood loading. According to the ANOVA model (Table 7.1), brood loading varies significantly with nest site (SITE) and seasonal time of hatch (TIME) (early-, mid-, and late-season nests were defined respectively as those hatching before, within, or after 0.5 standard deviations of the mean Julian hatch date). Brood loading does not vary significantly with other nest parameters, such as year of nest occurrence (YEAR), number of attempted broods per nest site within a given year (ATTEMPT), volume of nest material (DEPTH) and

Table 7.1. Multiway ANOVA model of brood loading (number of parasites per brood). The model tested is: Loading = Year Attempt Depth Hatch Brood Fledge Site Time (model parameters are defined in the text). For the entire model, $F = 9.63$; $R^2 = 0.32$; $df = 14\ 286$; $p < 0.0001$

Source	*df*	*F*
Year	4286	0.52
Attempt	1286	0.18
Depth	1286	2.32
Hatch	1286	1.83
Brood	1286	1.88
Fledge	1286	1.57
Site	2286	26.46*
Time	2286	12.08*

Asterisk indicates significant source of variation.
* $p < 0.0001$.

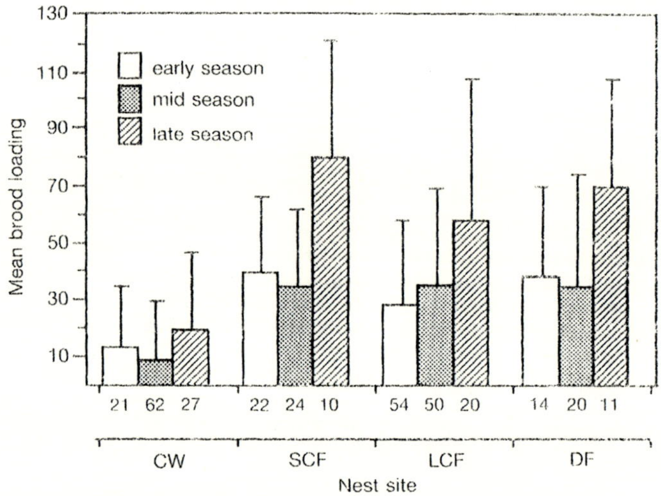

Fig. 7.1. Mean (± SE) brood loading in nests occurring at each site and time during the tree swallow breeding season. Number of nests in each site is indicated below the abscissa. See text for site descriptions.

number of nestlings that hatch (HATCH), survive to day 12 (BROOD), or fledge (FLEDGE).

Since SITE and TIME are the only significant factors in the ANOVA model, their independent effects on parasite loading can be examined without regard to the other nest parameters. At each site, brood loading was significantly higher among late season nests than among mid- or early-season nests (Fig. 7.1, Table 7.2).

Table 7.2. Comparison of brood loading for all sites in relation to time of the tree swallow breeding season

Times compared	Difference between median†	Relative brood loading
Early/mid	16.2	early = mid
Early/late	42.2*	early < late
Mid/late	58.4*	mid < late

† Absolute difference between median brood loading at times compared. Asterisk indicates a significant difference between medians at $p=0.05$, according to GT2 multiple range test.

Brood loading did not appear to be affected by nesting density within the range of inter-nest distances occurring in our study area. At each time during the tree swallow breeding season, brood loading was similar among nests in both the clumped small (SCF) and large (LCF) colonial field and dispersed field sites (DF). However, brood loading was significantly lower among the nests in the colonial wetland site (CW) (Table 7.3). Moreover, although a similar proportion of nests (83 per cent) was infested in each of the field sites ($\chi^2=1.663$, $df=2$, $p>0.05$), a significantly lower proportion of nests (37 per cent) were infested in the wetland site ($z=7.50$, $p< <0.001$).

Development and survival of nestlings

The Protocalliphoran ectoparasites had no evident effects upon their hosts within the range of parasite loading observed in this study. Multiway ANOVA models were devised (both including and excluding data from non-infested nests) for parameters of nest success, including

Table 7.3. Comparison of brood loading in relation to habitat and nest spacing

Sites compared†	Difference between median‡	Relative brood loading
CW/SCF	109.9*	CW < SCF
CW/LCF	82.6*	CW < LCF
SCF/LCF	27.3	SCF = LCF
SCF/DF	2.1	SCF = DF
LCF/DF	25.3	LCF = DF

† Nest sites defined in text.
‡ Absolute difference between median brood loading at the sites compared. Asterisk indicates a significant difference between medians at $p=0.05$, according to the GT2 multiple range test.

number and proportion of nestlings surviving to day 12, total and average weight of nestlings on day 12, and number and proportion of nestlings fledged. However, nestling loading was not a significant parameter in any model.

The minimum nestling loading presumed to result in significant detrimental effects on the host can be calculated as outlined by Gold and Dahlsten (1983). The calculations rely on estimates of host tolerance of blood loss and estimates of the average host blood loss which would be required to support the biomass of parasites found within the nest. These calculations indicate that the health and development of nestling tree swallows would be significantly impaired if nestling loading exceeds 15, and that the survival of host nestlings would be jeopardized if nestling loading exceeds 37. Although 56 (17 per cent) of our study nests experienced infestations presumed significant, and five (1 per cent) experienced infestations presumed critical, no debilitating effects attributable to parasitic exploitation were detected among these nests.

To test further for deleterious effects of parasitic exploitation, average nestling weights in 12-day-old broods were examined more closely. Average nestling weights are one measure of nestling development and are often indicative of the probability of survival (Ricklefs 1968; Moss and Camin 1970; Eschuis-van der Voet and Kluyver 1971). The ANOVA model indicates that average nestling weight depends on the year of nest occurrence (YEAR) and number of nest mates (average brood size). Accordingly, average nestling weights were adjusted to control for these factors by standardizing the weights in nests of each brood size and year to a mean of 0 and a standard deviation of 1. Adjusted weights were expressed subsequently as a number of standard deviations from the grand mean of the adjusted nestling weights. Standardized average nestling weights were then plotted for 50 nests that experienced low, presumably insignificant, infestations (less than 15 parasites per nestling) and for the 10 nests with the highest, presumably significant, infestation levels. (The 50 nests that experienced low infestations were chosen using a systematic random technique, by randomly choosing a first point and including every fifth point thereafter, from a data set ordered by increasing nestling loading.) Average nestling weights in the most heavily infested nests were not necessarily lower than those in nests with low infestation levels (Fig. 7.2). Considerable variation in average nestling weight is apparent even among uninfested nests.

For the same subset of nests, the proportion of nestlings fledged was plotted against nestling loading (Fig. 7.3). Fledging success was not necessarily lower for nests with high parasite loads. In fact, all highly infested nests fledged at least one nestling and the two most intensely

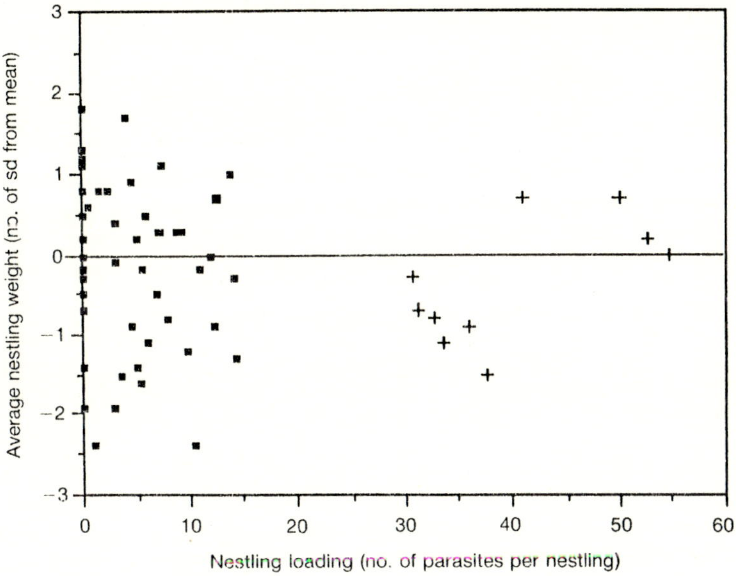

Fig. 7.2. Nestling weight in relation to severity of parasitism. Average nestling weight for each brood on day 12 is adjusted to control for years of nest occurrence and number of nest mates. It is expressed as a number of standard deviations from the grand mean of the adjusted nestling weights. Nestling loading is the average number of parasites per nestling within a brood. Data points represented by crosses are average nestling weights in the 10 most severely infested nests. Data points represented by squares are average nestling weights in 50 nests chosen at random from uninfested and lightly infested nests.

infested nests (54.7 and 52.5 parasites per nestling) fledged all of their young (three and two, respectively).

Experimental studies

Although large numbers of parasites were transferred among nests, these experimental manipulations were largely ineffective. Nests in which parasites were experimentally removed or added did not end up being consistently the least heavily or most heavily infested of all nests among both the experimental and control groups of nests (Kruskal–Wallis ANOVA, $H=0.86$, $p>0.05$).

Average feeding rates (number of parental visits per nestling per hour) of each tree swallow pair were calculated for the period after the parasite manipulations. Only broods of four to six nestlings were included. This analysis minimizes other effects on feeding rates, such as brood size, weather, and time of day. Average feeding rate after the redistribution of parasites was not significantly correlated with actual nestling loading,

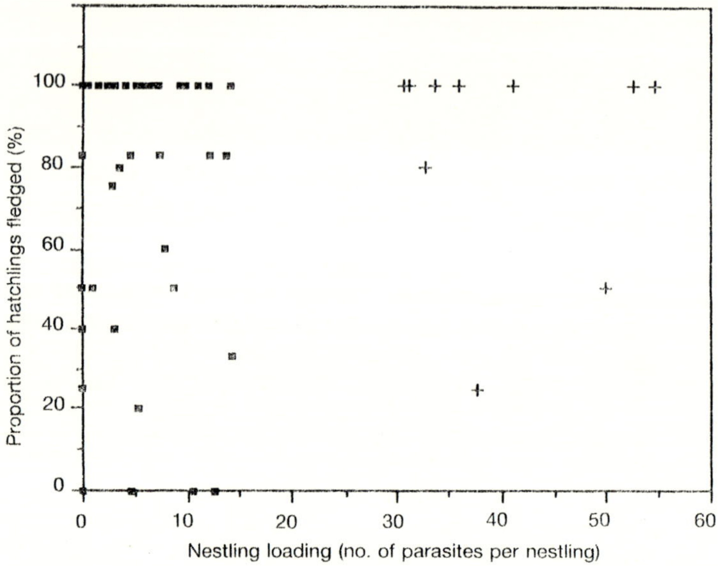

Fig. 7.3. Fledging efficiency in relation to severity of parasitism. Fledging efficiency is the ratio of nestlings fledged to nestlings hatched, expressed as a percentage. Nestling loading is the average number of parasites per nestling within a brood. Data points represented by crosses indicate fledging efficiency in the 10 most severely infested nests. Data points represented by squares indicate fledging efficiency in 50 nests chosen at random from uninfested and lightly infested nests.

within the range of nestling loads (0–17) observed in this experiment (Spearman Rank Correlation $r = -0.47$, $n = 9$, $p > 0.05$).

Discussion

Spatial and temporal patterns

Brood and nestling loading are highly variable. In our study, some tree-swallow broods hosted as many as 167 parasites, while others escaped infestation altogether. Within broods, individual nestlings hosted up to an average of 55 parasites each. Parasite loading in other populations may range higher still. Mason (1944) documented several cases of tree swallow nests that harboured more than 300 Protocalliphoran larvae.

Bennett (1957) suggests that infestations of *P. sialia* are usually most severe in nests located near open water. He concludes that this tendency reflects habitat preferences of adult *P. sialia* and also the metamorphosing larvae's need for relatively high ambient humidity. However, in our study, parasite loading in the wetland site was significantly lower than the field sites, indicating that infestations do not always conform to this trend. Bennett (1957) also suggests that Protocalliphoran infestation rates vary according to the degree of nest exposure to wind, rain, and light. However, since nest boxes in both our field and wetland sites were located in open areas, with the same orientation of the cavity entrance, this would not account for the lower parasite loads observed in the wetland site.

Parasite loading may also be influenced by the internal environment of the nest. This microenvironment is largely determined by nest composition, which in turn may reflect the habitat in which the nest is located. Variation in the degree of infestation among individual nests would be increased if ovipositing flies exhibit microhabitat (nest) preferences or if larvae exhibit microhabitat requirements which result in differential rates of oviposition and/or survival of the larvae. The distinct composition of nests in wetland sites may partially explain the relatively low levels of parasitism observed in the wetland site. Relative to nests in the field sites, most nests in the wetland site consisted of a smaller proportion of grass and larger proportions of both moss and pine needles. In contrast to chemically benign grasses which typify the nest material in field sites, the mosses and pine needles of nests in wetland sites may contain secondary compounds which deter oviposition by gravid females or impede larval development (Bennett 1957; Levin 1971; Wimberger 1984; Clark, Chapter 11, this volume). Tree swallow nests also vary with respect to the number of lining feathers they contain (Capreol 1983). Larvae often pupate amongst these feathers (personal observation) and Mason (1944) suggests, moreover, that feathers afford cover to ectoparasitic larvae or pupae. However, in a previous study on this population of tree swallows, the number of feathers was not significantly correlated with parasite loading (Capreol 1983), indicating that the feather lining neither favours nor deters infestations of *P. sialia*.

Parasite loads might be expected to increase with nest density due to the increased proximity of hosts which facilitates dispersal of parasites (Brown and Brown 1986; Møller 1987; Shields and Crook 1987). However, we found that parasite loads in dispersed nests were not significantly smaller than those in colonial nests, and also that parasite loads did not differ significantly among nests in the small and large colony field sites. The dispersion patterns of *P. sialia,* with a parasitic larva which does not disperse, but an adult which is potentially very mobile, might be expected to differ from those parasites which rely on direct movement or transportation on the host for dispersal from nest to nest. Also, the relatively large distances between nests of tree swallows (typically 20–40 m), compared to other colonial swallows where adjacent nests may be in direct contact, would probably affect the dispersion patterns of the parasite. The high parasite loads of solitary nests suggest that *P. sialia* adults are indeed highly mobile and can disperse over relatively long distances. This contrasts with Bennett's (1957) assertion that adult Protocalliphora are not highly mobile.

Our study shows a clear trend of increased Protocalliphoran parasite loads as the breeding season progresses, as has been found in other studies (Bennett 1957; Pinkowski 1977; Shields and Crook 1987). This pattern probably occurs because adult Protocalliphora become more active as ambient temperatures increase, and because reproductive flies become more abundant as the new generation matures. Furthermore, nests of potential hosts become more limited in number as the end of the season approaches, so oviposition becomes more concentrated within the remaining nests.

Parasite loading may be regulated in a density-dependent fashion. Our attempts to experimentally modify parasite loading were largely ineffective. The removal or addition of larvae may have respectively stimulated or inhibited further oviposition within the nest (Mason 1944). In addition, when Protocalliphoran larvae become crowded, there is increased larval competition for space and for feeding sites which may result in cannibalism (B.P. Smith, personal communication). Accordingly, the number of parasites in a nest should increase with increasing volume of nest material and/or brood size. Pinkowski (1977) found that, although eastern bluebird nests in man-made and natural cavities had equal probabilities of infestation, parasite loading was higher in man-made cavities which, being more spacious than natural ones, contained larger nests. In our study, relative nest volumes were measured as a function of depth of nesting material, since almost all nest boxes were of equal floor area. However, we found no significant relationship between nest depth and parasite loading. Furthermore, brood loading was not significantly correlated with brood size, which suggests that the larvae were not constrained by available food resources within the observed range in intensity of parasitism. Other population regulation mechanisms could include secondary parasitism by the wasp *Nasonia vitripennis* (Mason 1944; Jones and Turner 1987; personal observation), bacterial and viral infection, predation, and overwintering survivorship (Mason 1944).

Development and survival of nestlings

Larvae of *P. sialia* indisputably live at their hosts' expense, since nestlings that experience blood loss through parasitism must reallocate energy from maintenance and growth into tissue regeneration and repair, as well as into other physiological responses to parasitism. However, our study shows that heavily parasitized nestlings do not necessarily suffer unusually low growth rates or high mortality. Several instances of heavily infested, yet successful nests have been documented elsewhere as well (Rothschild and Clay 1952; Bennett 1957; Whitworth 1976; Pinkowski 1977; Gold and Dahlsten 1983).

Haufe (1987) suggests that the rationale for stable host–parasite associations is mutual adaptation of host and parasite species. Successful host species evolve physiological and/or behavioural tolerance of parasitic exploitation (Esch *et al.* 1975; Brown and Brown 1986; Emlen 1986; Haufe 1987). Moreover, parasitic species must evolve life history characteristics that exploit their host populations within the context of host availability (Loye 1985). It has previously been interpreted as maladaptive for sedentary parasites to kill their hosts (Noble and Noble 1982; Gold and Dahlsten 1983; Haufe 1987). However, reason would indicate that host nesting behaviour, arrival and migration patterns, and physiology respond to the pressures of nestling blood loss and reduced fitness.

The impact of parasites upon their hosts is often dependent upon the severity of infestation (Rothchild and Clay 1952; Pinkowski 1977; Wilson 1979). Typical infestations of *P. sialia* on tree swallows may be below the level at which nestlings are significantly impaired since limited withdrawal of blood does not result in a reduction of body weight or increased mortality among small birds (Stangel 1986). Density-dependent regulation within the complex nest community may commonly serve to restrict infestations to tolerable levels.

Brooks (1979) suggests that, through time, co-accommodation of parasitic and host species will result in a correlation of life-cycle stages and reduced pathogenicity. The growth of Protocalliphoran larvae and tree swallow nestlings is coordinated such that energetic demands imposed by the parasites are relatively low during the most critical stages of the nestlings' development. During the week following hatch, when nestlings are growing most rapidly (Quinney and Ankney 1985) and have correspondingly high energy requirements (Ricklefs 1968), larvae are absent from the nest or are relatively small. Furthermore, Procallophoran larvae most often complete development and stop feeding well before their hosts fledge (Bennett 1957), and anaemic nestlings, while still in the nest, may recover rapidly upon cessation of blood loss (Whitworth 1976).

It is also possible that deleterious effects of parasitism are somewhat averted, since tree swallow nestlings may have incidental tolerance of considerable blood loss. According to O'Connor's (1977) adaptive growth hypothesis, nestlings of many aerial insectivores have a flexible growth strategy which accommodates their dependence on fluctuating food supplies. Such nestlings tend either to store resources within their bodies for use during periods of food shortage, or to conserve existing resources by temporarily suspending growth and thereby diverting energy from biosynthesis into maintenance. If blood loss (energy loss) to parasites is in any way comparable to food shortage (energy shortage), plasticity of growth could serve to mitigate the immediate deleterious effects of parasitism.

The immediate costs of parasitism could be further offset by actions of the parent tree swallows. If parents of parasitized young increase their feeding rates in response to increased nestling demands, the augmented food supply may counteract the drain on nestling energy reserves (Mason 1944). Moreover, since host inactivity may facilitate parasite attachment, well-fed nestlings may experience lesser degrees of parasitism than hungry nestlings (Pinkowski 1977; Bortolotti 1985). However, our observations of broods with nestling loads ranging between 0 and 17 parasites per nestling indicate no significant correlation between the degree of parasitism and feeding rates by the parents under the normal weather conditions which prevailed during this study.

Subtle negative effects of parasitism may have been masked by the effects of factors that influence nesting success such as genetic character (Ricklefs 1968), sibling competition (Paynter 1954), quality of parental care (Austin and Low 1932; Ricklefs 1968; Quinney *et al.* 1986), and environmental conditions (Mason 1944; Ricklefs 1968). This possibility is substantiated by the wide range in average nestling weight observed in the non-parasitized tree swallow nests. Accordingly, parasitic exploitation should be acknowledged as only one of numerous stresses that nestlings may encounter (Whitworth 1976). Gold and Dahlsten (1983) point out that in many cases where nestling mortality has been attributed to Protocalliphoran ectoparasitism, conclusive evidence for the cause of death is lacking because of these complexities. Yet, in a previous 1-year study on part of the population of tree swallows that we studied, parasite loading was found to be correlated negatively with the number as well as the proportion of nestlings fledged (Capreol 1983). Mason (1944) suggests that parasitic exploitation may have severe effects on host nestlings under conditions of increased energy requirement and decreased energy supply, such as cold, wet, or windy conditions which reduce food availability. Capreol's (1983) results indicate that, in some years, parasitism by *P. sialia* might have a deleterious effect on growth and survival of tree swallow nestlings. However, the present study, covering a larger population over a 5-year period, indicates that, under most environmental conditions and over a wide range of parasite loading, parasitism by *P. sialia* has little, if any, effect on nestling tree swallows.

Given the apparent subtlety of effects in this and many avian parasite–host interactions, future research should concentrate on the sublethal effects of parasites as well as on the behavioural and physiological adaptations by which hosts minimize the impact of parasites. There is much scope for experimental manipulation of parasite–host relationships in species such as tree swallows which readily occupy nest boxes.

Acknowledgements

We are grateful to Barb Dowsley, Andy McClure, Tracey Wood, Janet Cox, Anne Davies, Laurie Robertson, Lisa Venier, Catherine Vardy, and Wally Rendell for their assistance in the field, and in sorting through nests, and to Andy Hurly and Kelvin Conrad for statistical advice. We also appreciate having the use of the facilities of the Queen's University Biological Station for this study. This project was supported by a grant from the Natural Sciences and Engineering Research Council of Canada to RJR.

References

Austin, O.L. Jr., and Low, S.H. (1932). Notes on the breeding of the tree swallow. *Bird banding* **3**, 39–44.

Bennett, G.F. (1957). Studies on the genus *Protocalliphora* (Diptera: Calliphoridae). Unpublished D.Phil. thesis, University of Toronto.

Bortolotti, G.R. (1985). Frequency of *Protocalliphora avium* (Diptera: Calliphoridae) infestations on bald eagles (*Haliaeetus leucocephalus*). *Canadian Journal of Zoology,* **63**, 165–8.

Boyd, E.M. (1951). The external parasites of birds: a review. *Wilson Bulletin,* **63**(4), 363–9.

Brooks, D.R. (1979). Testing the content and extent of host–parasite coevolution. *Systematic Zoology,* **28**, 299–307.

Brown, C.R. and Brown, M.B. (1986). Ectoparasitism as a cost of coloniality in cliff swallows (*Hirundo pyrrhonota*). *Ecology,* **67**, 1206–18.

Capreol, M. (1983). The adaptive significance of the feather lining in the tree swallow (*Tachycineta bicolor*) nest and its effect on reproductive success. Unpublished B.Sc thesis, Queen's University.

DeSteven, D. (1978). The influence of age on the breeding biology of the tree swallow (*Iridoprocne bicolor*). *Ibis,* **120**, 516–23.

Emlen, J.T. (1986). Responses of breeding cliff swallows to nidicolous parasite infestations. *Condor,* **88**, 10–111.

Esch, G.W., Whitfield-Gibbons, J., and Bourque, J. E. (1975). An analysis of the relationship between stress and parasitism. *American Midland Naturalist,* **93**(2), 339–53.

Esehuis-van der Voet, C.W. and Kluyver, H.N. (1971). Parasitism by *Protocalliphora azurea* (Fall). *Verh. K. Ned. Akad. Wet. Afd.* Natuurk de Tweede Reeks, **60**, 50–51. (Also in Gold and Dahlsten, (1983).)

Garrison, B.A., Vouchilas, C., and Stauffer, D.F. (1986). Nestling great crested flycatcher parasitized by larval fly (*Protocalliphora hirundo*). *Wilson Bulletin,* **98**(2), 321.

George, J.L. and Mitchell, R.T. (1948). Notes on two species of Calliphoridae (Diptera) parasitizing nestling birds. *Auk*, **65**, 549–52.

Gold, C.S. and Dahlsten, D.L. (1983). Effects of parasitic flies (*Protocalliphora* spp.) on nestlings of mountain and chestnut-backed, chickadees. *Wilson Bulletin*, **95**, 560–72.

Haufe, W.O. (1987). Host–parasite interaction of blood-feeding dipterans in health and productivity of mammals. *International Journal of Parasitology*, **17**(2), 607–14.

Herman, C.M. (1936). Ectoparasites and bird diseases. *Bird Banding*, **7**, 163–6.

Hicks, E.A. (1962). Checklist and bibliography on the occurrence of insects in birds' nests. Supplement I. *Iowa State College Journal of Science*, **36**, 233–398.

Hill, H.M. (1948). Raptorial hosts of Protocalliphora. *Condor*, **50**, 31.

Hill, H.M. and Work, T.H. (1947). Protocalliphoran larvae infesting nestling birds of prey. *Condor*, **49**, 74–5.

Johnson, C.W. (1930). Notes on *Protocalliphora* during the summer of 1930. *Bird Banding*, **1**, 169–73.

Jones, T.H. and Turner, B.D. (1987). The effect of host spatial distribution on patterns of parasitism by *Nasonia vitripennis*. *Entomologia Experimentalis et Applicata*, **44**, 169–75.

Knapton, R.W. (1984). Parental feeding of nestling Nashville Warblers: the effects of food type, brood size, nestling age, and time of day. *Wilson Bulletin*, **96**, 594–602.

Kuerzi, R.G. (1941). Life history studies of the tree swallow. *Proceedings of the Linnaean Society*, **52–53**, 1–52.

Levin, D.A. (1971). Plant phenolics: an ecological perspective. *American Naturalist*, **105**, 157–81.

Low, S.H. (1933). Further notes on the nesting of tree swallows. *Bird Banding*, **4**, 76–87.

Loye, J.E. (1985). Host-effects of feeding and survival of the polyphagus cliff swallow bug, *Oeciacus vicarius* (Hemiptera, Heteroptera: Cimicidae). *Bulletin of the Society of Vector Ecology*, **10**, 7–13.

Mason, E.A. (1944). Parasitism by *Protocalliphora* and management of cavity-nesting birds. *Journal of Wildlife Management*, **8**, 232–47.

McAtee, W.L. (1929). Further notes on insect inhabitants of bird houses. *Proceedings of the Entomological Society of Washington*, **31**, 105–11. (Also in Mason, E. A. (1944).)

Møller, A.P. (1987). Advantages and disadvantages of coloniality in the swallow *Hirundo rustica*. *Animal Behaviour*, **35**, 819–32.

Moss, W.W. and Camin, J.H. (1970). Nest parasitism, productivity, and clutch size in purple martins. *Science*, **168**, 1000–3.

Muldal, A., Gibbs, H.L., and Robertson, R.J. (1985). Preferred nest spacing of an obligate cavity-nesting bird, the tree swallow. *Condor*, **87**, 356–63.

Neff, J.A. (1945). Maggot infestation of nestling mourning doves. *Condor*, **47**, 73–6.

Noble, E.R. and Noble, G.A. (1982). *Parasitology*. Lea and Febiger, Philadelphia.

O'Connor, R.J. (1977). Growth strategies in nestling passerines. *Living Bird*, **16**, 209–38.

Paynter, R.A. Jr. (1954). Interrelations between clutch-size, brood-size, prefledging survival, and weight in Kent Island tree swallows. *Bird Banding*, **25**, 35–59, 102–11, 136–49.

Pinkowski, B.C. (1977). Blowfly parasitism of eastern bluebirds in natural and artificial nest sites. *Journal of Wildlife Management*, **41**, 272–6.

Plath, O.E. (1919). Parasitism on nestling birds by fly larvae. *Condor*, **21**, 30–8.

Quinney, T.E. and Ankney, C.D. (1985). Prey size selection by tree swallows. *Auk*, **102**, 245–50.

Quinney, T.E. and Ankney, C.D. (1986). Sources of variation in growth of tree swallows. *Auk*, **103**, 389–400.

Ricklefs, R.E. (1968). Patterns of growth in birds. *Ibis*, **110**(4), 419–48.

Robertson, R.J. and Gibbs, H.L. (1982). Superterritoriality in tree swallows: a re-examination. *Condor*, **84**, 313–16.

Robertson, R.J. and Stutchbury, B.J. (1988). Experimental evidence for sexually selected infanticide in tree swallows. *Animal Behaviour*, **36**, 749–53.

Rothschild, M. and Clay, T. (1952). *Fleas, flukes and cuckoos*. Philosophical Library, New York.

Shields, W.M. and Crook, J.R. (1987). Barn swallow coloniality: a net cost for group breeding in the Adirondacks? *Ecology*, **68**, 1373–86.

Stangel, P.W. (1986). Lack of effects from sampling blood from small birds. *Condor*, **88**, 244–5.

Stutchbury, B.J. and Robertson, R.J. (1987). Two methods of sexing adult tree swallows before they begin breeding. *Journal of Field Ornithology*, **58**, 236–42.

Stutchbury, B.J. and Robertson, R.J. (1988). Within-season and age-related patterns of reproductive performance in female tree swallows *Tachycineta bicolor*. *Canadian Journal of Zoology*, **66**, 827–34.

Wetherbee, K.B. (1932). Two pairs of tree swallows mated during two successive seasons. *Bird Banding*, **3**, 72–3.

Whitworth, T.L. (1976). Host and habitat preferences, life history, pathogenicity and population regulation in species of Protocalliphora Hough (Diptera: Calliphoridae). Unpublished Ph.D. thesis, Utah State University, Logan, Utah. (Also in Gold and Dahlsten (1983).)

Wilson, R.A. (1979). *An introduction to parasitology* (2nd edn). Camelot Press Ltd, London.

Wimberger, P.H. (1984). The use of green plant material in bird nests to avoid ectoparasites. *Auk*, **101**, 615–18.

Zach, R. and Mayoh, K.R. (1982). Weight and feather growth of nestling tree swallows. *Canadian Journal of Zoology*, **60**, 1080–90.

8 *Parasite communities in wet and dry forest subpopulations of the Hawaii common amakihi*

CHARLES VAN RIPER III

Introduction

Parasites can act in diverse ways upon a host population and in many instances the effects are subtle and often difficult to interpret. This is because, among other factors, differing environmental conditions and diverse host behavioural patterns have a profound influence upon parasite levels (Dogiel 1961; Bull 1964; Kennedy 1975; Holmes *et al.* 1977). Parasites can be a primary factor of population regulation, although the overall importance is probably more closely related to their role in increasing the susceptibility of the host to other mortality factors (Bradley 1972; Esch 1977; Forrester *et al.* 1978). This view is often extended to predict that, in times when food supplies are reduced, levels of parasite infection are greater (Esch *et al.* 1975). There are, however, few examples in which environmental variables and host behaviour in different habitats has been monitored so as to document the interrelationship of these differing variables upon parasite community structure.

Although some attention has been given to the parasites of poultry and introduced birds in Hawaii (Alicata 1969; Lewin and Holmes 1971; Smith and Guest 1974; Lewin and Mahrt 1983), little work has been done on the native Hawaiian birds. Warner (1968) speculated on the presumed role that introduced avian malaria and pox played in the extinction of the Hawaiian avifauna, and more recently van Riper *et al.* (1986) documented the impact that avian malaria is having on the extant endemic birds. Cid del Prado Vera *et al.* (1985) have identified nematodes, Baldwin (1948) cestodes, Goff (1980) ectoparasites, and Levine *et al.* (1980) new coccidia species from native birds. Van Riper and van Riper (1985) summarized all known avian parasites which have been reported in the land birds of Hawaii. But no information exists on parasite community structure in any native land bird species.

This study was concerned with a comparison of the effect of environmental and host behavioural factors upon parasite community

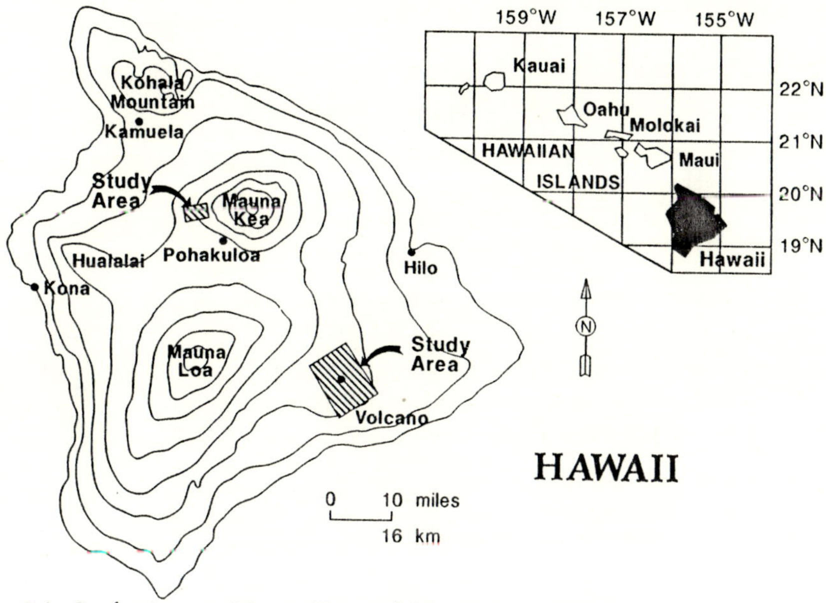

Fig. 8.1. Study sites on Mauna Kea and Mauna Loa, Hawaii.

structure in endemic common amakihi (*Hemignathus virens*) populations from the dry forest on Mauna Kea and wet forest on Mauna Loa, Hawaii. The questions that I attempted to answer were: (1) what parasites are present in subpopulations of a host species in different habitats on the same island? and (2) what are possible explanations for differing parasite community structures between habitats?

Methods

Study areas

This study was carried out on the island of Hawaii, on the southern slopes of Mauna Kea and Mauna Loa volcanoes (Fig. 8.1). I collected information on Mauna Kea at sites between 1980 and 2290 m elevation. This area is one of the last extensive dry forest ecosystems in the archipelago, typically a savannah, in places interrupted by bare lava, grassland, and scrub. The vegetation is dominated by two tree species, *Sophora chrysophylla* and *Myoporum sandwicense,* with a variety of shrub and groundcover plants (Hartt and Neal 1940; van Riper 1980). On Mauna Loa I sampled birds at sites from 1000 m elevation to the upper forest boundary at 2300 m elevation. The mesic vegetation on this volcano is dominated by ohia (*Metrosideros polymorpha*), koa (*Acacia koa*), and tree fern, (*Cibotium* spp.). Mueller-Dombois and Fosberg (1974) provide a detailed account of vegetation types in this region.

The bird

Hemignathus virens is one of the most common native birds on Mauna Kea (van Riper *et al.* 1978), while it is locally abundant on Mauna Loa (Scott *et al.* 1986). The species has been the subject of investigations concerning its distribution, breeding, ecology, and behaviour in the dry forest on Mauna Kea (Berger 1969; van Riper 1975, 1976, 1977, 1984, 1987; Kamil 1978; Kamil and van Riper 1982; Kern and van Riper 1984) and in the wet forest on Mauna Loa (Baldwin 1953; MacMillen 1974; Carpenter 1976; Pimm and Pimm 1982; Scott *et al.* 1986). The bird is a small (approximately 15 g) passerine of the endemic honeycreepers (Drepanidinae). Adult *H. virens* utilize considerable nectar during the breeding season, and through territoriality defend this resource (van Riper 1984). Their diet also includes insects and various fruits (Baldwin 1953).

Collection techniques

I collected *H. virens* parasites on Mauna Kea from 1972 through 1980, and on Mauna Loa from 1976 through 1980. Nests that had recently fledged young were placed in Berlese–Tullgren funnels for 48–60 hours and the fauna collected in 70 per cent ethanol with glycerine. From mist-netted *H. virens* toe clips were made and thin blood smears taken; slides were fixed for 30 s in absolute methyl alcohol, stained with Giemsa, then read under oil immersion at 1000 × until 30 000 red blood cells (RBC) had been surveyed. Faecal material collected from live-caught birds was transported in 2 per cent (w/v) $K_2Cr_2O_7$ solution, kept at room temperature for a week to allow any oocysts present to sporulate, stored in a refrigerator at 4°C, and later examined for coccidia.

Dead *H. virens* were either post-mortemed immediately or preserved in 70 per cent ethanol with glycerine. All post-mortemed birds were necropsied according to the technique outlined by van Riper and van Riper (1980). Ectoparasites were collected by shaking the bird in detergent and water for 2 min, removing the host, permitting the wash to stand for 1 h, breaking the surface tension with 95 per cent alcohol, then decanting; ectoparasites were preserved in 70 per cent alcohol. Feet and facial lesions were excised and a portion examined for mites before being preserved in 10 per cent formalin; the remainder was frozen and later examined for avian pox inclusion bodies by the USFWS Health Laboratory in Madison, Wisconsin. Impression smears of fresh organs were fixed in absolute methyl alcohol and stained with Giemsa. Cestodes were fixed in FAA, trematodes and nematodes in alcohol; all worms were preserved in 70 per cent ethanol with glycerine. Representative blood smears (no. 77567) have been deposited in the International Reference

Centre for Avian Hematozoa, St. Johns, Newfoundland, Canada; trematodes (no. 76484) at USNM at Beltsville, Maryland, USA; nematodes (no. 2127) at University of California Davis Nematode Collection; and all ectoparasites at the Bernice P. Bishop Museum, Honolulu, Hawaii, USA.

Results

Dry forest diseases

Nidicoles and Ectoparasites

Arthropods collected from 46 *H. virens* nests were associated with the nest as saprophages or as predators on other nest arthropods (Table 8.1).

Table 8.1. Acari collected from nests, body washes, and nasal cavities of *Hemignathus virens* on Mauna Kea and Mauna Loa, Hawaii

Order	Parasitiformes		
	Family	Rhinonyssidae:	*Ptilonyssus* n. sp.
		Macronyssidae:	*Ornithonyssus sylviarum* (Canestrini & Fanzago, 1877)
Order	Acariformes		
Suborder	Actinedida		
	Family	Cheyletidae:	*Cheyletus malaccensis* (Oudemans, 1903)
	Family	Harpyrhynchidae:	*Harpyrhynchus* sp.
Suborder	Acaridida		
	Family	Analgidae:	*Analges* n. sp.
	Family	Proctophyllodidae:	*Proctophyllodes* n. sp.
			Pterodectes n. sp.
	Family	Trouessartiidae:	*Trouessartia* n. sp.
	Family	Xolagidae:	*Ingrassiella* n. sp.

Only *Ornithonyssus sylviarum* (Parasitiformes: Macronyssidae) was parasitic to birds. There were, however, a number of parasitic acrina collected from body washes and nasal mites of the genus *Ptilonyssus* found in two birds (Table 8.1).

Avian pox

Avian pox lesions were observed on five (0.4 per cent) of 1250 mist-netted *H. virens*. Four birds had facial lesions, while the fifth had swellings on a leg and toes. One bird was a female, collected by hand on 27 March 1974 with a prominent keel; both opercula and the left eye were occluded by scabs and diphtheritic lesions extended into the oral cavity. This was during the height of the breeding season, yet no brood patch was evident. The bird died shortly after capture, as did one other individual with facial lesions.

Blood parasites

From mist-netted birds, 462 peripheral blood smears were taken and all were negative for parasites. None of the heart, liver, spleen, and lung impression smears of 27 post-mortemed birds contained blood parasites. It appears that haematozoa and other blood parasites are absent from the dry forest *H. virens* population on Mauna Kea.

Avian malaria did, however, develop in birds that were taken to lower elevations for behavioural studies. In 1972 three *H. virens* were removed from 2290 m elevation on Mauna Kea and transported to Kamuela, at approximately 760 m elevation on the southern slope of Kohala Mountain (Fig. 8.1). The birds were maintained in an outdoor aviary. One bird died at 32 days and the other two at 45 days, all from acute infections of *Plasmodium relictum*. In 1978, 15 birds were taken from the same location on Mauna Kea to Hawaii Volcanoes National Park at 1230 m elevation on the south-east slope of Mauna Loa. Six of these birds were challenged with malarial parasites, all became infected, and four died (see van Riper *et al.* 1986 for details of this experiment).

Gastrointestinal parasites

In December 1973 and January 1974, faecal samples collected from 14 mist-netted *H. virens* revealed three birds infected with *Isospora loxopis*. This coccidia was also found in three of the post-mortemed *H. virens*. All individuals exhibited light and, what appeared to be, non-pathogenic infections.

During post-mortem examinations of 27 birds, a variety of other gastrointestinal parasites were encountered (Table 8.2). The tapeworm, *Anonchottaenia brasilense*, was found in three adult birds and, in one individual, worms were so numerous that they formed a solid mass that could have influenced food passage through the intestinal tract. The most frequently encountered parasite, *Capillaria*, was present in 9 of the 27 birds, while *Viguiera hawaiiensis* occurred in six. The prevalence of *Capillaria* fell from 58 per cent in 1972, to 33 per cent in 1973, and in 1974 none were detected in any of the post-mortemed birds. One immature bird had three flukes (*Urotocus rossittensis*) in the bursa of Fabricius.

Table 8.2. Parasites found in *Hemignathus virens* from Mauna Loa mesic and Mauna Kea xeric forests on the island of Hawaii

Parasite	Mauna Loa			Mauna Kea		
	Number infected	(*N*)	%	Number infected	(*N*)	%
*Poxvirus**	20	(518)	3.9	5	(1250)	0.4
*Plasmodium**	38	(518)	7.3	0	(462)	—
Anonchotaenia	4	(15)	26.6	3	(27)	11.1
Viguiera	8	(15)	53.3	6	(27)	22.2
*Capillaria**	0	(15)	—	9	(27)	33.3
Urotocus	1	(15)	6.6	1	(27)	3.7
Isospora	3	(24)	12.5	6	(41)	14.6

* Significant differences ($p \leqslant 0.05$) of parasite numbers between habitats using chi-square corrected analysis.

Wet forest diseases

Nidicoles and ectoparasites

Thirty-three *H. virens* were body washed for ectoparasites and several genera of parasitic acarina were present (Table 8.1). All birds were parasitized by at least one genus, but parasite loads were often quite low. Two of 15 necropsied *H. virens* harboured nasal mites of the genus *Ptilonyssus* (van Riper and van Riper 1985). Three *H. virens* nests from Mauna Loa yielded eight genera of arthropods, all of which were saprophagous or predators on other nest arthropods except for one that contained *Ornithonyssus sylviarum* (Goff 1980).

Avian pox

Lesions, characteristic of the avian pox virus, occurred on 20 (3.9 per cent) of 518 *H. virens* captured by mist-net, while 52 additional birds (10 per cent) had digits missing or scars, apparently the result of an earlier infection. The majority of infected birds ($n = 19$) had lesions on the legs and feet, with facial lesions on one individual. Avian pox virus contributed to the death of at least eight of the 14 recently dead or moribund *H. virens* found on Mauna Loa during this study. The incidence of lesions was greatly influenced by elevation in that birds from the mid-elevational ranges had the greatest prevalence. This is apparently the result of the absence of high vector (mosquito) densities in the upper forest areas (Goff and van Riper 1980), where *H. virens* is much more abundant.

Blood parasites

From 518 *H. virens* sampled in the wet forests of Mauna Loa, 38 (7 per cent) were found positive for *Plasmodium relictum*. As with avian pox, malaria prevalence was influenced by elevation. Birds from 900 to 1500 m had a much higher prevalence (17.2 per cent) than did birds above 1500 m elevation (2.1 per cent). The Mauna Loa *H. virens* population is, however, one of the most resistant of the native passerines to malarial parasite infection (van Riper *et al.* 1986).

Gastrointestinal parasites

Of 24 faecal samples from *H. virens,* three adult males were found positive for *Isospora loxopis*. As in the dry forest areas, prevalence and intensity were not high. None of the post-mortemed birds had detectable coccidia infections.

A total of 15 post-mortem examinations were done on birds from the wet forest areas. Most parasite species found in the dry forest were also present, but some occurred in significantly different prevalences (Table 8.2). *Anonchottaenia brasilense* had a higher prevalence in *H. virens* adults and young from mesic forest habitat. *Capillaria* was absent in the *H. virens* that I sampled from Mauna Loa, although it is found in other native birds in that habitat (van Riper and van Riper 1985). It appears that *Viguiera hawaiiensis* is the most abundant gastrointestinal parasite in the Mauna Loa subpopulation. One young bird parasitized by the fluke (*Urotocus rossittensis*) had the bursa of Fabricius completely filled so that this organ occupied a major portion of the posterior abdominal cavity.

Discussion

The parasite communities of xeric and mesic forest *H. virens* subpopulations on Hawaii differ in several respects. A few parasites are unique to either the wet or dry areas, but most occur with regular freqency in both habitats. The parasites that occur in both habitats include nest and body ectoparasites, and most of the gastrointestinal parasites. Malarial and pox virus parasites predominate in the mesic forest, corresponding to greater vector (mosquito) abundance in this habitat.

Of the nest fauna that I collected, most were of minor parasitic importance. Only the northern fowl mite (*Ornithonyssus sylviarum*) feeds directly on host blood. It occurred in single nests from the wet and dry forests, but this does not necessarily indicate a low incidence. These Macronyssid mites share time between nest and host and rapidly leave the nest following fledging of the young and can, therefore, be easily missed if the nest is not immediately collected. A number of feather mites of the superfamily Analgoidea were found, but due to the complexity of their taxonomy complete identifications are not yet available from either the dry or wet forest birds. These mites have little impact on a host because they are saprophagous rather than parasitic, feeding on feather fragments, desquamated skin scales, and oil secretions from the host (Krantz 1978).

The *H. virens* parasitic helminth communities on Hawaii are quite simple and, when worm groups occur, they are represented by a single species. Several major groups are absent, for example, the Acanthocephala. The most numerous helminth parasites were gastrointestinal worms and each species occupied a different attachment site. Where several parasitic species occur within the same host, their distribution may be restricted by competition with other species (Holmes 1961; MacKenzie and Gibson 1970). However, this does not appear to be the case with *H. virens* in Hawaii because of the low number of potential gastrointestinal parasites. In all birds with *Capillaria,* the site of attachment was the upper third of the intestinal tract, while *Viguiera hawaiiensis* was found only in the proventriculus and gizzard. However, the cestode site of attachment was also the upper third of the small intestine. But as Kennedy (1975) has pointed out, conclusive evidence for interspecific competition for sites can only be proven by a comparison of the distribution of the species when they occur alone and together.

Several of these gastrointestinal parasites can have severe effects on their hosts. For example, *Viguiera* causes extreme erosion of the tissue in the proventriculus and gizzard (Cid del Prado Vera *et al.* 1985), and *Capillaria* can cause extensive damage to the intestinal lining such that the absorption capabilities may be greatly decreased (Wehr 1971). But most of the infections that we observed were not severe, and occurred singly in individuals.

Capillaria was the only gastrointestinal parasite that occurred in one habitat and not the other, as I found it exclusively in the dry forest. I am confident that if more birds were analysed it would occur in the mesic forest because it is found there in other avian species (Cid del Prado Vera *et al.* 1985; van Riper and van Riper 1985). The intermediate host of this parasite is unknown in Hawaii so speculation is difficult concerning the life cycle. The incidence of this helminth went from 58 per cent in 1972 to 0 per cent in 1974 in the dry forest, suggesting a dynamic population cycle.

The overall annual helminth infection rate in the dry forest was positively, but not significantly, correlated ($r = 0.425$) with the Mauna

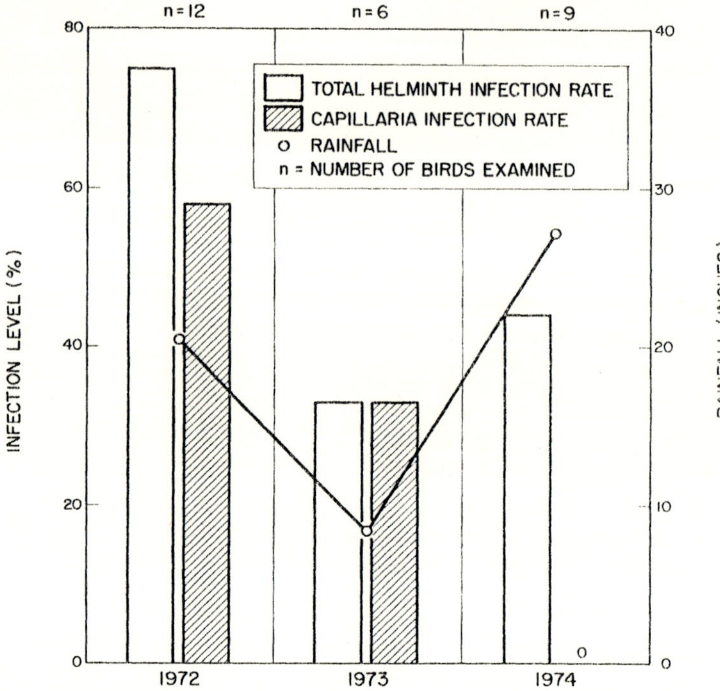

Fig. 8.2. Helminth infection rates as related to yearly rainfall on Mauna Kea, Hawaii.

Kea rainfall pattern (Fig. 8.2) This relationship did not hold true for the Mauna Loa birds probably because of the smaller host sample size and the more consistent pattern of interyear precipitation (Bridges 1981). There are several possible explanations for the increased gastrointestinal parasite loads of *H. virens* during wetter years in xeric habitat. The first is that, with increased food resources, the birds are able to physically support heavier parasite loads. If this is the case, it would suggest that the host-parasite system in *H. virens* is finely tuned and in dynamic equilibrium, being able to respond rapidly to changing environmental conditions. An alternative, and probably more likely, explanation is that in years of high rainfall there is a greater production of insect vectors (e.g. Conant 1985) and, because of the increased probability of encountering intermediate hosts, parasite loading in *H. virens* increases.

Climate differences certainly contribute to differences in *H. virens* parasite community structure, in that the much reduced precipitation of the xeric habitat precludes mosquito reproduction, and therefore a number of the vector-borne parasites are either absent or at extremely low prevalances. In the rainforest areas where mosquitoes occur, avian malaria is one of the most important diseases and continually exerts selection pressure on the entire host subpopulation. The lack of movement and exchange of individual hosts between subpopulations (Baldwin 1953; van Riper 1987), has resulted in differences in immunogenic capabilities to this vector borne parasite (van Riper *et al.* 1986).

Malarial parasites are sometimes very localized (Garnham 1966), and this appears to be the case in Hawaii. None of the *H. virens* on the higher slopes of Mauna Kea were found to harbour blood parasites. These birds are, however, quite susceptible as evidenced by individuals that were transferred to lower elevations on Kohala Mountain and contracted malaria, and by the greater than 65 per cent death rate of birds that van Riper *et al.* (1986) challenged with malarial parasites on Mauna Loa. Warner (1968), after moving native land birds to lower elevations on Kauai, also found that they became infected with malarial parasites. There are, however, native populations of *H. virens* that do survive at lower elevations. I have regularly observed individuals below 200 m elevation in Manoa Valley and Makiki Heights (Oahu), and on Hawaii in the Kiholo and lower Hilina Pali areas. Berger (1981) found *H. virens* at 76 m elevation in the Malama-Ki Forest Reserve on Hawaii. The extant *H. virens* populations that presently exist within vector range on Mauna Loa have a 7 per cent prevalence rate, and challenges with malarial parasites by van Riper *et al.* (1986) revealed that this species has developed immunogenic mechanisms of resistance to this introduced parasite. They postulated that this immunity developed because *H. virens* is sedentary and the mesic forest population at lower elevations was continually exposed to the malarial parasite, whereas the xeric forest subpopulation occurs above the mosquito distribution zone, and therefore rarely comes into contact with the parasite.

One disease that is present in both the forests of Mauna Kea and Mauna Loa, albeit at low levels in the former, is the avian pox virus. Several references to 'avian pox-like' lesions in endemic avifauna of Hawaii are found in the early literature (Perkins 1893; Henshaw 1902; Amadon 1950; Munro 1944). More recently Warner (1968) appears to have proven the susceptibility of some native and introduced birds to this virus. The disease was probably introduced to Hawaii through the importation of domestic chickens (*Gallus gallus*). Mosquitoes have been implicated as the primary mechanical carrier of avian pox infections, but the virus may also be spread directly by contact of susceptible birds with infected ones, by contact with contaminated objects, or by bird mites (Karstad 1971). Manifestation of diphtheritic pox lesions around the oral cavity are not frequently observed in the wild, and are more likely to be pathogenic to the bird than are lesions on the leg or foot. The lack of mosquitoes in the higher xeric forests of Mauna Kea may be why the incidence of avian pox (0.5 per cent) was so low. It would appear that in the absence of mosquitoes, avian pox is probably present in the endemic birds of Mauna Kea either because of direct contact with a contaminated object, possibly by transovarial transmission of the virus *Ornithonyssus sylviarum*, or probably by contracting the disease at lower elevations during postbreeding movements. Four of the five xeric forest *H. virens* found with avian pox had the more debilitating diphtheritic form, suggesting a high susceptibility of this host subpopulation to *Poxvirus avium*.

The prevalence of avian pox was much higher (14 per cent) in the wet forest areas of Mauna Loa, undoubtedly the result of increased vector densities (Goff and van Riper 1980). But only one bird had diptheritic facial lesions while 19 others had the more common skin form with leg and/or toe lesions. The small number of the more debilitating facial lesions detected in the wet forest birds might indicate a greater resistance to *Poxvirus avium* than in the less frequently exposed dry forest subpopulation. This virus was present in over half (57 per cent) of the moribund or dead mesic forest *H. virens* found during this study, suggesting that, although it does not appear by itself to be a major mortality factor, in combination with other diseases it can contribute to mortality in this native host species. This is much the same situation that D. Forrester (personal communication) found in Florida wild turkeys, where avian pox alone had little impact but, when combined with other active diseases, resulted in higher host mortality.

In conclusion, the different parasite communities of *H. virens* subpopulations on the island of Hawaii appear to be the result of several interacting factors. Climatic differences certainly contribute, in that the much reduced precipation of the xeric habitat precludes mosquito reproduction, and therefore parasites borne by this vector are either absent (malaria) or at extremely low prevalances (avian pox). The lack of host movement and concomitant genetic exchange between subpopulations has resulted in immunogenic differences to the malarial parasite on Hawaii (van Riper *et al.* 1986), and possibly to *Poxvirus avium*. Those parasites (other than avian malaria) that occur in both habitats as single species appear not to be pathogenic, but additivity or complementarity could be a problem.

Acknowledgements

I am grateful to A. R. Maggenti, N. D. Levine, A. O. Bush, and M. L. Goff for identification of parasites. S. G. van Riper assisted with collection of field data and made valuable comments on various drafts of this chapter. Financial support was provided by the World Wildlife Fund Grant US-35, the Frank M. Chapman Memorial Fund, Earthwatch, and Contract CX 8000 7 0009 from the National Park Service to the Cooperative National Park Resources Studies Units at the University of Hawaii and the University of California, Davis.

References

Alicata, J.E. (1969). *Parasites of man and animals in Hawaii.* S. Karger, Basel.

Amadon, D.O. (1950). The Hawaiian honeycreepers (Aves, Drepaniidae). *Bulletin of the American Museum of Natural History*, **95**, 151–262.

Baldwin, P.H. (1948). Discovery of tapeworm parasites in three genera of Hawaiian honey suckers. *Elepaio,* **9**, 29.

Baldwin, P.H. (1953). Annual cycle, environment and evolution in the Hawaiian honeycreepers (Aves: Drepaniidae). *University of California Publications in Zoology,* **52**, 285–398.

Berger, A.J. (1969). The breeding season of the Hawaii Amakihi. *Occasional Papers of Bishop Museum* **24**, 1–8.

Berger, A.J. (1981). *Hawaiian birdlife* (2nd ed. University Press of Hawaii, Honolulu.

Bradley, D.J. (1972). Regulation of parasite populations: a general theory of the epidemiology and control of parasite infections. *Transactions of the Royal Society of Tropical Medicine and Hygiene,* **66**, 697–708.

Bridges, K.W. (1981). Temporal relationships of island biota: introduction. In *Island ecosystems: biological organization in selected Hawaiian communities* (ed. D. Mueller-Dombois, K.W. Bridges, and H.L. Carson), pp. 367–91. Hutchinson Ross, Stroudsburgh, Pennsylvania.

Bull, P.C. (1964). Ecology of helminth parasites of the wild rabbit (*Oryctolagus cuniculus*) in New Zealand. *Bulletin of the New Zealand Department of Science and Industrial Research,* **158**, 1–147.

Carpenter, F.L. (1976). Plant-pollinator interactions in Hawaii: pollination energetics of *Metrosideros collina* (Myrtaceae). *Ecology,* **57**, 1125–44.

Cid del Prado Vera, I., Maggenti, A.R. and van Riper, C. III. (1985). New species of Spiruridae (Nemata: Spirurida) from Hawaiian honeycreepers (Passeriformes: Drepanididae), the Japanese White-eye (Passeriformes: Zosteropidae) and a new species of Acuariidae (Nemata: Spirurida) from the Japanese white-eye collected on the island of Hawaii. *Proceedings of the Helminthological Society, Washington,* **52**, 247–59.

Conant, M. (1975). *Seasonal abundances of the mamane moth, its nuclear polyhedrosis virus, and its parasites.* US IBP Island Ecosystems IRP Technical Report, no. 64. University of Hawaii, Honolulu.

Dogiel, V.A. (1961). Ecology of the parasites of freshwater fishes. In *Parasitology of fishes* (ed. V.A. Dogiel, G.K. Petrushevski, and Y.I. Polyanski), pp. 1–47. Oliver and Boyd, Edinburgh.

Esch. G.W. (ed.) (1977). *Regulation of parasite populations.* Academic Press, New York.

Esch, G.W., Gibbons, J.W., and Bourque, J.E. (1975). An analysis of the relationship between stress and parasitism. *American Midland Naturalist,* **93**, 339–53.

Forrester, D.J., Carpenter, J.W., and Blankinship, D.R. (1978). Coccidia of whooping cranes. *Journal of Wildlife Diseases,* **14**, 24–7.

Garnham, P.C.C. (1966). *Malaria parasites and other Haemosporidia*. Blackwell Scientific, Oxford.

Goff, M.L. (1980). *Mites (Chelicerata: Acari) parasitic on birds in Hawaii Volcanoes National Park*. Cooperative Parks Studies Unit, University of Hawaii Technical Report, no. 29, Honolulu.

Goff, M.L. and van Riper, C. III (1980). Distribution of mosquitoes (Diptera: Culicidae) on the east flank of Mauna Loa Volcano, Hawaii. *Pacific Insects*, **22**, 178–88.

Hartt, C.E. and Neal, M.C. The plant ecology of Mauna Kea, Hawaii. *Ecology*, **21**, 237–66.

Henshaw, H.W. (1902). *Birds of the Hawaiian possessions with notes on their habits*. T.G. Thrum, Honolulu.

Holmes, J.C. (1961). Effects of concurrent infections on *Hymenolepis diminuta* (Cestoda) and *Moniliformis dubis* (Acanthocephala). I. General effects and comparison with crowding. *Journal of Parasitology*, **47**, 209–16.

Holmes, J.C., Hobbs, R.P., and Leong, T.S. (1977). Populations in perspective: community organization and regulation of parasite populations. In *Regulation of parasite populations* (ed. G.W. Esch), pp. 209–45. Academic Press, New York.

Kamil, A.C. (1978). Systematic foraging by a nectar-feeding bird, the amakihi (*Loxops virens*). *Journal of Comparative Physiology and Psychology*, **92**, 388–96.

Kamil, A.C. and van Riper, C. III (1982). Within-territory division of foraging space by male and female Amakihi (*Loxops virens*). *Condor*, **84**, 117–19.

Karstad, L. (1971). Pox. In *Infectious and parasitic diseases of wild birds*. (ed. J.W. Davis, R.C. Anderson, L. Karstad, and D.O. Trainer), pp. 34–42. Iowa State University Press, Ames.

Kennedy, C.R. (1975). *Ecological animal parasitology*. John Wiley, New York.

Kern, M.D. and van Riper, C. III (1984). Altitudinal variations in nests of the Hawaiian honeycreeper *Hemignathus virens virens*. *Condor*, **86**, 443–54.

Krantz, G.W. (1978). *A manual of acarology*. Oregon State University Press, Corvallis, Oregon.

Levine, N.D., van Riper, S., and van Riper, C. III (1980). Five new species of *Isospora* from Hawaiian birds. *Journal of Protozoology*, **27**, 258–9.

Lewin, V. and Holmes, J.C. (1971). Helminths from the exotic game birds of the Puuwaawaa Ranch, Hawaii. *Pacific Science*, **25**, 372–81.

Lewin, V. and Mahrt, J.L. (1983). Parasites of Kalij pheasants (*Lophura leucomelana*) on the island of Hawaii. *Pacific Science*, **37**, 81–3.

MacKenzie, K. and Gibson, D. (1970). Ecological studies on some parasites of plaice (*Pleuronectes platessa* L.) and flounder (*Platichthys flesus* L.). *Symposia of the British Society of Parasitology*, **8**, 1–42.

MacMillen, R.E. (1974). Bioenergetics of Hawaiian honeycreepers: the amakihi (*Loxops virens*) and the anianiau (*L. Parva*). *Condor*, **76**, 62–9.

Mueller-Dombois, D. and Fosberg, F.R. (1974). *Vegetation map of Hawaii Volcanoes National Park*. University of Hawaii Cooperative National Park Resources Studies Unit Technical Report, no. 4. Honolulu.

Munro, G.C. (1944). *Birds of Hawaii*. Bridgeway Press, Rutland, Vermont.

Perkins, R.C.L. (1893). Notes on collecting in Kona, Hawaii. *Ibis*, **1893**, 101–14.

Pimm, S.L. and Pimm, J.W. (1982). Resource use, competition, and resource availability in Hawaiian honeycreepers. *Ecology*, **63**, 1468–80.

Scott, J.M., Mountainspring, S., Ramsey, F.L., and Kepler, C.B. (1986). *Forest bird communities of the Hawaiian Islands: their dynamics, ecology, and conservation*, Cooper Ornithological Society Studies in Avian Biology, no. 9, 1–43.

Smith, H.E. and Guest, S.J. (1974). *A survey of internal parasites of birds on the western slopes of Diamond Head, Oahu, Hawaii 1972–1973.* US IBP Island Ecosystems IRP Technical Report, no. 37. University of Hawaii, Honolulu.

van Riper, C. III (1975). *Parasites of the Hawaii Amakihi (Loxops virens virens).* US IBP Island Ecosystems IRP Technical Report, no. 62, University of Hawaii, Honolulu.

van Riper, C. III (1976). Superimposition of Amakihi nest on one of an Elepaio. *Condor,* **78**, 267–9.

van Riper, C. III (1977). The use of sheep wool in nest construction by Hawaiian birds. *Auk,* **94**, 646–51.

van Riper, C. III (1980). The phenology of the dryland forest of Mauna Kea, Hawaii, and the impact of recent environmental perturbations. *Biotropica,* **12**, 282–91.

van Riper, C. III (1984). The influence of nectar resources on nesting success and movement patterns of the Hawaii Amakihi (*Hemignathus virens*). *Auk,* **101**, 38–46.

van Riper, C. III (1987). Breeding ecology of the Hawaii common amakihi. *Condor,* **89**, 85–102.

van Riper, C. III and van Riper, S.G. (1980). A necropsy procedure for sampling disease in wild bird populations. *Condor,* **82**, 85–98.

van Riper, C. III, Scott, J.M., and Woodside, D.H. (1978). Distribution and abundance patterns of the Palila on Mauna Kea, Hawaii. *Auk,* **95**, 518–27.

van Riper, C. III, van Riper, S.G., Goff, M.L. and Laird, M. (1986). Epizootiology and ecological significance of malaria in Hawaiian land birds. *Ecological Monographs,* **56**, 327–44.

van Riper, S.G. and van Riper, C. III (1985). A summary of known parasites and diseases from the avifauna of the Hawaiian Islands. In *The preservation and management of terrestrial Hawaiian ecosystems* (ed. C. Stone and J.M. Scott) pp. 298–374. University Press of Hawaii, Honolulu.

Warner, R.E. (1968). The role of introduced diseases in the extinction of the endemic Hawaiian avifauna. *Condor,* **70**, 101–20.

Wehr, E.E. (1971). Nematodes. In *Infectious and parasitic diseases of wild birds* (ed. J.W. Davis, R.C. Anderson, L. Karstad, and D. O. Trainer), pp. 185–233. Iowa State University Press, Ames.

9 *Analysis and ecological correlates of tick burdens in a New Guinea avifauna*

MELINDA PRUETT-JONES and STEPHEN PRUETT-JONES

Introduction

Ticks (Arachnida: Acari) are nearly ubiquitous ectoparasites of terrestrial vertebrates. Although taxonomically conservative, with approximately 800 species in 20 genera in three families (Ixodidae, Argasidae, and Nuttalliellidae; Hoogstraal 1978), ticks are geographically widely distributed, being found on all continents, and most islands (Hoogstraal 1973). They infest most orders and families of mammals, reptiles, and birds and are vectors for a number of pathogens which are known to adversely affect the survival and reproduction of both human and non-human hosts (Hoogstraal 1966; Wilde 1978). As vectors of human diseases, ticks rank second only to mosquitoes (Obenchain and Galun 1982). With respect to bird populations, effects of tick infestations on reproduction include mortality of young, nest failure, and nest desertion (Duffy 1983, and references therein).

In this chapter we present data on the ecology of tick burdens in a tropical forest avifauna in Papua New Guinea. Specifically, we examine how tick infestations vary across species, individuals within a given species, and with seasons and altitude. We address tick–host relationships but, in contrast to standard treatments of this issue (e.g. Fairchild *et al.* 1966; Sonenshine and Stout 1970; Pegrem *et al.* 1981), we focus on a quantitative comparison of tick burdens across species and attempt to elucidate the proximate causes for such variation.

The ticks of New Guinea are poorly known. Doss *et al.* (1979) list 35 and Hoogstraal (1982) 31 species as recorded from the island. From the latter survey, the genus *Ixodes* is the most diverse, with 11 species. Seven of the 31 species appear to have been introduced to New Guinea from domestic animals. Of the native species, seven are endemic to New Guinea and another seven are endemic to both New Guinea and Australia (Hoogstraal 1982). A number of tick species, including some of those we studied, are still undescribed from New Guinea.

Study area

The majority of birds examined in this study were trapped over an approximately 1100-ha area from 1300 to 2200 m altitude on the south-western flank of Mount Missim (7°16′S, 146°47′E), Kuper Range, Morobe Province, Papua New Guinea. The area consisted of continuous, mature, montane rain forest and was undisturbed except for occasional hunting by local people. Small, natural gaps in the forest due to tree-fall or land slides were common but otherwise there were no habitat breaks or ecotones. The forest at 1300 m altitude adjoined cultivated areas of the Wau valley. Above the study area, mature rain forest continued to the summit of Mount Missim at 2860 m altitude. Tree-species diversity on a 6-ha plot at 1600 m altitude equalled 1826 adult trees of 168 species (Pratt 1983). This diversity is comparable to other montane forests on New Guinea (Paijmans 1976).

During 1982, the only year for which we have continuous weather data, rainfall totalled 196, 186, and 220 cm at 1600, 1850, and 2100 m altitude, respectively. Rain fell every week of the year (Fig. 9.1), but December and January were substantially wetter than other months. On Mount Missim, as across New Guinea generally, May to October represents the dry and November to April, the wet season (McAlpine *et al.* 1983). Weekly minimum and maximum temperatures on Mount Missim during 1982 at 1850 m altitude ranged from 9 to 15°C and 16 to 23°C, respectively (Fig. 9.1).

We also include data on tick burdens gathered during occasional work on Mount Missim up to 2400 m altitude as well as several records from birds netted in areas of second growth and disturbed vegetation on the grounds of Wau Ecology Institute (1150 m altitude).

The avifauna

There are approximately 150 bird species on Mount Missim above 1300 m altitude (Beehler 1978, 1982, 1983; Pratt 1983). Individual species show distinct patterns of distribution and abundance across altitude (Beehler 1978, 1982; Pruett-Jones and Pruett-Jones 1986) and few species are found across the entire study area. Numerically, the avifauna is dominated by fruit pigeons (Columbidae), parrots (Psittacidae), honeyeaters (Meliphagidae), and birds of paradise (Paradisaeidae). The species for which we have data comprise the majority of the common breeding species across the study area. We were unable to capture and examine for ticks species of nomadic fruit doves which were unpredictably present in any given year, and the exclusively canopy-dwelling species.

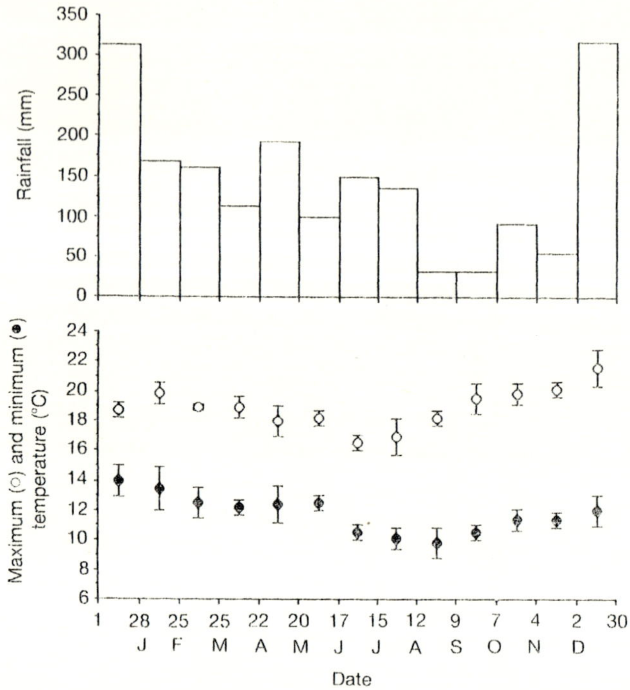

Fig. 9.1. Seasonality during 1982 of rainfall (upper) and minimum and maximum temperatures (lower) at 1850 m altitude on Mount Missim. Weather data were gathered each week but are summarized here in intervals of 4 weeks' duration. The bar graph for rainfall shows total rainfall recorded during each interval. The temperature plot shows mean and range (width of the vertical bars) of minimum and maximum temperatures recorded during each interval.

All of the species we studied share similar aspects of ecology in that they appear to reside in the forest throughout the year and do not migrate. Some species may show altitudinal movements, if not across seasons then with the age of the individual (Diamond 1972; Pruett-Jones and Pruett-Jones 1986).

Methods

We worked on Mount Missim from 1980 to 1987. The field work was continuous during the period September 1981 to December 1983, whereas in 1980 and then from 1985 to 1987 it was seasonal, from September to December each year. No data were gathered in 1984.

All birds except the megapode (*Aepypodius arfakianus*) were trapped in mist nets set along prominent ridge lines or, to a lesser extent, on side slopes and along water courses. The nets were set singly or as a tier of two together but canopy and tiers of four nets on a pulley system were occasionally used. Nets were opened during daylight hours, from 05.30 to 18.30, rarely during the night, and were checked at 30–60 min. intervals. All birds examined were free-flying and no nestlings were included.

Upon capture, birds were first banded and then examined for the presence of ticks. The examination was visual, lasted approximately 1 min, and consisted of scanning the bird's body while moving or blowing back feathers so that the skin could be seen. Initially, during the first 2 years of study, all body parts of each bird were checked. Without exception, however, ticks were only found attached to the bird's head, primarily above the eyes, on either side of the gape, and on the chin. After 1982, casual wholebody examinations were occasionally made, but intensive scanning of each individual was confined to the bird's head region. Engorged ticks were readily visible, and small nymphs could be seen once the bird's skin was exposed. Observers and assistants examining birds were frequently checked to make sure that consistent techniques were being used. Ticks were not, as a matter of protocol, removed from the birds, although about half of the ticks seen were collected for identification.

Many banded birds were recaptured and examined on multiple occasions (see 'Results'). Information on the period of attachment for the tick species involved (nymphal and immature *Ixodes*) is not available, but in general, *Ixodes* remain on their host between 1 and 2 weeks (J. Cairns, personal communication). As a conservative measure, we chose 1 month as the time between captures necessary for the second capture to be considered an independent sample; thus data from recaptures within the same month were deleted.

For each bird species, we summarized the following aspects of morphology and ecology:

- body size (total body length from Beehler *et al.* 1986);
- male showiness, reflecting showiness and elaborateness of plumage and appearance (a subset of 79 of the species studied here were scored for showiness in conjunction with a separate study of blood parasites and sexual selection in this avifauna; see Pruett-Jones *et al.* in press);
- sexual chromatism (species were categorized as either sexually dichromatic or monochromatic);
- foraging strata, reflecting the strata within the forest where the species foraged. Four levels were considered:
  (1) ground (<0.5 m);

(2) understorey (0.5–2 m);

(3) lower canopy (2–10 m);

(4) upper canopy (>10 m); mean scores were given to species foraging over more than one level;

- foraging levels, the number of different forest strata the species utilized;
- diet. The categories used were:

(1) nectarivorous;

(2) frugivorous;

(3) insectivorous, including other invertebrates;

(4) predatory (species eating terrestrial vertebrates);

(5) omnivorous (combination of nectarivorous, frugivorous, and insectivorous);

(6) general predators (combination of insectivorous and predatory); list modified from Bennett 1986 as cited in Read 1987 and unpublished manuscript.

We refer to tick prevalence as the proportion of individuals infested by at least one tick, relative intensity as the average number of ticks recorded across all individuals examined, and mean intensity as the average number of ticks recorded across all individuals that were infested (cf. Margolis *et al.* 1982; Duffy and Campos De Duffy 1986).

Statistical analysis

Because of the large number of species for which few individuals were examined (see below) and which exhibited 0 prevalence values, standard statistical analyses were not possible for interspecific comparisons. We took two separate approaches as a solution to this problem. First, for most comparisons, we only considered those species for which adequate samples were taken (more than 10 individuals examined; see below), and both parametric and non-parametric statistical methods were employed. Analyses using this data set were conducted across species and, in an effort to reduce potential phylogenetic artefacts (Pagel and Harvey 1988), within and among families. Second, in analysing trends where it was desirable to include data from all species, we calculated the measures of tick infestation across all individuals sampled rather than across species. This was done in comparing passerines versus non-passerines and

Table 9.1. Tick burdens across species and individuals of non-passerine and passerine birds on Mount Missim

Group	No. of species	No. (%) infested	No. of individuals	No. (%) infested
Non-passerines	30	1 (0.3)	161	1 (0.6)
Passerines	85	51 (60.0)	2335	384 (16.4)
Total	115	52 (45.2)	2496	385 (15.4)

examining seasonal and altitudinal trends in tick burdens. Our use of contingency-table analysis in these comparisons is done with the assumption that the variation across species in numbers of individuals sampled does not covary with the variables of interest (i.e. altitude and season). All probability values reported are for two-tailed tests.

Results

All ticks found on birds were nymphs or immatures. In an initial assessment, Hoogstraal (personal communication) identified them to the genus *Ixodes* and suggested there were four to five species represented. Hoogstraal tentatively assigned many of the specimens to *Ixodes cordifer* and *I. confusus* but exact species identification was not completed, nor can it be with the current material. The specimens were lodged in the Harry Hoogstraal collection at Smithsonian Institution.

The following identified species of ticks were also collected on Mount Missim from the indicated hosts or substrate: *I. cordifer*, on the opossum *Phalanger vestitus* and on vegetation; *I. luxuriosus*, on the opossums *P. vestitus* and *Dactylonax palpator*, and on the wallaby *Thylogale* sp.; and *Rhipicephalus sanguineus*, on a domestic dog.

Distribution of ticks across species and individuals

We made 2527 examinations of 2230 different individuals of 115 species in 27 families in nine orders of birds (see Table 9.6 at end of Chapter). Of 297 recaptures of 207 individuals, 31 were within the same month of the initial capture and were deleted. The data set thus consists of 2496 records from 2230 individuals.

Tick burdens were low, and only 15.4 per cent of all individuals were infested (Table 9.1). Non-passerines suffered significantly lower levels of infestation (0.6 per cent prevalence) than did passerines (16.4 per cent prevalence; $\chi^2 = 28.9$; $p < 0.001$; Table 9.1). Considering all individuals sampled, relative intensity averaged 0.21 ticks per individual ($n = 2496$;

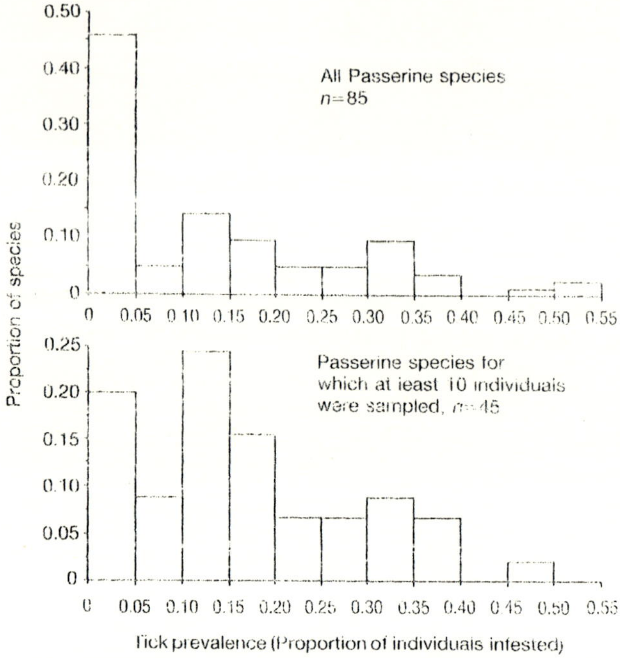

Fig. 9.2. Frequency distribution of tick prevalences across all passerine species combined (upper) and those species for which at least 10 individuals were sampled (lower).

range = 0–8). In passerines, a total of 2263 (96.9 per cent) of the 2335 individuals examined had either no ticks or just one tick. Because of the extremely low tick burdens in non-passerines, they are excluded, unless otherwise stated, from all analyses that follow.

Across passerine species, tick prevalence was significantly correlated with the number of individuals sampled ($n=85$, $r_s=0.502$, $Z=4.60$, $p<0.0001$). This is due, in part, to the large number of species with 0 prevalence values (see Table 9.6). Given low levels of tick infestation and potential sampling errors, large sample sizes are required to accurately determine species' specific tick burdens. Of the 45 species in our sample for which 10 or more individuals were examined, 41 (91.1 per cent) showed some level of tick infestation (Fig. 9.2). This contrasts with the 40 species for which one to nine individuals were examined, of which just 10 (25.0 per cent) showed infestations; these were deleted from subsequent analyses. With respect to the remaining 45 species, there was no correlation between sampling effort and tick prevalence ($r_s=-0.233$, $Z=-1.542$, $p>0.10$).

Species in which more individuals were infested (larger prevalence values) also showed higher mean intensities (across species, $df=40$, $R^2=0.085$, $F=4.714$, $p=0.036$ (Fig 9.3); across families, $df=12$,

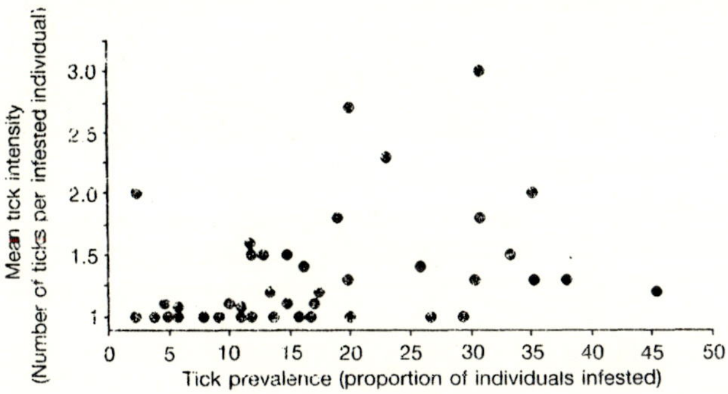

Fig. 9.3 Tick prevalence versus mean tick intensity across passerine species ($n=41$) for which at least 10 individuals were sampled.

$R^2=0.390$, $F=8.606$, $p=0.0134$). This suggests that, as the probability of infestation increases, so does the level of infestation.

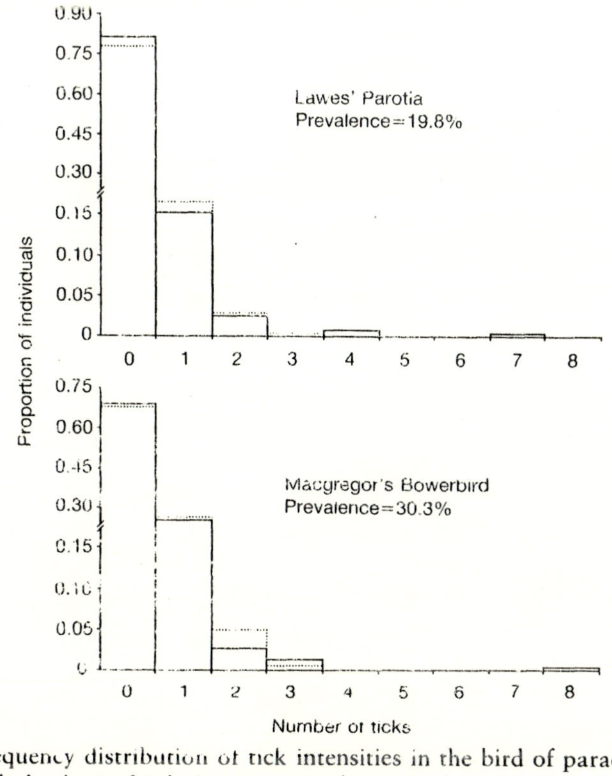

Fig. 9.4. Frequency distribution of tick intensities in the bird of paradise, Lawes' Parotia, and the bowerbird, Macgregor's bowerbird. Solid lines represent the observed frequency and the dotted lines, the expected frequency assuming a Poisson process. In both cases, the observed distributions did not differ significantly from the expected Poisson distributions. See text.

Table 9.2. Sexual differences in tick burdens for species of bird of paradise (Lawes' parotia) and bowerbird (Macgregor's bowerbird) from Mount Missim

Age/sex	No. of individuals	Prevalence*	Relative intensity†	Mean intensity†
Lawes' parotia				
Immature male	32	0.22	0.25 ± 0.51	1.1 ± 0.38
Adult male	89	0.16	0.17 ± 0.41	1.1 ± 0.27
Adult female	108	0.22	0.26 ± 0.52	1.2 ± 0.38
Statistical probability‡		> 0.10	0.389	0.719
Macgregor's bowerbird				
Immature male	58	0.21	0.21 ± 0.41	1.0 ± 0
Adult male	142	0.33	0.45 ± 0.91	1.4 ± 1.13
Adult female	26	0.15	0.27 ± 0.72	1.8 ± 0.96
Statistical Probability‡		> 0.10	0.116	0.375

* Proportion of individuals infested.
† The mean ± SD number of ticks infesting all individuals sampled (relative intensity) and just those individuals infested (mean intensity).
‡ Probabilities that the differences across age and sex classes are significantly different from random. The values are from a chi-square analysis for prevalence values and an ANOVA for mean and relative intensities.

To examine the distribution of ticks across individuals, we compared the observed frequencies of tick intensities with expected frequencies assuming a Poisson distribution for all passerines combined and then separately for two intensively studied species, the bird of paradise Lawes' parotia (*Parotia lawesii*) and Macgregor's bowerbird (*Amblyornis macgregoriae*) (Fig. 9.4). The distribution of ticks across all passerines differed significantly from the Poisson ($\chi^2 = 39.5$, $p < 0.001$) and the variance to mean ratio (1.85) indicated a clumped dispersion ($t = 29.01$, $p < 0.001$). The observed distribution also differed significantly from that of a negative binomial ($\chi^2 = 43.06$, $p < 0.001$). In contrast, within Lawes' parotia and Macgregor's bowerbird, ticks were randomly distributed across individuals (neither distribution differed significantly from the Poisson; for Lawes' parotia, $\chi^2 = 2.06$, $p > 0.10$; for Macgregor's bowerbird, $\chi^2 = 1.10$, $p > 0.50$).

Data from other long-term studies on Mount Missim (Pruett-Jones and Pruett-Jones 1982, 1990; Beehler and Pruett-Jones 1983) yielded accurate information on the age and sex of most individual birds of paradise and bowerbirds examined. Tick burdens were equivalent for immature males, adult males, and adult females for both Lawes' parotia and Macgregor's bowerbird (Table 9.2). Similarly, considering all 10 species of birds of paradise together, there were no significant differences

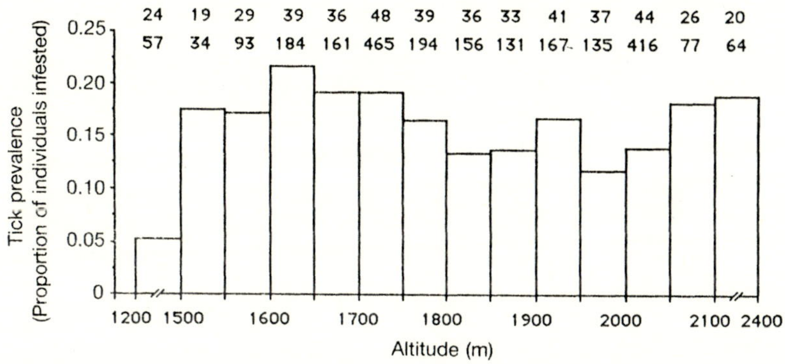

Fig. 9.5. Tick prevalence in passerines across altitude. The data from all years are combined. Above each bar are listed the number of species (upper) and individuals (lower) examined in each altitudinal interval.

in tick burdens across age and sex classes of the hosts ($p > 0.10$ in each comparison).

Ecology of tick infestations

Tick prevalence in passerines did not vary systematically across altitude, with the exception that below 1500 m altitude there was an apparent drop in prevalence (Fig. 9.5). Our sampling in that area was limited. Also, five of the 24 species examined below 1500 m did not occur above that altitude.

Across seasons, lower tick prevalences were observed during the dry season (May to October, $\bar{x} = 14.6$ per cent) than in the wet season (November to April, $\bar{x} = 16.9$ per cent; Fig. 9.6), yet the differences were not significant ($t = 0.974$, $p = 0.337$). For the annual period in which we have the most consistent records, September to December, tick prevalences showed similar trends across the seven years of the study (Table 9.3). Higher prevalence values during 1980 and 1981 were most likely due to our focus during netting those seasons on *P. lawesii* and *A. macgregoriae,* which both showed higher tick burdens relative to other species (see Table 9.6).

Tick burdens were correlated with aspects of foraging behaviour across species (Table 9.2). Considering relative tick intensities, there were significant differences in intensities across foraging strata (ANOVA, $df = 44$, $F = 4.627$, $p = 0.0013$). Species with a strata score of 1.5 (those foraging both on the ground and in the understorey) had significantly higher intensities than species foraging at higher levels as well as those foraging exclusively on the ground (strata score of 1.0; ANOVA, $p < 0.05$

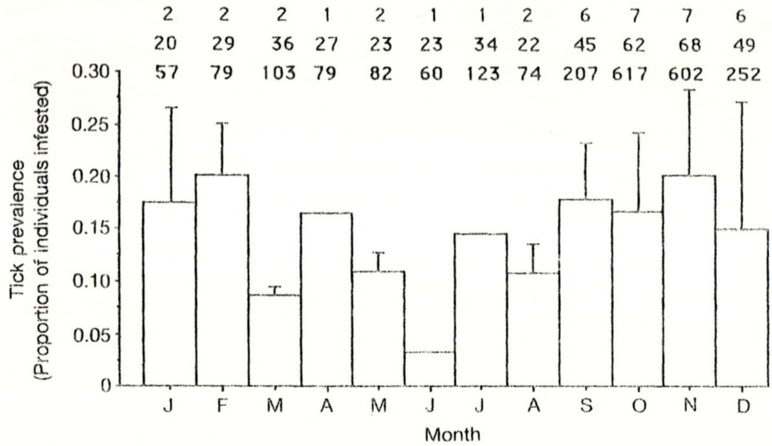

Fig. 9.6. Tick prevalence in passerines across months. Shown are mean prevalence values for all individuals examined each month, with the vertical bar representing one-half of the range of monthly prevalence values across years. Above each bar are listed the number of years (upper) in which birds were examined during the respective month, and the number of species (middle) and individuals (lower) examined.

Table 9.3. Yearly variation in tick burdens in passerine birds on Mount Missim*

Variable	1980	1981	1982	1983	1985	1986	1987
No. of species examined	26	36	41	45	55	52	64
No. of individuals examined	135	215	188	205	328	315	292
Proportion of individuals infested	26.7	24.2	19.2	15.6	16.8	13.6	15.4

* Only records from September to December are included. No data were obtained in 1984.

in each comparison). The decrease in relative tick intensity associated with foraging in higher strata was significant for species suffering relatively high levels of intensity (greater than the median value of 0.18 ticks per individual; $n = 23$, $r_s = -0.464$, $Z = -2.176$, $p < 0.03$; Fig. 9.7), but not all species combined (Table 9.4). Mean tick intensity, in contrast, did show a significant negative correlation with increasing foraging strata across all species (Table 9.4).

The correlations between tick burdens and foraging strata and foraging levels observed across species were not significant when analysed across families (Table 9.4). These relationships were similarly not significant among species within families (just considering species for which at least 10 individuals were examined, and families for which at least three species were sampled; $p > 0.05$ in each comparison).

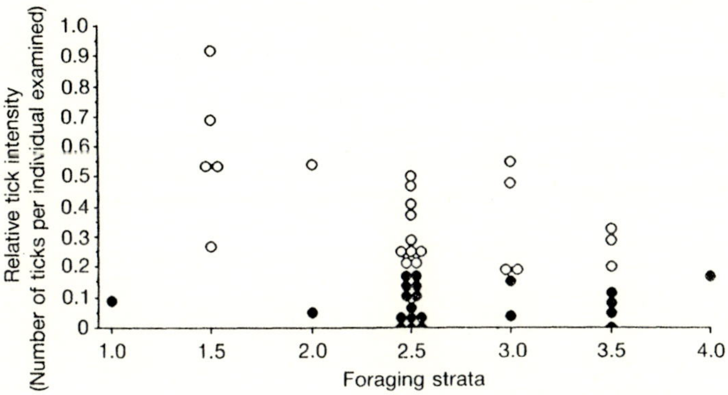

Fig. 9.7. Relative tick intensity in passerines as it relates to the forest strata where species forage. Only species for which at least 10 individuals were sampled are included ($n=45$). The open circles represent species ($n=23$) above the median level of intensity. See 'Methods' for description of foraging strata, and text for discussion of statistics.

Table 9.4. Correlations between tick burdens and ecology of passerine species on Mount Missim*

Measure of infestation	Variable	Species level†		Family level‡	
		r_s	Z	r_s	Z
Prevalence	Body size§	0.070	0.453	0.036	0.124
	Foraging strata	−0.130	−0.864	−0.056	−0.193
	Foraging levels	0.307	2.039¶	0.422	1.461
Relative intensity	Body size§	0.073	0.472	−0.117	−0.406
	Foraging strata	−0.232	−1.54	−0.105	−0.363
	Foraging levels	0.306	2.029¶	0.354	1.228
Mean intensity	Body size§	0.135	0.841	−0.363	−1.258
	Foraging strata	−0.371	−2.349¶	−0.167	−0.579
	Foraging levels	0.298	1.883	0.536	1.855

* Only species for which at least 10 individuals were sampled are included.
† Sample size is 45 species.
‡ Sample size is 13 families.
§ Two species of long-tailed birds of paradise (*Epimachus meyeri*) and *Astrapia stephaniae*) were excluded from this analysis; $n=43$.
¶ $p<0.05$.

Table 9.5. Tick burdens across passerine species according to diet*

Dietary group	Number of species	Tick prevalence Mean	Tick prevalence SD	Tick intensity† Mean	Tick intensity† SD
Insectivores	24	14.4	10.9	0.24	0.24
Frugivores	5	23.5	16.0	0.30	0.22
Omnivores	16	16.6	10.7	0.21	0.16

* Only species for which at least 10 individuals were sampled are included.
† Relative intensity.

Body size in passerines did not correlate with any measure of tick infestation (Table 9.4) and the differences in tick burdens across dietary groups were similarly not significant (Table 9.5; ANOVA, $df=44$, $F=1.31$, $p=0.2806$ for tick prevalence; $df=44$, $F=0.367$, $p=0.6948$ for relative intensity).

Across species ($n=41$), tick burdens were independent of blood parasitaemia (blood parasite prevalence: $r_s=0.239$, $p>0.10$, and $r_s=0.176$, $p>0.20$ for tick prevalence and relative intensity, respectively; blood parasite intensity: $r_s=0.249$, $p>0.10$, and $r_s=0.145$, $p>0.20$ for tick prevalence and relative intensity, respectively) and male showiness ($r_s=0.126$, $p>0.40$, and $r_s=0.025$, $p>0.50$ for tick prevalence and relative intensity, respectively). Comparing species with respect to sexual chromatism yielded similar results to those examining male showiness specifically: sexually dichromatic and monochromatic species suffered, on average, similar levels of tick infestation ($t=-0.104$, $p=0.918$; $t=0.115$, $p=0.909$; and $t=0.303$, $p=0.763$ for tick prevalence, relative intensity, and mean intensity, respectively).

Comparing passerine and non-passerine species as two groups, non-passerines were significantly larger ($t=5.18$, $p=0.0001$) and foraged over significantly fewer levels in the forest ($t=3.73$, $p=0.0003$), but otherwise used equivalent forest strata ($t=0.16$, $p=0.877$) as passerine species.

Discussion

Prior to this study, host–tick relationships from New Guinea were known for only 19 species of birds in six orders (Hoogstraal 1982). In that summary, adult ticks were listed from just three non-passerine species: a double-wattled cassowary (*Casuarius casuarius*), the common coucal (*Centropus phasianinus*), and the dusky moorhen (*Gallinula tenebrosa*). Ticks infesting these species were *Amblyomma papuanum*, *Haemaphysalis bancrofti*, and *H. novaeguineae*, respectively. All records of ticks collected from passerines were of nymphs and immature ticks. The passerine genera listed by Hoogstraal (1982)—*Sericornis*, *Rhipidura*,

Gerygone, Poecilodryas, Pachycephala, Pitohui, Zosterops, Meliphaga, Amblyornis, Cnemophilus, and *Paradigalla*—were each infested with *I. cordifer*. All but one of these genera (*Paradigalla*) were included in our study. Hoogstraal (personal communication) believed that, separate from *I. cordifer* and *I. confusus* which were tentatively identified from Mount Missim, at least two new tick species were collected during our study.

Given the total absence of adult ticks from all passerines examined in this study, the tick species involved clearly use at least two hosts during their life cycle. This is consistent with the general life-history of *Ixodes* sp. The hosts of the adult ticks are most likely mammals, which would entail marsupials, bats, or rodents. Snakes, as well as monitor lizards, were uncommon on Mount Missim within the altitude where we worked, but they may also serve as hosts for adult ticks.

Before interpreting the results of this study, it is important to note that our method of examining birds for ticks was subject to error. In a visual examination, it would be easy to miss larval and immature ticks. Given the presumed low tick burdens, the statistical impact of missing one tick could be significant. Thus, while we consider the data on tick burdens as reported in Table 9.6 reliable at the species level for those for which at least 10 individuals were sampled, we must consider the data approximate for all others. Also, although this potential error may reduce the reliability of species-specific values of tick burdens, the probability of error should be similar across species.

Tick burdens appeared to be low across all species, with a maximum prevalence in species for which at least 10 individuals were sampled of 45.4 per cent in the magnificant bird of paradise (*Cicinnurus magnificus*). Most individuals of all species were infested with zero to one tick. High levels of intensity (to a maximum of eight ticks in *A. macgregoriae*) were, however, occasionally observed and produced an overall contagious distribution of intensities across all individuals (see 'Results'). Judging from the random distribution of ticks across individuals of Lawes' parotia and Macgregor's bowerbird (Fig. 9.4), the species for which we had the largest samples, we suspect that random distributions probably prevail in most species. Cross-species contagious distributions can result from a combination of independent Poisson distributions (Johnson and Kotz 1969).

Previous studies of the distribution of ticks across individuals (Crofton 1971*a,b*; Schmid and Robinson 1972; Randolph 1975; see also May 1985) have shown contagious distributions that are approximated by a negative binomial distribution. Contagious distributions are characterized by most individuals having low infestations, but with a few

Table 9.6. Tick burdens across bird species on Mount Missim, Papua New Guinea. Taxonomy follows that of Beehler *et al.* (1986)

Family/species	No. of birds examined	Birds with ticks		Relative intensity
		No.	%	
NON-PASSERINES				
Accipitridae				
Accipiter cirrhocephalus	2	0	0	0
Megapodidae				
Aepypodius arfakianus	1	0	0	0
Rallidae				
Rallicula forbesi	4	0	0	0
Columbidae				
Macropygia amboinensis	11	0	0	0
Macropygia nigrirostris	[illegible]	0	0	0
Gallicolumba rufigula	6	0*	0	0
Gallicolumba beccarii	9	0	0	0
Gallicolumba jobiensis	1	0	0	0
Ptilinopus magnificus	1	0	0	0
Ptilinopus superbus	12	0	0	0
Ptilinopus rivoli	13	0	0	0
Ducula chalconata	2	0	0	0
Psittacidae				
Trichoglossus haematodus	1	0	0	0
Charmosyna wilhelminae	2	0	0	0
Charmosyna pulchella	1	0	0	0
Charmosyna papou	7	0	0	0
Neopsittacus musschenbroekii	1	0	0	0
Neopsittacus pullicauda	2	0	0	0
Micropsitta bruijnii	2	0	0	0
Psittacella brehmii	5	0	0	0
Psittacella madaraszi	3	0	0	0
Alisterus chloropterus	4	0	0	0
Cuculidae				
Cacomantis variolosus	1	0	0	0
Cacomantis castaneiventris	45	0	0	0
Chrysococcyx meyerii	3	0	0	0
Caliechthrus leucolophus	2	0	0	0
Aegothelidae				
Aegotheles albertisi	4	0	0	0
Alcedinidae				
Tanysiptera nympha	1	0	0	0
Halcyon sancta	1	0	0	0
Halcyon megarhyncha	9	1	11.1	0.11
PASSERINES				
Campephagidae				
Coracina caerulogrisea	1	0	0	0
Coracina melaena	[illegible]	0	0	0
Turdidae				
Zoothera dauma	2	0	0	0
Orthonychidae				

Table 9.6. *continued*

Family/species	No. of birds examined	Birds with ticks No.	%	Relative intensity
Ptilorrhoa leucosticta	22	2	9.1	0.09
Ptilorrhoa castanonotus	1	0	0	0
Ifrita kowaldi	1	0	0	0
Sylviidae				
Phylloscopus trivirgatus	2	0	0	0
Acanthizidae				
Crateroscelis murina	15	4	26.7	0.27
Crateroscelis nigrorufa	13	3	23.1	0.54
Crateroscelis robusta	13	4	30.8	0.92
Sericornis spilodera	1	0	0	0
Sericornis nouhuysi	26	7	26.9	0.38
Sericornis perspicillatus	31	5	16.1	0.23
Sericornis papuensis	6	2	33.3	1.00
Sericornis arfakianus	2	0	0	0
Gerygone cinerea	1	0	0	0
Rhipiduridae				
Rhipidura brachyrhyncha	3	0	0	0
Rhipidura atra	92	2	2.2	0.02
Rhipidura albolimbata	23	0	0	0
Rhipidura rufiventris	1	0	0	0
Myiagridae				
Monarcha axillaris	27	4	14.8	0.22
Machaerirhynchus nigripectus	8	0	0	0
Eopsaltriidae				
Microeca papuana	46	5	10.9	0.11
Tregellasia leucops	23	4	17.4	0.22
Poecilodryas albispecularis	77	27	35.1	0.69
Poecilodryas albonotata	10	2	20.0	0.20
Amalocichla incerta	4	0	0	0
Peneothello sigillatus	2	1	50.0	0.50
Peneothello cyanus	17	6	35.3	0.47
Pachycephalopsis poliosoma	35	7	20.0	0.54
Pachycephalidae				
Pachycare flavogrisea	3	1	33.3	0.33
Rhagologus leucostigma	80	8	10.0	0.11
Pachycephala soror	98	5	5.1	0.05
Pachycephala schlegelii	42	1	2.4	0.05
Pachycephala modesta	1	0	0	0
Pachycephala rufinucha	22	1	4.6	0.05
Colluricincla megarhyncha	34	4	11.8	0.18
Pitohui dicrous	6	1	16.7	0.67
Pitohui nigrescens	26	0	0	0
Eulacestoma nigropectus	6	2	33.3	0.33
Climacteridae				
Cormobates placens	43	5	11.6	0.19
Dicaeidae				
Melanocharis nigra	3	0	0	0
Melanocharis longicauda	6	0	0	0
Melanocharsis versteri	38	6	15.8	0.16

Table 9.6. *continued*

Family/species	No. of birds examined	Birds with ticks No.	Birds with ticks %	Relative intensity
Melanocharsis striativentris	75	11	14.7	0.16
Rhamphocharis crassirostris	26	1	3.8	0.04
Oreocharis arfaki	3	0	0	0
Zosteropidae				
Zosterops atrifrons	1	0	0	0
Zosterops novaeguineae	1	0	0	0
Meliphagidae				
Timeliopsis fulvigula	19	0	0	0
Melilestes megarhycha	29	11	37.9	0.48
Toxorhamphus poliopterus	225	38	16.9	0.18
Oedistoma iliolophus	38	3	7.9	0.08
Myzomela adolphinae	1	0	0	0
Myzomela rosenbergii	12	2	16.7	0.17
Meliphaga orientalis	8	2	25.0	0.25
Meliphaga albonotata	2	0	0	0
Meliphaga aruensis	1	0	0	0
Lichenostomus subfrenatus	5	1	20.0	0.40
Xanthotis polygramma	1	0	0	0
Pycnopygius cinereus	3	0	0	0
Ptiloprora plumbea	4	0	0	0
Ptiloprora meekiana	2	0	0	0
Ptiloprora guisea	47	6	12.8	0.19
Melidectes belfordi	41	2	4.9	0.05
Melidectes torquatus	21	4	19.0	0.33
Melipotes fumigatus	63	7	11.1	0.11
Estrildidae				
Erythrura trichroa	22	3	13.6	0.14
Erythrura papuana	58	3	5.2	0.05
Oriolidae				
Oriolus szalayi	1	0	0	0
Dicruridae				
Chaetorhynchus papuensis	26	3	11.5	0.15
Dicrurus hottentottus	1	0	0	0
Grallinidae				
Grallina bruijnii	1	0	0	0
Ptilonorhynchidae				
Ailuroedus melanotis	26	8	30.8	0.54
Amblyornis macgregoriae	261	79	30.3	0.38
Paradisaeidae				
Cnemophilus loriae	18	6	33.3	0.50
Manucodia keraudrenii	3	0	0	0
Ptiloris magnificus	3	1	33.3	0.33
Epimachus albertisi	9	1	11.1	0.11
Epimachus meyeri	17	5	29.4	0.29
Astrapia stephaniae	13	0	0	0
Lophorina superba	45	6	13.3	0.16
Parotia lawesii	272	54	19.8	0.26
Cicinnurus magnificus	11	5	45.4	0.55
Paradisaea rudolphi	6	3	50.0	0.67

individuals exhibiting very high infestations. High infestations could reflect low immunity in some individuals, reduced health or body condition causing sluggishness and increased encounter rates with ticks, or random processes through some individuals inhabitating areas with high tick densities.

Our observation of a random distribution of tick intensities suggests that individuals encounter ticks randomly or perhaps that individuals do not behave in ways to avoid encountering ticks. Nevertheless, the finding that mean tick intensity increases with tick prevalence (Fig. 9.3) suggests that the factors, ecological or immunological, that make species more vulnerable to tick infestation also make individuals vulnerable to greater levels of infestation. A similar result was found with respect to prevalence and intensity of blood parasites in this avifauna (Pruett-Jones *et al.* in press).

Non-passerines appear to be significantly less infested with ticks than passerines (Table 9.1). Assuming these trends are real and not an artefact of the generally small samples for most non-passerines, they warrant explanation. Although the non-passerine species utilized fewer levels in the forest than did passerines (see 'Results'), the forest strata used were similar and we doubt that differences in foraging behaviour between the two groups can explain the order of magnitude difference in tick burdens between them (Table 9.1). Non-passerines were also larger than passerines, which should otherwise increase exposure to ticks. We believe the most likely explanation is that non-passerines, being longer-lived than passerines, acquire a greater immunity to ticks and thus suffer lower infestations. Acquired immunity to ticks is well known (Trager 1939; Riek 1962; Roberts 1968) but does not necessarily occur for all species (Randolph 1975*a*). Given that such immunity is dependent on multiple exposures of hosts to ticks, long-lived species may potentially experience more exposures and acquire a greater degree of immunity than shorter-lived species. This may be particularly true when tick densities are low as our data suggest may be the case on Mount Missim. This idea has not, to our knowledge, been tested nor have species' specific differences in immunity been quantified in field situations. It is important to note that non-passerines also suffered lower intensities of blood protozoa and microfilarial worms than did passerines on Mount Missim, differences attributed to greater immunity to blood parasites in non-passerines (Pruett-Jones *et al.* in press).

Two aspects of ecology, where species foraged in the forest column and the vertical range of their movements, appeared to significantly affect tick infestations in passerines (Table 9.4, Fig. 9.7). The correlations were not, however, significant across families, or across species within families, and we cannot discount the possibility that the significance of the relationships was due to artefacts of phylogeny within the set of species studied. We consider the most compelling result relating foraging behaviour to tick burdens to be the significant difference in relative intensity between species foraging both on the ground and in the understorey (strata score = 1.5) and all other species (see 'Results'). These results are to some extent expected in that nymphal and immature ticks remain on the ground and in low vegetation until a host is encountered. It is surprising, nonetheless, that the one exclusively ground foraging species in our sample (spotted jewel babbler, *Ptilorrhoa leucosticta*) had such a low relative tick intensity (0.09; see Table 9.6 and Fig. 9.7). These data might suggest that ticks seeking hosts may be more abundant in low vegetation than they are on the ground. It is a general interpretation that host species' differences in tick burdens relate to those differences in their biology that influence encounter rates with ticks (Randolph 1975). Such factors include body size (Milne 1947), home range (Randolph 1975), nesting substrate (Duffy 1983; Duffy and Campos De Duffy 1986), as well as foraging behaviour. If we assume equivalent levels of immunity against ticks across individuals, which may be valid within taxa but, as discussed above, is probably not valid across taxa, those host species that spend greater amounts of time in areas where ticks reside should experience higher levels of infestation. For species which inhabit similar habitats, or in the case of forest birds the same levels within the forest, a major factor affecting infestations may be the manner of foraging or the total area over which individuals move and forage.

Interpretation of the seasonal and altitudinal patterns of tick prevalence (Figs 9.5 and 9.6) is difficult because the ticks involved were not identified to species. The distributions recorded may thus be representative of each respective tick species, or they may represent the combination of disparate species-specific patterns. The seasonal trends in tick prevalence, with more host individuals infested during the wet than in the dry season (Fig. 9.5) are likely to be representative of each tick species and is a commonly observed pattern (cf. Hoogstraal 1978; Pegrem *et al.* 1981). Whether each species of tick is distributed across the entire altitudinal range of the study area is harder to evaluate. For example, Guoshi *et al.* (1984) found distinct altitudinal zonation in ticks in montane habitat from 300 to 2400 m altitude in Xinjiang, China. Distinct spatial patterning of tick species, when even the same host species is infested, has also been observed (Bull *et al.* 1981; Bull 1986). Given the importance of altitudinal zonation and segregation in avian populations in New Guinea (Diamond 1972), we suspect that the tick species also show altitudinal zonation.

Answers to the questions raised in this study will ultimately require species' identifications of the ticks. Our findings show that ticks are widespread, if uncommon, parasites of birds on Mount Missim. Non-random aspects of the ecology of passerine species, specifically their foraging habits, appear to influence the level of infestation by ticks, but individuals within species may encounter ticks randomly. From our perspective, the most interesting question is whether the observed low levels of infestation reflect widespread immunity in the host species or some aspect of the biology of the ticks themselves, for example low population densities. This study has described the pattern of tick burdens across species in a tropical forest avifauna and shown how these patterns vary according to ecology. A full understanding of tick–host interactions in this system will require additional data on both immunological responses to ticks and patterns of disease transmission within the host species.

Acknowledgements

Our work during 1980–83 was funded by the National Geographic Society, New York Zoological Society, Chem-tronics Inc and Dan Brimm, Papua New Guinea Biological Foundation, Frank M. Chapman Memorial Fund, Joseph Henry Fund, Flora and Fauna Preservation Society, Carl Koford Memorial Fund, and Harry Hoogstraal. The work during 1985–87 was supported by NSF Grant BSR 8416000 to Jack Bradbury and SPJ. Recreational Equipment Inc, North Face, and Ski Hut generously provided items of camping equipment. This work would not have been possible without the much appreciated assistance from the following volunteers: D. Alexander, M. Alfano, J. Archer, K. Bellchambers, J. Benshemesh, J. Burt, G. Dandliker, J. Dearnaley, S. Edwards, S. Fretz, B. Hastings, A. Hunt, E. Hacker, S. Johnston, M. Kowalski, K. Kristensen, R. Magrath, D. McNiven, B. Nelson, M. Norman, P. Pugh, J. Scott, J. Starks, S. Stirrat, M. Schultz, M. Shields, A. Thomas, B. Traill, D. Westcott, and M. Willett; and the following Papua New Guinea men: Buka, Geri, Herean, Iling, Mendon, Mungem, Sapawe, and most of all Mr Diro Manawe. We were greatly helped during our stay in Papua New Guinea by A. Allison, B. Beehler, T. Pratt, A. Safford, H. Sakulas, the Eltham, Fraser, Harvey-Hall, Violaris, and Winn families, New Guinea Goldfields Ltd. Pty., and the staffs at Wau Ecology Institute and Museum of Vertebrate Zoology. The national and Morobe provincial governments allowed us to carry out this study. We wish to thank J. Cairns, D. Clayton, J. Kethley, G. Needham, and R. Robbins for advice, F. Pitelka and J. Bradbury for support, J. Diamond for scoring the species for forest strata, and H. McClure, C. van Riper III, and an anonymous reviewer for comments on an earlier draft of the chapter. This chapter is dedicated to the memory of Harry Hoogstraal.

References

Beehler, B.M. (1978). *Upland birds of northeastern New Guinea.* Wau Ecology Institute, Wau, Papua New Guinea.

Beehler, B.M. (1982). Ecological structuring of forest bird communities in New Guinea. In *New Guinea biogeography* (ed. J.L. Gressitt), pp. 837–60. W. Junk, The Hague.

Beehler, B.M. (1983). The behavioral ecology of four birds of paradise. Unpublished Ph.D. thesis, Princeton University, New Jersey.

Beehler, B. and Pruett-Jones, S. G. (1983). Display dispersion and diet of birds of paradise: a comparison of nine species. *Behavioral Ecology and Sociobiology*, **13**, 229–38.

Beehler, B.M., Pratt, T.K., and Zimmerman, D.A. (1986). *Birds of New Guinea.* Princeton University Press, Princeton, New Jersey.

Bennett, P.M. (1986). Compative studies of morphology, life history and ecology among birds. Unpublished Ph.D. thesis, University of Sussex, UK.

Bull, C.M. (1986). Ticks on reptiles, a difference of scale. In *Morphology, physiology, and behavioral biology of ticks* (ed. J. R. Sauer and J.A. Hair), pp. 391–405. John Wiley, New York.

Bull, C.M., Sharrad, R.D., and Petney, T.N. (1981). Parapatric boundaries between Australian reptile ticks. *Proceedings of the Ecological Society of Australia*, **11**, 95–107.

Crofton, H.D. (1971*a*). A quantitative approach to parasitism. *Parasitology*, **62**, 179–93.

Crofton, H.D. (1971*b*). A model of host–parasite relationships. *Parasitology*, **63**, 343–64.

Diamond, J.M. (1972). *Avifauna of the Eastern Highlands of New Guinea.* Nuttal Ornithological Club, Cambridge, Massachusetts.

Doss, M.A., Farr, M.M., Roach, K.F., and Anastos, G. (1979). *Ticks and tickborne diseases IV. Geographical distribution of ticks.* Index-Catalogue of Medical and Veterinary Zoology, Special Publication no. 3. United States Department of Agriculture, Washington, DC.

Duffy, D.C. (1983). The ecology of tick parasitism on densely nestling Peruvian seabirds. *Ecology*, **64**, 110–19.

Duffy, D.C. and Campos De Duffy, M.J. (1986). Tick parasitism at nesting colonies of blue-footed boobies in Peru and Galapagos. *Condor*, **88**, 242–4.

Fairchild, G.B., Kohls, G.M., and Tipton, V.J. (1966). The ticks of Panama (Acarina: Ixodoidea). In *Ectoparasites of Panama* (ed. B.L. Wenzel and V.J. Tipton), pp. 167–219. Field Museum of Natural History, Chicago, Illinois.

Guoshi, C., Guangyuan, C., Guannan, S., Daomao, P., and Huanzhang, X. (1984). Occurrence of some ticks at different altitudes in Xinjiang, China. In *Acarology VI*, Vol. 2 (ed. G.A. Griffiths and C.E. Bowman), 1189–92. John Wiley, New York.

Hoogstraal, H. (1966). Ticks in relation to human diseases caused by viruses. *Annual Review of Entomology*, **11**, 261–308.

Hoogstraal, H. (1973). Ticks. In *Parasites of laboratory animals* (ed. R. J. Flynn), pp. 398–424. Iowa State University Press, Ames.

Hoogstraal, H. (1978). Biology of ticks. In *Tickborne diseases and their vectors* (ed. J.K.H. Wilde), pp. 3–14. University of Edinburgh Press, Edinburgh.

Hoogstraal, H. (1982). Ticks (Acari: Ixodoidea): a preliminary study. In *New Guinea biogeography* (ed. J.L. Gressitt), pp. 537–44. W. Junk, The Hague.

Johnson, N.L. and Kotz, S. (1969). *Discrete distributions*. John Wiley, New York.

Margolis, L., Esch, G.W., Holmes, J.C., Kuris, A.M., and Schad, G.A. (1982). The use of ecological terms in parasitology. *Journal of Parasitology*, **68**, 131–3.

May, R.M. (1985). Host–parasite associations: their population biology and population genetics. In *Ecology and genetics of host–parasite interactions* (ed. D. Rollinson and R.M. Anderson), pp. 243–62. Academic Press, New York.

McAlpine, J.R., Keig, G., and Falls, R. (1983). *Climate of Papua New Guinea*. CSIRO and Australian National University, Canberra.

Milne, A. (1947). The ecology of the sheep tick, *Ixodes ricinus* L. The infestations of hill sheep. *Parasitology*, **38**, 34–50.

Obenchain, F.D. and Galun, R. (ed.) (1982). *Physiology of ticks*. Pergamon, New York.

Pagel, M.D. and Harvey, P.H. (1988). Recent developments in the analysis of comparative data. *Quarterly Review of Biology*, **63**, 413–40.

Paijman, K. (1976). Vegetation. In *New Guinea Vegetation* (ed. K. Paijmans), pp. 23–104. CSIRO and Australian National University, Canberra.

Pegrem, R.G., Hoogstraal, H., and Wassef, H.Y. (1981). Ticks (Acari: Ixodoidea) of Ethiopia. I. Distribution, ecology, and host relationships of species infesting livestock. *Bulletin of Entomological Research*, **71**, 339–59.

Pratt, T.K. (1983). Seed dispersal in a montane forest in Papua New Guinea. Unpublished D.Phil. thesis, Rutgers University, New Brunswick.

Pruett-Jones, M.A. and Pruett-Jones, S.G. (1982). Spacing and distribution of bowers in Macgregor's bowerbird (*Amblyornis macgregoriae*). *Behavioral Ecology and Sociobiology*, **11**, 25–32.

Pruett-Jones, S.G. and Pruett-Jones, M.A. (1986). Altitudinal distribution and seasonal activity patterns of birds of paradise. *National Geographic Research*, **2**, 87–105.

Pruett-Jones, S.G. and Pruett-Jones, M.A. (1990). Sexual selection through female choice in Lawes' Parotia, a lek-breeding bird of paradise. *Evolution*, **44**, 486–501.

Pruett-Jones, S.G., Pruett-Jones, M.A., and Jones, H.I. (1990). Parasites and sexual selection in birds of paradise. *American Zoologist*, **30**, 287–98.

Pruett-Jones, S.G., Pruett-Jones, M.A., and Jones, H.I. (in press). Parasites and sexual selection in a New Guinea avifauna. *Current Ornithology*.

Randolph, S.G. (1975). Patterns of distribution of the tick *Ixodes trianguliceps* Birula on its hosts. *Journal of Animal Ecology*, **44**, 451–74.

Read, A.F. (1987). Comparative evidence supports the Hamilton and Zuk hypothesis on parasites and sexual selection. *Nature*, **327**, 68–70.

Riek, R.F. (1962). Studies on the reactions of animals to infestation with ticks. VI. Resistance of cattle to infestation with the tick *Boophilus microplus* (Canestrini). *Australian Journal of Agricultural Research*, **13**, 532–50.

Roberts, J.A. (1968). Resistance of cattle to the tick *Boophilus microplus* (Canestrini). II. Stages at the life cycle of the parasite against which resistance is manifest. *Journal of Parasitology*, **54**, 667–73.

Trager, W. (1939). Acquired immunity to ticks. *Journal of Parasitology*, **25**, 57–81.

Schmid, W.D. and Robinson, E.J. (1972). The pattern of a host–parasite distribution. *Journal of Parasitology*, **58**, 907–10.

Sonenshine, D.E. and Stout, I.J. (1970). A contribution to the ecology of ticks infesting wild birds and rabbits in the Virginia–North Carolina Piedmont (Acarina: Ixodidae). *Journal of Medical Entomology*, **6**, 645–54.

Wilde, J.K.H. (ed.) (1978). *Tickborne diseases and their vectors*. University of Edinburgh Press, Edinburgh.

Part III
Behaviour

10 *Effects of avian host age and arbovirus infection on mosquito attraction and blood-feeding success*

THOMAS W. SCOTT and JOHN D. EDMAN

Introduction

In this chapter we review the role of birds in arthropod-borne virus (arbovirus) transmission cycles with emphasis on interactions between avian hosts and mosquito vectors. Our discussion begins with an introduction to the role of birds in arbovirus transmission and is followed by a summary of contemporary thoughts on attraction of mosquitoes to birds and on avian defensive behaviour toward mosquitoes that are attempting to blood feed. The last section of the chapter addresses research that we conducted to determine how avian host age and virus infection status affect the attraction and blood-feeding success of mosquitoes.

The role of birds in mosquito-borne virus transmission cycles

The principal role of vertebrates in arbovirus transmission cycles is to amplify virus in their bodies and to infect arthropod vectors that imbibe their viraemic blood (Scott 1988). Transmission cycles are maintained by a continuous vertebrate–vector–vertebrate exchange of virus; vectors imbibe infected blood, become infected during an extrinsic incubation period, and then biologically transmit virus to a different, susceptible host during refeeding. Vertebrates, therefore, play an important role in the transmission and amplification of arboviruses: a single infected host can infect numerous vectors. During times when transmission does not occur—that is, in winter in temperate regions and the dry season in the tropics—birds may participate in mechanisms of virus persistence. For most arboviruses, however, these processes are poorly understood.

Birds and mammals are the principal vertebrate sources of virus-infected blood for vectors, although blood-sucking arthropods feed on all four classes of terrestrial vertebrates (Downes 1971; Hocking 1971; Scott 1988). Only infrequently or with inconclusive evidence have reptiles or amphibians been associated with transmission of an arbovirus. Of the more than 500 arboviruses currently recognized, 79 have been recovered from birds (Karabatsos 1985). This is a conservative estimate of bird involvement in arbovirus transmission cycles. Isolation of a virus from a vertebrate does not necessarily mean that a host is important in enzootic transmission. Moreover, little is known about the natural history of many arboviruses. For some viruses there is no information on vertebrate hosts.

The contribution of birds as vertebrate hosts to arbovirus transmission was first recognized during the late 1930s and 1940s. During 1938, eastern equine encephalomyelitis (EEE) virus was recovered from sick and dead ring-necked pheasants (*Phasianus colchius*) in Connecticut and New Jersey (Tyzzer *et al.* 1938; Beaudette and Hudson 1945). That same year, Ten Broeck (1938) proposed that birds might be the vertebrate host of EEE virus. Birds were implicated as vertebrate hosts of St. Louis encephalitis (SLE) virus in 1940 (Howitt and Van Herick 1941; McLean and Bowen 1980). Early studies focused on chickens as arbovirus hosts because of their close association to human and equine infections (Hammon *et al.* 1941) and because, in early studies on mosquito feeding habits, chickens were found to be a common host (Bang and Reeves 1942; Reeves and Hammon 1944). Other researchers, however, began to implicate wild birds (Cox *et al.* 1941; Sellards *et al.* 1941), and during the early 1940s a series of investigations by Reeves and Hammon (1962) showed that in California songbirds are the vertebrate hosts of western equine encephalomyelitis (WEE) and SLE viruses. During June 1950, Kissling *et al.* (1951) recovered the first isolate of EEE virus from an enzootic avain host—a common grackle (*Quiscalus quiscula*) captured in Louisiana. These studies formed the foundation for subsequent investigations of avian involvement in arbovirus amplification and transmission (Kissling 1958; Stamm 1958, 1966; Reeves and Hammon 1962; Downs 1963; McClure 1963; Stamm and Newman 1963; Work 1963; Theiler and Downs 1973; McLean and Scott 1979; McLean and Bowen 1980; Scott 1988).

Although most of the early work on avian hosts of arboviruses was done in the United States, it is now clear that birds are involved in virus transmission cycles throughout the world (Scott 1988). These viruses, like their avian hosts, are more diverse in tropical than temperate regions (Karabatsos 1985; also see Zuk, Chapter 16, this volume).

The contribution of individual birds, bird species, or groups of avian species to arbovirus transmission cycles varies for different viruses and for the same virus in different locations. McLean and Bowen (1980) and

Scott (1988) summarize the terms used to describe a vertebrate host's contribution to arbovirus transmission. In general, an important host, avian or otherwise, possesses certain key characteristics. They are abundant, accessible to vectors in time and space, attractive to vectors, and receptive to vector feeding. They respond to infection with a viraemia of sufficient magnitude and duration to regularly infect vectors. They have a low probability of mortality from infection. Their life history is such that, at times of active transmission, immunologically susceptible animals are entering the host population.

The response of a bird to virus infection can be influenced by several factors (Sawyer and Lloyd 1931; Webster 1933; Nathanson 1980; Brinton *et al.* 1984; Brinton 1986; Monath 1986). Intrinsic variables include species-specific variation, genetics, age, health, and immune status. Extrinsic factors are the type or strain of virus inoculated, its virulence characteristics, and the dose inoculated.

The infection process typically begins with the bite of an infected arthropod (Nathanson 1980; Monath 1985; Shope 1985). Some replication can occur in tissues at the site of the vector bite. Virus then enters the lymphatic system and replicates in regional lymph nodes, after which it is released into the circulatory system where replication can take place in extraneural tissues. The result is a viraemia that is usually brief, 2–5 days in duration (Scott 1988). In some animals the viral infection may cause morbidity or mortality, but most infections in enzootic avian hosts are terminated by an immune response without detectable disease. Immune birds are protected from future infection by the same virus, or in some cases closely related viruses (Nathanson 1980; Vorndam 1980). In addition to acquired immunity, some female birds passively transfer antibodies to their progeny through the egg (Scott 1988).

An exception to the pattern of acquired protective immunity is immune enhancement, which has been most thoroughly studied for humans infected with dengue virus (Halstead 1980*a*,*b*; Monath 1986). One study with house sparrows (*Passer domesticus*) demonstrated viraemic enhancement (Ludwig *et al.* 1986). Nestlings who received passive antibodies from their mothers and were subsequently infected with SLE virus had viraemias that were greater in duration and magnitude than nestlings who did not receive maternal antibodies.

In addition to amplifying virus and infecting vectors, birds can contribute to arbovirus transmission in at least three other ways.

1. For certain viruses, transmission can occur from one bird to another without the bite of an infected arthropod vector. Eastern equine encephalomyelitis virus is transmitted this way by pecking or cannibalism among penned, exotic gallinaceous birds (Scott and Weaver

1989). The result can be extensive mortality among flocks. This mode of transmission has not been demonstrated for free-ranging birds.

2. Virus can be dispersed by flying, viraemic birds; this could expand a focus of transmission activity or modify the geographical distribution of a virus. Movement of infected avian hosts was suspected for ciconiiform birds infected with Japanese encephalitis virus (Scherer *et al.* 1959) and was demonstrated for passerine birds infected with EEE virus or Highlands J virus (Stamm and Newman 1963; Lord and Calisher 1970).
3. Birds may participate in the overwintering of arboviruses. The most likely role for vertebrates in overwintering mechanisms is annual reintroduction of virus from areas of continuous transmission or chronic latent infections with relapsing viraemias (Reeves 1974). Although there is some circumstantial evidence supporting these hypotheses, it is doubtful, based on available data, that birds regularly reintroduce virus during migration, and, if relapsing viraemias are important, they are more likely to occur in reptiles and amphibians than in birds (Scott 1988).

Attraction of mosquitoes to birds

General understanding of mosquito attraction

Since the pioneering work of Rudolfs (1922), many studies have been carried out on the mechanisms of mosquito attraction to their vertebrate hosts. Early investigations focused on a single mosquito species, *Aedes aegypti*, because it was easy to work with in the laboratory, was colonized and responded well in small, experimental enclosures under visible light (Brown 1966; Wright 1975; Bar-Zeev *et al.* 1977; Khan 1977).

Although the behaviour of this urban vector of yellow fever and dengue viruses is different from that of other mosquito species, much of the early literature on mosquito host-seeking behaviour is based on *Ae. aegytpi* and implies that there is an olfactory basis for host orientation to which all mosquito species conform (Brown 1966). Differences in host preference were attributed to subtle differences in the chemosensory sensitivity of various mosquito species and chemical emanations of various hosts (Smith *et al.* 1970; Price *et al.* 1979). Despite the lack of definitive data to support this proposition, the idea is still popular today.

Laboratory studies on mosquito attraction have examined only short-range orientation and landing stimuli (Gillies 1980). Visual factors, such as host size, colour, and shape, but not movement, were found to influence attraction and landing (Sippell and Brown 1953; Brown 1954; Wright and Kellogg 1964). The relative role of olfactory versus visual cues, however, has been difficult to assess even for *Ae. aegypti*. Data on this topic are ambiguous, and interpretation of results is limited by our incomplete understanding of the sensory physiology of mosquitoes.

The general view that chemical cues provide more information to mosquitoes for host discrimination than visual cues led to research emphasis on the isolation of specific host attractants, which usually were from humans (Smith *et al.* 1970; Khan 1977). The observation that certain individual hosts are more attractive to mosquitoes than others (Brouwer 1960; Khan 1977) was the primary reason researchers began to look for 'unique chemicals' that account for individual differences in the attractiveness of hosts to mosquitoes.

Carbon dioxide, heat, and moisture were identified early as attractants for almost all blood-feeding diptera (Khan 1977). Carbon dioxide was recognized as a particularly strong cue (Reeves 1953), although whether it functions as an attractant or a stimulant is still controversial (Gillies, 1980). Recognition of the universal role of CO_2, as well as heat and moisture, encouraged researchers to continue to search for additional chemicals. Breath, blood, sweat, and urine were all fractioned and tested. Unfortunately, little insightful information was generated from these analyses (Hocking 1971). Lactic acid and certain other acids, esters of acids, and steroids were found to have attraction properties but only in combination with CO_2, and the attractiveness of these compounds was never equal to that of an intact host (Smith *et al.* 1970; Khan 1977). In fact, field trials using lactic acid to enhance CO_2 attraction were disappointing (Peters, personal communication).

Receptors that are stimulated by lactic acid and become unresponsive when a mosquito is not in a biting mode have been found on *Ae. aegypti* (Davis 1984). However, these receptors also responded to other chemicals. In fact, in the laboratory, *Ae. aegypti* and many other mosquito species are attracted to and feed on a wide variety of hosts including many that they seldom, if ever, feed on in nature. This finding suggests that results from field studies showing restricted mosquito feeding patterns do not necessarily indicate that a species has a restricted capacity to locate and feed on certain hosts. Other factors, such as daily activity patterns of mosquitoes and their hosts or host defensive behaviour, may influence the blood-feeding patterns of a given mosquito species.

Only field studies can take into account the full range of factors affecting mosquito attraction to hosts—from activation of host seeking; to searching flight; through long-, mid-, and short-range orientation; to approach; and finally to landing on the host. Early field studies on mosquito attraction focused on the behaviour of anopheline vectors of human malaria and used large mammals, such as humans and cattle, as experimental hosts (Ribbands 1949; Learman 1955; Reid 1961; Service 1964). Later, when investigators became interested in the attraction of nocturnal mosquitoes to smaller, often cryptic hosts such as birds, they were influenced by literature dealing with large mammals and day-biting mosquitoes (Williams 1963; Gillies and Wilkes 1974). Studies with large mammals were complicated by the simultaneous presentation of both visual and olfactory cues. Results did demonstrate, however, that mosquitoes tend to fly upwind or crosswind in search of odours and that, once they encounter an odour plume, they follow it anemotactically to its source (Gillies 1972, 1980). The zigzag or casting nature of orientation flight is caused by the discontinuity of wind-borne odour plumes and the propensity of these insects to turn back into the plume when their flight takes them out of the odour trail (Kennedy 1983).

One important finding from this kind of investigation was that large hosts often attract mosquitoes from a greater distance (about 30–50 m maximum) than does CO_2 alone (Gillies and Wilkes 1969, 1970, 1972). This and other evidence suggested that CO_2 is not merely a flight stimulant but also attracts mosquitoes once they have begun to fly toward a host (Gillies 1980). It was also shown that: (1) not all mosquito species are attracted to CO_2 (Edman 1979; Price *et al.* 1979; Gillies 1980); (2) a pulsed CO_2 release may be required for sustained orientation; and (3) changes in CO_2 concentration are more important than the actual concentrations (Gillies 1980).

Attraction to birds

Interest in mosquito–bird interactions began at the start of this century when the role of the tropical house mosquito, *Culex quinquefasciatus* (=*fatigans*), in the transmission of avian malaria was recognized. *Cx. quinquefasciatus* was examined in detail because at the time it was the only experimental vector available for studying human malaria. Unfortunately, these early investigations did not include field studies to investigate the role of mosquito–bird interactions in the natural transmission of avian malaria. Significant interest in this kind of interaction did not emerge until the 1950s when birds were recognized as natural reservoirs for the enzootic transmission of several arboviruses that cause disease in humans and their domestic animals (Dow *et al.* 1957; Blackmore and Dow 1958; Stamm 1966; Reeves 1971).

The northern house mosquito, *Cx. pipiens*, was among the first mosquitoes to be recognized as a specialist that fed primarily on birds. Studies during the 1950s with laboratory colonies of this species showed that it was reluctant to bite any host other than a bird (Ekis 1971). At about the same time, EEE virus was recovered for the first time from its enzootic vector, *Culiseta melanura* (Chamberlain *et al.* 1951). Like *Cx. pipiens*, this mosquito was shown to feed almost exclusively on birds and rarely to attack humans or other mammals (Hayes 1961; Scott and Weaver 1989). *Cx. restuans* and *Cs. morsitans* were examined and similarly found to be ornithophilic. Blood feeding of mosquitoes on birds in the field was subsequently documented, and our understanding of the phenomenon was refined based on studies with baited traps (Means 1968; Wright and DeFoliart 1970) and serological identification of mosquito blood meals (Joseph and Bickley 1969; Tempelis 1975; Morris and Zimmerman 1981; Nasci and Edman 1981*a*). Eventually, species that feed primarily on birds—usually in the genus *Culex*—were recognized from all regions of the world (Williams 1963; Service 1971; Snow and Boreham 1973, 1978; Snow 1983).

Serological studies of mosquito blood meals showed that three important arbovirus vectors in North America (*Cx. tarsalis*, *Cx. quinquefasciatus*, and *Cx. nigripalpus*) commonly feed on both avian and mammalian hosts (Tempelis 1975). In other parts of the world, *Culex* vectors of arboviruses (e.g., *Cx. annulirostris* in Australia; Kay *et al.* 1979), have similarly been found to feed on a variety of hosts including birds. These discoveries support experiments that compared the relative attractiveness of various hosts to mosquitoes that are apparently opportunistic in their feeding behaviour (Dow *et al.* 1957). Later, it was suggested that the feeding plasticity and seasonal shifts in feeding, from birds to mammals, may be responsible for the epidemic transmission of WEE and SLE viruses to mammalian populations. Conversely, the rarity with which EEE virus infections occur in mammals is in part attributable to the ornithophilic specificity of *Cs. melanura* (Edman *et al.* 1985).

Several important ecological generalizations concerning bird-feeding mosquitoes can be drawn from the literature.

- Bird-feeding species tend to quest for hosts during the dark part of the night. Host seeking begins near the end of evening twilight when birds have settled on their roosts for the night.
- Bird-feeding mosquitoes bite throughout the night but rarely attack their hosts during the daytime, even when encouraged to do so.
- These mosquitoes spend their daylight hours in moist, heavily shaded, and usually forested areas. Their nocturnal host-seeking behaviour normally occurs in these same sylvan habitats.
- Bird-feeding mosquitoes commonly seek hosts in the forest canopy, whereas their mammal-feeding counterparts rarely seek hosts in treetops, at least at night. Mammal feeders usually seek hosts in campestral or transitional habitats during crepuscular periods.

There is also growing evidence that diurnal blood feeders (e.g., *Ae. aegypti*) tend to 'perch' in elevated locations frequented by hosts and wait until they are in some way stimulated before commencing their orientation flight toward the host (Edman and Spielman 1988). Moreover, they rely heavily on visual cues for locating hosts. In contrast, crepuscular and nocturnal species rely less on visual cues and embark on searching flights spontaneously when changing light levels or some other environmental cue stimulates the beginning of rhythmic, diel activity periods (Edman and Spielman 1988).

The generalizations above do not describe some unique situations. For example, high densities of salt-marsh or freshwater *Aedes* mosquitoes can result in opportunistic attacks on nearly all warm-blooded animals, including nesting or feeding birds, regardless of time of day or type of habitat (Corbet and Downe 1966; Day and Curtis 1983; Edman *et al.* 1984; Anderson and Fortner 1988).

Snow (1983) suggested that nocturnal mosquitoes tend to be more host-specific than diurnal or crepuscular ones because nocturnal mosquitoes must rely more on chemical cues to locate their hosts. In fact, the most host-specific species (i.e., certain bird, fish, and amphibian feeders) are nocturnal, but not all nocturnal species are highly host-specific. It is noteworthy that species which are almost exclusively bird feeders engorge principally on passerines. Perching birds constitute about 80 per cent of all birds in many habitats and tend to be smaller than the birds in other orders. In contrast, generalist bird feeders engorge on a wider variety of species, including larger birds. Generalists also host seek in a greater diversity of habitats (Edman and Spielman 1988).

It is reasonable to suggest that the host-attraction process for generalist bird-feeding mosquitoes differs from that for specialist bird feeders in some fundamental way (Gillies 1972). Gillies and Wilkes (1974) failed to demonstrate any long-range orientation toward chickens, pigeons, or CO_2 with an ornithophilic African species, *Cx. decans*. *Cx. decans* was captured in pigeon-baited traps situated in adjacent tree foliage (5.33 m high) and, to a much lesser extent, at the base of the tree trunk (0.91 m). In contrast, both Gillies and Wilkes (1974) and Edman (1979) demonstrated upwind orientation toward birds and CO_2 by generalist *Culex* species. In fact, CO_2-baited baffle traps are used successfully to monitor and collect generalist bird feeders for arbovirus isolation, but these traps are not useful with specialist species (Service 1976). Among the possible interpretations of these data are that specialists seek hosts at higher elevations, orient to different cues, or orient in different ways with regard to the spatial distribution of their hosts than do generalists.

Some mosquitoes may use CO_2 from trees as well as from roosting birds to locate their hosts. Trees produce CO_2 at night. Concentrations in foliage (0.6–1.0 per cent) are significantly higher than atmospheric levels (0.3–0.4 per cent). The CO_2 in air expired by vertebrates is about 4–5 per cent. Because concentration changes as small as 0.1 per cent are detectable by *Cx. tarsalis*, they conceivably could use both sources to orient to the roosting habitat in trees and to the roosting birds themselves (Gillies 1980).

In the tree canopy during early evening, the warm odour plume from a roosting bird should rise upward, and CO_2, which is heavier than air, should tend to descend. This suggests that mosquito orientation might originate from either above or below the roosting bird rather than horizontally as occurs with hosts situated on the ground. Such vertical orientation behaviour has never been demonstrated, however, and specialist bird feeders, such as *Cs. melanura*, are attracted to bird-baited traps situated on the forest floor. In a mature forest, most foliage is in the canopy and in the understorey; sparse cover lies in between at the trunk level. Thus, a search strategy that includes both canopy and understorey-roosting birds would seem adaptive. The flight patterns of *Cs. melanura* suggest that this may in fact occur (Nasci and Edman 1981*b*).

Differences in the attraction of ornithophilic mosquitoes to birds of different species, ages, and health status has been examined in the field on only a few occasions (Dow *et al.* 1957; Blackmore and Dow 1958; Means 1968; Joseph and Bickley 1969; Mahon and Gibbs 1982). Young animals appear to be less attractive than adults in both laboratory (LaPointe 1982) and field situations (Dow *et al.* 1957; Blackmore and Dow 1958), and there is some evidence that virus-infected birds may be more attractive to mosquitoes than uninfected conspecifics (Mahon and Gibbs 1982).

Based on available data, we cannot describe a biologically relevant scenario for mosquito attraction to vertebrate hosts. This applies to data on day-active species like *Ae. aegypti* and to the sparse but more anomalous data on several other mosquito species, including both specialists that feed on birds and generalists. The result is that we know far more about which vertebrates particular mosquitoes feed on than why or how they select these hosts.

Avian defensive behaviour to blood-feeding mosquitoes

General understanding of vertebrate defensive behaviour

Grooming behaviour by small mammals, lizards, and birds can affect both the number and distribution of permanent or semipermanent ectoparasites such as fleas, lice, and ticks on a host (Nelson *et al.* 1975;

Edman and Spielman 1988; Clayton, Chapter 14, this volume). Until recently, observations on the effect of temporary or rapid blood-feeding arthropods, like mosquitoes, on host defensive behaviour were restricted to anecdotal reports (Edman *et al.* 1985). Recent experimental approaches emphasized avian hosts (Edman and Kale 1971; Webber and Edman 1972; Edman *et al.* 1974, 1985; Edman and Scott 1987; Scott *et al.* 1988), but several small mammals have also been examined in detail (Waage and Nondo 1982; Day and Edman 1984; Isaacs 1986; Walker and Edman 1986).

Avian defensive behaviour

Interspecific variation

Mosquito feeding success on birds has been studied with a variety of avian species, representing eight different orders (Edman and Kale 1971; Edman *et al.* 1974). However, careful analysis of defensive behaviour toward mosquitoes by different birds has only been carried out with seven species of herons (Webber and Edman 1972), three passerine species, one columbiform species, and one galliform species (Edman *et al.* 1974). Consistent species-specific differences in defensive behaviour were found among healthy birds of the same age. This was particularly well documented among the ciconiiforms, whose defensive patterns appear to be related to their strategy of foraging for food; that is, passive foragers were more tolerant of mosquito attack than were active foragers (Edman *et al.* 1984). Behaviours that were directed against mosquitoes attacking the legs and feet were particularly effective in preventing engorgement (Webber and Edman 1972). Defensive species also killed significant numbers of mosquitoes and interrupted more blood meals than did less defensive species (Edman *et al.* 1985). To some extent, larger avian species were more tolerant of mosquitoes than were smaller species, but this correlation was much stronger among the mammalian species studied than among the birds (Edman and Scott 1987).

Intraspecific variation

Within the same avian species, birds can show considerable behavioural variation to mosquito attack. Intraspecific behavioural variation has been attributed to differences in experience, age, health, and social status (Edman *et al.* 1985; Edman and Scott 1987; Edman and Spielman 1988; Scott 1988; in this volume: Kirkpatrick *et al.*, Chapter 18; Johnson and Boyce, Chapter 20; Saumier *et al.*, Chapter 15). Even the same individual may suddenly change its behavioural response to mosquito attack for no apparent reason (Isaacs 1986).

In general, young nestling birds lack the defensive capabilities of juvenile and adult birds (Kale *et al.* 1972). The young of precocial gallinaceous species are the exceptions; they are more defensive than are juvenile and adult conspecifics (Edman *et al.* 1974). Older nestlings of altricial night herons were also somewhat more defensive than were the highly tolerant adult birds (Kale *et al.* 1972). Nestling house sparrows appeared to develop their defensive behaviour during days 14–18 post-hatching so that fledgelings were nearly as defensive as adult birds. Juvenile house sparrows that were held for 60 days in a mosquito-free environment were only slightly less defensive than control sparrows exposed to mosquito attack each night (Edman and Scott 1987).

Experience is not required for efficient defensive behaviour to occur; birds seem to learn quickly to be defensive. However, previously exposed birds caused more mosquitoes to interrupt blood meals than did naïve cohorts, which suggests an acquired sensitivity or resistance (Edman and Scott 1987).

Effect of mosquito density

For all avian hosts examined, the frequency of defensive behaviour toward mosquitoes increases with the number of attacking mosquitoes (Edman *et al.* 1972). This response results in a density-dependent reduction in the frequency of mosquito feeding success. For example, feeding success on the cattle egret (*Bubulcus ibis*) ranged from a low of 3 per cent when 800 mosquitoes were present to a high of 44 per cent when only 25 were present (Edman *et al.* 1972). Partial blood meals also varied with density from 0 per cent when 25 mosquitoes were present to 13 per cent when 800 mosquitoes were present. Successful feeding on house sparrows increased from less than 5 per cent to near 50 per cent when the density was decreased from 80 to 5 mosquitoes per night (Edman and Scott 1987). The relationships between mosquito density and avian defensive behaviour were linear with distinct annoyance thresholds, that is, above a particular mosquito density, defensive behaviour of birds increased significantly.

In nature, extremely high levels of annoyance may cause birds to take more radical steps than merely accelerating the frequency of their grooming. Temporary abandonment of roosting, feeding, or nesting sites may occur. High densities of both *Argus* ticks (Duffy 1983; Chapter 13, this volume) and salt-marsh mosquitoes (Anderson and Fortner 1988) have been credited with causing seabirds to abandon nests with eggs. Colonial roosting and nesting on offshore islands by some birds may in part be a behavioural adaptation to avoid high densities of biting mosquitoes during the night in estuaries, a place where these birds feed during the day.

Behavioural repertoires and their frequencies

On a fully feathered bird, mosquito mouth parts can penetrate only naturally exposed areas, such as the legs, feet, base of the bill, and fleshy region around the eyes. Some roosting birds can protect these areas by covering their head with scapulars or protecting a large portion of their feet and legs by pulling their legs up into their abdominal feathers. This tactic minimizes the surface area on which mosquitoes can probe and successfully feed.

We have observed that roosting birds respond readily to the wing-beat sound from approaching mosquitoes and even attempt to catch hovering mosquitoes in the air (Edman, unpublished data). It might be adaptive for bird-feeding mosquitoes to land on the perch near their roosting host and walk on to the sleeping target. Such behaviour has been observed in *Chrysops* sp., which feed on sleeping monkeys (Duke 1972).

The specific kinds of defensive movements, their intensity, and their frequency are major factors that lead to differential feeding success of mosquitoes on different birds. The behavioural repertoires and the intensity of these defensive movements seem to be species-specific. However, almost all herons observed displayed a full range of defensive movements (Webber and Edman 1972). The intensity of some movements, such as the head shake and foot stamp, were less vigorously performed by more tolerant species. Herons that successfully prevented most mosquitoes from engorging did so in large part because of the frequency with which they performed certain movements that were particularly effective in disturbing mosquitoes, for example, the foot peck, foot stamp, and foot slap. The frequency of key defensive behaviours among herons exposed to 300 *Culex* ranged upward from 2500 movements per hour for defensive species to under 600 per hour for more tolerant species.

Marked frequency differences among passerine species were also observed. Rufous-sided towhees (*Pipilo erythrophthalmus*) and house sparrows relied mainly on head shaking and foot shaking or stamping to dissuade attacking mosquitoes (Edman *et al.* 1974). They performed these defensive movements about 3500 times per hour compared with 40–50 times per hour when no mosquitoes were present. In contrast, the defensive strategy of the Carolina wren (*Thryothorus ludovicianus*) consisted almost entirely of covering its exposed areas with its feathers. This strategy appeared to be less efficient than the more active defences of the towhee and sparrow.

It is clear that major quantitative and qualitative differences in defensive behaviour exist even among closely related species of birds. Thus, the role that different species play as hosts for mosquitoes and reservoirs for arboviruses may vary dramatically. Generalizations are difficult to make, and research should take into account both intra- and interspecific differences in defensive behaviour.

Research with house sparrows and North American mosquito vectors

The information above summarizes contemporary understanding of mosquito–bird interactions in arbovirus transmission cycles. It provides the knowledge base for our experiments with house sparrows, arboviruses, and mosquito vectors.

It should now be clear that a fundamental component of arbovirus transmission is the host upon which a vector feeds. Understanding the dynamics of this interaction is critical for understanding or accurately simulating arbovirus transmission (Scott *et al.* 1983). Anything favouring mosquitoes feeding on a particular age group in the avian host population or on infected birds rather than uninfected ones would increase the probability of virus transmission, amplification, and survival by increasing the probability of mosquito infection.

Our hypothesis for the research discussed below was that arbovirus mosquito vectors feed preferentially on virus-infected birds. To test this prediction, we designed experiments that examined mechanisms of enzootic vector–host interaction. There are at least three ways that vertebrate virus infection could affect mosquito feeding: (1) by increasing the attractiveness of vertebrates to mosquitoes (Mahon and Gibbs 1982); (2) by reducing the defensive behaviour of hosts toward blood-feeding mosquitoes (Day and Edman 1983*a*,*b*; Scott *et al.* 1988); and (3) by modifying blood chemistry so that probing mosquitoes can quickly locate and imbibe host blood (Rossignol 1988). We chose to study mosquito attraction to birds and avian defensive behaviour. Except perhaps for viruses that are transmitted transovarially or venereally among vectors, a mosquito must first find and then probe its mouthparts into a susceptible host to transmit a virus or blood feed on a viraemic host to contract an infection. After our studies began, we realized that avian host age, without regard to infection status, might have a more important influence on the interactions between mosquitoes and sparrows than we had originally considered.

The response of house sparrows to virus infection

At the beginning of our project, it was necessary to define the viraemic responses of experimentally infected birds (Scott *et al.* 1988). In subsequent experiments we would then be able to determine if the timing of the viraemia overlapped with increased attraction or reduced defensive behaviour. This would have occured if modification of attraction and defensive behaviour caused by virus infection were to make a meaningful difference in virus transmission.

Throughout our attraction and defensive behaviour experiments we used birds, viruses, and mosquitoes that would be associated naturally in North American transmission cycles. House sparrows were collected in central Maryland, the virus strains had low passage histories, and the mosquito species were known vectors of the virus being studied. Birds were inoculated intramuscularly with 10^{3-4} 50 per cent tissue culture infective doses ($TCID_{50}$) of SLE, WEE, or EEE virus or 0.1 ml of diluent for controls. During viraemia studies, birds were bled at 24-h intervals from the jugular vein. During experiments on mosquito attraction and feeding success, sparrows were bled only once to minimize stress resulting from handling and bleeding. The single bleeding confirmed the infection status of birds. The duration of viraemias for attraction and defensive experiments was estimated based on results of viraemia studies.

Following infection with SLE, WEE, or EEE virus, nestling sparrows responded with viraemias of greater magnitude and duration than adults. Moreover, none of the nestlings infected with an alphavirus—WEE or EEE viruses—survived; they became ill and died 24–48 h after inoculation. None of the adults or nestlings infected with SLE virus showed signs of virus-associated morbidity or mortality. Based on these findings, we predicted that increased sparrow attraction or reduced sparrow antimosquito behaviour would be greatest for nestlings infected with one of the alphaviruses.

Effects of house sparrow age and virus infection on mosquito attraction

We used an olfactometer designed by Feinsod and Spielman (1979) that measures passive, vertical movement of host-related cues to determine if arbovirus infection increases the attractiveness of house sparrows to mosquitoes (Scott *et al.* 1990). Twenty-four hours after birds had been inoculated with virus or diluent, they were exposed to 10 mosquitoes for

10 min in the olfactometer. The number of mosquitoes moving toward the bird—or into the middle chamber of the olfactometer—was used as the measure of relative attractiveness. This process was repeated at 24-h intervals for the duration of the experiment, which in most cases lasted as long as the sparrow viraemia for that virus. Each bird was exposed to at least three groups of mosquitoes during each day of experimentation. During each trial, infected and control birds were examined simultaneously. Preliminary trials demonstrated that *Cx. quinquefasciatus* and *Cx. tarsalis* mosquitoes would respond when an adult sparrow was in the olfactometer and would not respond when the avian host was removed.

Host-seeking behaviour of mosquitoes is a complex phenomenon, and consequently, it has been difficult to study (Hocking 1971; Edman and Spielman 1988; Scott 1988). Our approach of using an olfactometer and laboratory-reared mosquitoes had limitations that suggest cautious interpretation of results. Nevertheless, several conclusions can be drawn from our experiments.

Infection of adult or nestling house sparrows with SLE or WEE virus did not make individual birds more attractive to *Cx. quinquefasciatus* or *Cx. tarsalis* mosquitoes, respectively. This result failed to support our hypothesis that virus infection would increase host attractiveness.

We did note, however, that during these trials adult sparrows seemed to attract more mosquitoes than did nestlings. Three subsequent experiments showed that sparrow age was positively associated with increased attraction of *Cx. quinquefasciatus*. In two experiments, nestlings were less attractive than adults, and there was a trend for increased mosquito attraction as nestling age increased.

In our final experiment on attraction, we studied sparrows in groups designed to simulate aggregations of birds that would be exposed to mosquitoes at a nest. Seven treatments were examined: individual birds (young nestling, old nestling, and adult) as an internal control for comparison with previous experiments, three young nestlings, three old nestlings, three young nestlings plus one adult, and three old nestlings plus one adult. Results from this experiment demonstrated that the effect of sparrow age was not changed by adding more birds of the same age. Attraction of *Cx. quinquefasciatus* to a group of sparrows was not simply the sum of the attractiveness of all birds in the group. Only when one adult was added to a group of old nestlings was attraction significantly greater than that observed for an individual nestling or group of nestlings of the same age. There was no difference in mosquito attraction at close range between individual nestlings and groups of nestlings.

Effects of virus infection on house sparrow antimosquito behaviour

To determine the effect of virus infection on avian antimosquito behaviour, we exposed individual house sparrows to mosquitoes in 1-foot-square screened cages and recorded the number of mosquitoes that imbibed blood (Scott *et al.* 1988). Unless otherwise noted, sparrows were exposed to 10 mosquitoes during a 2-h simulated twilight period. Infected sparrows and noninfected control birds were examined simultaneously at 24–h intervals for the duration of the experiment. In preliminary studies, we observed, as Waage and Nondo (1982) had with rabbits, that sparrows could learn quickly to become defensive behaviourally to host-seeking mosquitoes. Consequently, we designed our experiments with one habituation session before data were recorded; this reduced relatively high feeding rates during the initial exposure.

Our first two experiments were conducted with adult sparrows. Birds that had been habituated to mosquito feeding were exposed to mosquitoes: for 2 days before they were inoculated with virus or diluent, throughout the duration of their estimated viraemia, and for 2 days after the viraemia was estimated to have terminated. In each experiment we compared mosquito feeding rates on birds in one of four treatment categories: (1) virus inoculated and unrestrained; (2) control and unrestrained; (3) virus inoculated and restrained; and (4) control and restrained. We expected that, if virus infection modifies sparrow antimosquito behaviour, mosquito feeding rates would be high on both restrained and infected birds but low on unrestrained controls.

Results from experiments with adult sparrows did not support that prediction. Feeding success of *Cx. quinquefasciatus* on unrestrained SLE virus-infected adult sparrows was not high compared to unrestrained controls. Birds in both treatments allowed relatively few mosquitoes to feed on them. Conversely, infected and control birds that were restrained allowed relatively large numbers of mosquitoes to imbibe blood. Similar results were obtained when WEE virus and *Cx. tarsalis* were studied. These data agree with previous studies (Edman *et al.* 1974) that concluded that house sparrows are behaviourally defensive toward blood-feeding mosquitoes, but they do not support the hypothesis that virus infection causes birds to become behaviourally lethargic, thus making it easier for mosquitoes to obtain a blood meal.

Using a similar protocol, the above experiments were repeated with nestling sparrows to determine if bird age would affect the outcome. In these trials, habituated nestlings were exposed to mosquitoes for 1 day before they were inoculated and also for 2–3 days after inoculation. There were no postviraemia trials with nestlings.

Fig. 10.1. The ontogeny of house sparrow defensive behaviour to host-seeking *Culex quinquefasciatus.*

Feeding rates of *Cx. quinquefasciatus* on unrestrained nestlings inoculated with either SLE or EEE virus were high compared to adults that had received the same treatment. However, feeding rates on uninfected controls were also high. In a fifth experiment, we increased the mosquito density to 20 *Cx. quinquefasciatus* and inoculated nestlings with SLE virus. Again there was no difference in mosquito feeding success on infected versus noninfected birds. We concluded that, although nestling sparrows were more susceptible to clinical illness especially following alphavirus inoculation, they did not exhibit the antimosquito behaviour that adults did. Instead, all nestlings acted as if they were restrained.

A noteworthy observation during our experiments with nestling sparrows was the ontogeny of defensive behaviour (Edman and Scott 1987). At approximately 15–18 days of age, the time period when house sparrows fledge from their nest (Weaver 1942), the defensive behaviour of uninfected birds increased markedly (Fig. 10.1). This finding suggests that infection of sparrows at the appropriate age, when they are still susceptible to overt disease and ready to fledge, might reduce defensive behaviour and increase mosquito feeding success in an epidemiologically meaningful way.

Conclusions from experimental studies

Table 10.1 summarizes results from our studies on the effects of avian host age and arbovirus infection on mosquito attraction and blood-feeding success. Nestling house sparrows were more susceptible to debilitating disease from virus infection than adults, but they were not as attractive to mosquitoes as adult conspecifics, and they had not yet

Table 10.1. A summary of the effects of house sparrow age and arbovirus infection on the attraction and blood-feeding success of mosquitoes

	Sparrow age		
	Nestling	Fledgling	Adult
Viraemia	High titre, long duration	High titre, long duration	Low titre, short duration
Disease			
Alphavirus	Overt	Overt	None
Flavivirus	None	None	None
Attractiveness to mosquitoes	Low	Increasing	High
Antimosquito behaviour	None	Developing	Well developed

developed antimosquito behaviour. Fledgling sparrows were still susceptible to overt illness, and their attractiveness to mosquitoes was increasing, although it was less than that of adults. Birds 15--18 days of age were beginning to develop mosquito defensive behaviour; this age might constitute the critical period for the ontogeny of antimosquito behaviour by house sparrows. Adult birds showed no signs of illness following experimental infection. Adults attracted more mosquitoes than any other age group, and they were highly defensive to host-seeking mosquitoes.

After integrating results from our study with other published information on mosquito attraction and avian antimosquito behaviour we speculate that the following scenario occurs during the transmission of SLE and WEE viruses. Mosquitoes are attracted to a nest of passerine birds primarily by the incubating adult. However, because adult birds have well-developed physical and behavioural defences, persistent mosquitoes feed most often on the altricial nestlings. Nestlings have a poorly developed feather barrier and little or no antimosquito behaviour, which makes them more susceptible to mosquito feeding (Blackmore and Dow 1958; Edman and Spielman 1988; Scott 1988; Scott *et al.* 1988). Six to 24 hours—depending on the virus—after being bitten by an infected mosquito, the young birds become viraemic (Scott and Olson 1986; Scott *et al.* 1988). Nestlings respond with a viraemia that is significantly greater in duration and magnitude than that of adults. They are, therefore, more infective to mosquitoes than are older conspecifics. Susceptible mosquitoes that feed on viraemic nestlings are again attracted by the incubating adult; we assume that virus infection does not increase attractiveness. Because of inferior defences, mosquitoes have a higher probability of feeding on the infected nestling than on the uninfected adult. The results of this interaction are that: (1) nestling sparrows contribute more to the maintenance of SLE and WEE virus transmission cycles than adults; and (2) more mosquitoes become infected than if all vertebrate-host age groups contributed equally as a source of blood for host-seeking mosquitoes.

Under some circumstances, fledgeling birds might make an especially important contribution to virus amplification. In a synchronous breeding species with a large number of young birds leaving the nest at the same time, infected fledgelings could infect an unusually large number of mosquitoes. Infected fledgelings would be quite infective and attractive to mosquitoes. Because they are susceptible to debilitating alphavirus-associated illness at the same time that they are developing antimosquito behaviours, infected fledgelings might become behaviourally lethargic and allow more mosquitoes to feed on them than uninfected cohorts or adults. This could contribute to rapid amplification of virus among mosquitoes and birds. The result would be epizootic transmission. The low probability of these events all happening at the same time might help explain the erratic nature of epidemics for some arboviruses.

Assumptions of our speculation could be tested in field studies by observing mosquitoes feeding on reproductively active birds and by developing assays to detect proteins in mosquito blood meals that would identify whether the source of blood was an adult or immature host (Day and Edman 1982). Attraction experiments would as usual be difficult to interpret, but could include manipulation of the number and age of birds present at a nest. Studies with experimentally infected vertebrates should only be done in a laboratory where virus can be safely contained.

Acknowledgements

This research was supported by the National Institutes of Health (grants AI20675, AI22119, AI24989, and AI26787) and the Maryland Agricultural Experiment Station. Judith Grumstrup-Scott edited the manuscript.

References

Anderson, D.J. and Fortner, S. (1988). Waved albatross egg neglect and associated mosquito ectoparasitism. *Condor*, **90**, 727–9.

Bang, F.B. and Reeves, W.C. (1942). Mosquitoes and encephalitis in the Yakima Valley, Washington. III. Feeding habits of *Culex tarsalis* Coq. a mosquito host of the viruses of western and St. Louis encephalitis. *Journal of Infectious Diseases*, **70**, 272–4.

Bar-Zeev, M.H., Maibach, I., and Khan, A.A. (1977). Studies on the attraction of *Aedes aegypti* (Diptera: Culicidae) to man. *Journal of Medical Entomology*, **14**, 113–20.

Beaudette, F.R. and Hudson, C.B. (1945). Additional outbreaks of equine encephalomyelitis in New Jersey pheasants. *Journal of the American Veterinary Medicine Association*, **57**, 384.

Blackmore, J.S. and Dow, R.P. (1958). Differential feeding of *Culex tarsalis*, on nestling and adult birds. *Mosquito News*, **18**, 15–17.

Brinton, M.A. (1986). Replication of flaviviruses. In *The togaviridae and flaviviridae* (ed. S. Schlesinger and M.J. Schlesinger) pp. 327–74. Plenum Press, New York.

Brinton, M.A., Blank, K.J., and Nathanson, N. (1984). Host genes that influence susceptibility to viral diseases. In *Concepts in viral pathogenesis* (ed. A.L. Notkins and M.B.A. Oldstone), pp. 71–8. Springer-Verlag, New York.

Brouwer, R. (1960). Variations in human body odour as a cause of individual differences of attraction for malaria mosquitoes. *Tropical and Geographical Medicine*, **12**, 186–192.

Brown, A.W.A. (1954). Studies on the responses of the fémale *Aedes* mosquito. VI. The attractiveness of coloured clothes to Canadian species. *Bulletin of Entomological Research*, **45**, 67–78.

Brown, A.W.A. (1966). The attraction of mosquitoes to hosts. *Journal of the American Medical Association*, **196**, 159–62.

Chamberlain, R.W., Rubin, H., Kissling, R.E., and Eidson, M.E. (1951). Recovery of virus of eastern equine encephalomyelitis from a mosquito, *Culiseta melanura* (Coquillett). *Society of Experimental Biology and Medicine*, **77**, 396–7.

Corbet, P.S. and Downe, A.E.R. (1966). Natural hosts of mosquitoes in northern Ellesmere Island. *Arctic*, **19**, 153–61.

Cox, H.R., Jellison, W.L., and Hughes, L.E. (1941). Isolation of western equine encephalomyelitis virus from a naturally infected prairie chicken. *Public Health Reports*, **56**, 1905–6.

Davis, E.E. (1984). Regulation of sensitivity in the peripheral chemoreceptor systems for host-seeking behaviour by a haemolymph-borne factor in *Aedes aegypti*. *Journal of Insect Physiology*, **30**, 179–83.

Day, J.F. and Curtis, G.A. (1983). Opportunistic blood-feeding on egg-laying sea turtles by saltmarsh mosquitoes (Diptera: Culicidae). *Florida Entomologist*, **66**, 359–60.

Day, J.F. and Edman, J.D. (1982). Age of murine hosts determined by binding of estradiol to mosquito blood meals. *Journal of Medical Entomology*, **19**, 357–60.

Day, J.F. and Edman, J.D. (1983*a*). Feeding patterns of mosquitoes (Diptera: Culicidae) simultaneously exposed to malarious and healthy mice, including a method for separating blood meals from conspecific hosts. *Journal of Medical Entomology*, **20**, 120–7.

Day, J.F. and Edman, J.D. (1983*b*). Malaria renders mice susceptible to mosquito feeding when gametocytes are most infective. *Journal of Parasitology*, **69**, 163–70.

Day, J.F. and Edman, J.D. (1984). Mosquito engorgement on normally defensive hosts depends on host activity patterns. *Journal of Medical Entomology*, **21**, 732–40.

Dow, R.P., Reeves, W.C., and Bellamy, R.E. (1957). Field test of avian host preferences of *Culex tarsalis* Coq. *American Journal of Tropical Medicine and Hygiene*, **6**, 294–303.

Downes, J.A. (1971). The ecology of blood-sucking Diptera: An evolutionary perspective. In *Ecology and physiology of parasites* (ed. A. M. Fallis), pp. 232–58. University of Toronto Press, Toronto.

Downs, W.G. (1963). Birds in relation to arthropod-borne viruses in Trinidad. *Proceedings of the Ornithological Congress*, **13**, 581–90.

Duffy, D.C. (1983). The ecology of tick parasitism on densely nesting Peruvian seabirds. *Ecology*, **64**, 110–19.

Duke, B.O.L. (1972). Behavioral aspects of the life cycle of Loa. In *Behavioral aspects of parasite transmission*. Supplement 1 of the *Zoology Journal of the Linnean Society*, Vol. 51. (ed. E. Canning and C. Wright), pp. 97–107. Academic Press, London.

Edman, J.D. (1979). Orientation of some Florida mosquitoes (Diptera: Culicidae) toward small vertebrates and carbon dioxide in the field. *Journal of Medical Entomology*, **15**, 292–6.

Edman, J.D. and Kale II, H.W. (1971). Host behavior: its influence on the feeding success of mosquitoes. *Annals of the Entomological Society of America*, **64**, 513–16.

Edman, J.D. and Scott, T.W. (1987). Host defensive behaviour and the feeding success of mosquitoes. *Insect Science and its Application*, **8**, 617–22.

Edman, J.D. and Spielman, A. (1988). Blood-feeding by vectors: physiology, ecology behavior and vertebrate defense. In *The arboviruses: epidemiology and ecology* (ed. T.P. Monath), Vol. 1, pp. 153–90. CRC Press, Boca Raton, Florida.

Edman, J.D., Webber, L.A., and Kale II, H.W. (1972). Effect of mosquito density on the interrelationship of host behavior and mosquito feeding success. *American Journal of Tropical Medicine and Hygiene*, **21**, 487–91.

Edman, J.D., Webber, L.A., and Schmid, A.A. (1974). Effect of host defenses on the feeding pattern of *Culex nigripalpus* when offered a choice of blood sources. *Journal of Parasitology*, **60**, 874–83.

Edman, J.D., Day, J.F., and Walker, E.D. (1984). Field confirmation of laboratory observations on the differential antimosquito behavior of herons. *Condor*, **86**, 91–2.

Edman, J.D., Day, J.F., and Walker, E.D. (1985). Vector–host interplay: Factors affecting disease transmission. In *Ecology of mosquitoes: Proceedings of a workshop* (ed. L.P. Lounibos, J.R. Rey, and J.H. Frank), pp. 273–85. University of Florida Press, Gainesville.

Ekis, G. (1971). Host acceptance of *Culex pipiens* L. (Culicidae: Diptera). *New York Entomological Society* **79**, 191–5.

Feinsod, F.M. and Spielman, A. (1979). An olfactometer for measuring host-seeking behavior of female *Aedes aegypti* (Diptera: Culicidae). *Journal of Medical Entomology*, **15**, 282–5.

Gillies, M.T. (1972). Some aspects of mosquito behaviour in relation to the transmission of parasites. In *Behavioral aspects of parasite transmission*, Supplement 1 of the *Zoology Journal of the Linnean Society*, Vol. 51 (ed. E. Canning and C. Wright), pp. 69–81. Academic Press, London.

Gillies, M.T. (1980). The role of carbon dioxide in host-finding for mosquitoes (Diptera: Culicidae): a review. *Bulletin of Entomological Research* **70**, 525–32.

Gillies, M.T. and Wilkes, T.J. (1969). A comparison of the range of attraction of animal baits and of carbon dioxide for some West African mosquitoes. *Bulletin of Entomological Research*, **59**, 441–56.

Gillies, M.T. and Wilkes, T.J. (1970). The range of attraction of single baits for some West African mosquitoes. *Bulletin of Entomological Research*, **60**, 225–35.

Gillies, M.T. and Wilkes, T.J. (1972). The range of attraction of animal baits and carbon dioxide for mosquitoes. Studies in a freshwater area of West Africa. *Bulletin of Entomological Research*, **61**, 389–404.

Gillies, M.T. and Wilkes, T.J. (1974). The range of attraction of birds as baits for some West African mosquitoes (Diptera: Culicidae). *Bulletin of Entomological Research*, **63**, 573–81.

Halstead, S.B. (1980*a*). Immunopathological parameters of togavirus disease syndromes. In *The togaviruses: biology, structure, replication* (ed. R.W. Schlesinger), pp. 107–74. Academic Press, New York.

Halstead, S.B. (1980*b*). Dengue haemorrhagic fever—a public health problem and a field for research. *Bulletin of the World Health Organisation*, **58**, 1–21.

Hammon, W.M., Gray, J.A., Evans, F.C., and Izumi, E.M. (1941). Western equine and St. Louis encephalitis antibodies in the sera of mammals and birds from an endemic area. *Science*, **94**, 305–7.

Hayes, R.O. (1961). Host preferences of *Culiseta melanura* and allied mosquitoes. *Mosquito News*, **21**, 179–87.

Hocking, B. (1971). Blood-sucking behaviour of terrestrial arthropods. *Annual Review of Entomology*, **16**, 1–26.

Howitt, B.F. and Van Herick, W. (1941). Neutralizing antibodies against St. Louis and Western equine encephalitis viruses in horses and fowl. *Proceedings of the Society of Biology and Medicine*, **48**, 247–50.

Isaacs, C. (1986). Anti-mosquito behavior of *Peromyscus leucopus* (white-footed mouse). Unpublished Ph.D. thesis, University of New Hampshire, Durham.

Joseph, S.R. and Bickley, W.E. (1969). *Culiseta melanura (Coquillett) on the Eastern Shore of Maryland (Diptera: Culicidae)*. University of Maryland Agricultural Experimentation Station Bulletin, no. A-161.

Kale, H.W., Edman, J.D., and Webber, L.A. (1972). Effect of behavior and age of individual ciconiiform birds on mosquito feeding success. *Mosquito News*, **32**, 343–50.

Karabatsos, N. (1985). *International catalogue of arboviruses I. Including certain other viruses of vertebrates* 3rd edn. American Society of Tropical Medicine and Hygiene, San Antonio, Texas.

Kay, B.H., Boreham, P.F.L., and Williams, G.M. (1979). Host preferences and feeding patterns of mosquitoes (Diptera: Culicidae) at Kowanyama, Cape York Peninsula, northern Queensland. *Bulletin of Entomological Research*, **69**, 441–57.

Kennedy, J.S. (1983). Zigzagging and casting as a programmed response to wind-borne odour: a review. *Physiological Entomology*, **8**, 109–20.

Khan, A.A. (1977). Mosquito attractants and repellants. In *Chemical control of insect behavior: theory and application* (ed. H.H. Shorey and J.J. McKelvey, Jr.), pp. 305–25. John Wiley, New York.

Kissling, R.E. (1958). Host relationships of the arthropod-borne encephalitides. *Annals of the New York Academy of Science*, **70**, 320–7.

Kissling, R.E., Rubin, H., Chamberlain, R.W., and Eidson, M.E. (1951). Recovery of eastern equine encephalomyelitis from the blood of a purple grackle. *Proceedings of the Society of Experimental Biology and Medicine*, **77**, 398–9.

LaPointe, D. (1982). Effect of host age on mosquito attraction, feeding and fecundity. Unpublished M.S. thesis, University of Massachusetts, Amherst.

Learman, J.J. (1955). The host-seeking behaviour of the malaria mosquito *Anopheles maculipennis* atroparvus. *Acta Leidensi*, **25**, 1–144.

Lord, R.D. and Calisher, C.H. (1970). Further evidence of southward transport of arboviruses by migratory birds. *American Journal of Epidemiology*, **92**, 73–8.

Ludwig, G.V., Cook, R.S., McLean, R.G., Francy, D.B., and Beaty, B.J. (1986). Viremic enhancement due to transovarially acquired antibodies to St. Louis encephalitis virus in birds. *Journal of Wildlife Disease*, **22**, 326–34.

Mahon, R. and Gibbs, A. (1982). Arbovirus infected hens attract more mosquitoes. In *Viral diseases in South-east Asia and the western Pacific* (ed. J.S. Mackenzie), pp. 502–4. Academic Press, Sydney.

McClure, H.E. (1963). Birds and the epidemiology of Japanese encephalitis. *Proceedings of the International Ornithological Congress*, **13**, 604–10.

McLean, R.G. and Bowen, G.S. (1980). Vertebrate hosts. In *St Louis encephalitis* (ed. T.P. Monath) pp. 381–450. American Public Health Association, Washington, D.C.

McLean, R.G. and Scott, T.W. (1979). Avian hosts of St. Louis encephalitis virus. In *Proceedings of the 8th Bird Control Seminar* (ed. W.B. Jackson) pp. 143–55. Bowling Green State University, Bowling Green, Ohio.

Means, R.G. (1968). Host preferences of mosquitoes (Diptera: Culicidae) in Suffolk County, N. Y. *Annals of the Entomological Society of America*, **61**, 116–20.

Monath, T.P. (1985). Flaviviruses. In *Virology* (ed. B.N. Fields), pp. 955–1004. Raven Press, New York.

Monath, T.P. (1986). Pathobiology of flaviviruses. In *The togaviridae and flaviviridae* (ed. S. Schlesinger and M.J. Schlesinger), pp. 955–1004. Plenum Press, New York.

Morris, C.D. and Zimmerman, R.H. (1981). Epizootiology of eastern equine encephalomyelitis virus in upstate New York, USA. *Journal of Medical Entomology*, **18**, 313–16.

Nasci, R.S. and Edman, J.D. (1981*a*). Blood-feeding patterns of *Culiseta melanura* (Diptera: Culicidae) and associated sylvan mosquitoes in southeastern Massachusetts eastern equine encephalitis enzootic foci. *Journal of Medical Entomology*, **18**, 493–500.

Nasci, R.S. and Edman, J.D. (1981*b*). Vertical and temporal flight activity of the mosquito *Culiseta melanura* (Diptera: Culicidae) in southeastern Massachusetts. *Journal of Medical Entomology*, **18**, 501–4.

Nathanson, N. (1980). Pathogenesis. In *St. Louis encephalitis* (ed. T.P. Monath), pp. 201–36. American Public Health Association, Washington, D.C.

Nelson, W.A., Keirans, J.E., Bell, J.F., and Clifford, C.M. (1975). Host–ectoparasite relationships. *Journal of Medical Entomology*, **12**, 143–66.

Price, G.D., Smith, N., and Carlson, D.A. (1979). The attraction of female mosquitoes (*Anopheles quadrimaculatus* Say) to stored human emanations in conjunction with adjusted levels of relative humidity, temperature and carbon dioxide. *Journal of Chemical Ecology*, **3**, 383–95.

Reeves, W.C. (1953). Quantitative field studies on a carbon dioxide chemotropism of mosquitoes. *American Journal of Tropical Medicine and Hygiene*, **2**, 325–31.

Reeves, W.C. (1971). Mosquito vector and vertebrate host interaction: the key to maintenance of certain arboviruses. In *Ecology and physiology of parasites* (ed. A. M. Fallis), pp. 223–31. University of Toronto Press.

Reeves, W.C. (1974). Overwintering of arboviruses. *Progress in Medical Virology*, **17**, 193–220.

Reeves, W.C. and Hammon, W.McD. (1944). Feeding habits of the proven and possible mosquito vectors of western equine and St. Louis encephalitis in the Yakima Valley, Washington. *American Journal of Tropical Medicine*, **24**, 131–4.

Reeves, W.C. and Hammon, W.McD. (1962). *Epidemiology of the arthropod-borne viral encephalitides in Kern County, California, 1943–1952, Vol. 4.* University of California Press, Berkeley, California.

Reid, J.A. (1961). The attraction of mosquitoes by human or animal baits in relation to the transmission of disease. *Bulletin of Entomological Research*, **52**, 43–62.

Ribbands, C.K. (1949). Studies on the attractiveness of human populations to *Anopheles. Bulletin of Entomological Research*, **40**, 227–37.

Rossignol, P.A. (1988). Parasite modification of mosquito probing behavior. In *Proceedings of a symposium: the role of vector–host interactions in disease transmission*, Miscellaneous Publication of the Entomological Society of America, no. 68 (ed. T.W. Scott and J. Grumstrup-Scott), Entomological Society of America, College Park, Maryland.

Rudolfs, W. (1922). Chemotropism of mosquitoes. *Bulletin of the New Jersey Agricultural Experiment Station* **367**, 1–23.

Sawyer, W.A. and Lloyd, W. (1931). The use of mice in tests of immunity against yellow fever. *Journal of Experimental Medicine*, **54**, 533.

Scherer, W., Buescher, F.E.L., and McClure, H.E. (1959). Ecologic studies of Japanese encephalitis. V. Avian factors. *American Journal of Tropical Medicine and Hygiene*, **8**, 689–97.

Scott, T.W. (1988). Vertebrate host ecology. In *The arboviruses: epidemiology and ecology* (ed. T.P. Monath), pp. 257–80. CRC Press, Boca Raton, Florida.

Scott, T.W. and Olson, J.G. (1986). Detection of eastern equine encephalomyelitis viral antigen in avian sera by enzyme immunoassay: a laboratory study. *American Journal of Tropical Medicine and Hygiene*, **35**, 611–18.

Scott, T.W. and Weaver, S.C. (1989). Eastern equine encephalomyelitis virus: epidemiology and evolution of mosquito transmission. *Advances in Virus Research*. **37**, 277–328.

Scott, T.W., McLean, R.G., Francy, D.B., and Card, C.S. (1983). A simulation model for the vector–host transmission system of a mosquito-borne avian virus, Turlock (Bunyaviridae). *Journal of Medical Entomology*, **20**, 625–40.

Scott, T.W., Edman, J.D., Lorenz, L.H., and Hubbard, J.L. (1988). Effects of disease on vertebrates' ability behaviorally to repel host-seeking mosquitoes. In *Proceedings of a symposium: the role of vector–host interactions in disease transmission*, Miscellaneous Publication of the Entomological Society of America, no. 68. (ed. T.W. Scott and J. Grumstrup-Scott), pp. 9–17. Entomological Society of America, College Park, Maryland.

Scott, T.W., Lorenz, L.H., and Edman, J.D. (1990). The effect of house sparrow age and arbovirus infection on attraction of mosquitoes. *Journal of Medical Entomology*. (In press.)

Sellards, A.W., Tyzzer, E.E., and Bennett, B.L. (1941). The infection of birds with the virus of equine encephalomyelitis. *American Journal of Hygiene*, **33**, 61–8.

Service, M.W. (1964). The attraction of mosquitoes by animal baits in the northern Guinea savannah of Nigeria. *Journal of the Entomological Society of South Africa*, **27**, 29–36.

Service, M.W. (1971). Feeding behaviour and host preferences of British mosquitoes. *Bulletin of Entomological Research*, **60**, 653–61.

Service, M.W. (1976). *Mosquito ecology: field sampling methods*. Allied Science Publishing, London.

Shope, R.E. (1985). Alphaviruses. In *Virology* (ed. B.N. Fields), pp. 931–53. Raven Press, New York.

Sippell, W.I. and Brown, A.W.A. (1953). Studies on the responses of the female *Aedes* mosquito. V. The role of visual factors. *Bulletin of Entomological Research*, **43**, 567–74.

Smith, C.N., Smith, N., Gouck, H.K., Weidhaas, D.E., Gilbert, I.H., Mayer, M.S., Smittle, B.J., and Hofbaueer, A. (1970). L-Lactric acid as a factor in the attraction of *Aedes aegypti* (Diptera: Culicidae) to human hosts. *Annals of the Entomological Society of America*, **65**, 605–12.

Snow, W.F. (1983). The attractiveness of some birds and mammals for mosquitoes in the Gambia, West Africa. *Annals of Tropical Medicine and Parasitology*, **77**, 641–51.

Snow, W.F. and Boreham, P.F.L. (1973). The feeding habits of some West African *Culex* (Dipt., Culicidae) mosquitoes. *Bulletin of Entomological Research*, **62**, 517–26.

Snow, W.F. and Boreham, P.F.L. (1978). The host-feeding patterns of some culicine mosquitoes (Diptera: Culicidae) in the Gambia. *Bulletin of Entomological Research*, **68**, 695–706.

Stamm, D.D. (1958). Studies on the ecology of equine encephalomyelitis. *American Journal of Public Health*, **48**, 328–35.

Stamm, D.D. (1963). Susceptibility of bird populations to eastern, western, and St. Louis encephalitis viruses. *Proceedings of the International Ornithological Congress*, **13**, 591–603.

Stamm, D.D. (1966). Relationships of birds and arboviruses. *Auk*, **83**, 84–97.

Stamm, D.D. and Newman, R.J. (1963). Evidence of southward transport of arboviruses from the U.S. by migratory birds. *Microbiology*, **11**, 123–33.

Tempelis, C.H. (1975). Host-feeding patterns of mosquitoes, with a review of advances in analysis of blood meals by serology. *Journal of Medical Entomology*, **11**, 635–53.

Ten Broeck, C. (1938). Birds as possible carriers of the virus of equine encephalomyelitis. *Archives of Pathology*, **25**, 759.

Theiler, M. and Downs, W.G. (1973). *The arthropod-borne viruses of vertebrates.* Yale University Press, New Haven, Connecticut.

Tyzzer, E.E., Sawyer, A.W., and Bennett, B.L. (1938). Occurrence in nature of 'equine encephalomyelitis' in the ring-necked pheasant. *Science*, **88**, 505–6.

Vorndam, A.V. (1980). Immunization. In *Encephalitis* (ed. T.P. Monath), pp. 623–35. American Public Health Association, Washington, D.C.

Waage, J.K. and Nondo, J. (1982). Host behavior and mosquito feeding success: an experimental study. *Transactions of the Royal Society of Tropical Medicine and Hygiene*, **76**, 119–23.

Walker, E.D. and Edman, J.D. (1986). Influence of defensive behavior of eastern chipmunks and gray squirrels (Rodentia: Sciuridae) on feeding success of *Aedes triseriatus* (Diptera: Culicidae). *Journal of Medical Entomology*, **23**, 1–7.

Weaver, R.L. (1942). Growth and development of English sparrows. *Wilson Bulletin*, **54**, 183–91.

Webber, L.A. and Edman, J.D. (1972). Anti-mosquito behaviour of ciconiiform birds. *Animal Behaviour*, **20**, 228–32.

Webster, L.T. (1933). Inherited and acquired factors in resistance to infection. I. Development of resistance and susceptible lines of mice through selective breeding. *Journal of Experimental Medicine*, **57**, 793.

Williams, M.C. (1963). Studies on mosquitoes (Diptera: Culicidae) biting birds using twenty-four-hour catches in the Entebbe area, Uganda. *Bulletin of Entomological Research*, **54**, 407–23.

Work, T.H. (1963). Virology in the biology of birds. *Proceedings of the International Ornithological Congress*, **13**, 570–80.

Wright, R.E. and DeFoliart, G.R. (1970). Associations of Wisconsin mosquitoes and woodland vertebrate hosts. *Annals of the Entomological Society of America*, **63**, 777–86.

Wright, R.H. (1975). Why the mosquito repellents repel. *Scientific American*, **233**, 104–11.

Wright, R.H. and Kellogg, F.E. (1964). Host size as a factor in the attraction of malaria mosquitoes. *Nature*, **202**, 321–2.

11 *The nest protection hypothesis: the adaptive use of plant secondary compounds by European starlings*

LARRY CLARK

Introduction

Some species of birds build their nests exclusively out of fresh vegetation; others may place only sprigs of fresh plant material into an otherwise dry nest matrix (Collias and Collias 1984). Patterns for the use of green plants as nest material have periodically attracted the attention of ornithologists and led to speculation as to why birds would go to great lengths to procure fresh plants (Gibb 1950; Jacobs *et al.* 1978). Among the hypotheses accounting for the use of green plants as nest material are:

1. *The aesthetic hypothesis.* Weaving flowers or green vegetation into an otherwise dry grass or twig matrix somehow fulfils an artistic sense of proportion of the breeding bird (Welty 1972).
2. *The crypsis hypothesis.* Green plants may serve to camouflage the nest from visually orienting predators (Skutch 1976).
3. *The water loss hypothesis.* Fresh vegetation may reduce desiccation of eggs when parents are not in attendance by raising the vapour pressure in the nest environment (Taverner 1933; Sengupta 1968). This latter hypothesis seems an unlikely explanation because many of the species characterized by nests with fresh vegetation nest in cavities where vapour deficits are low. Also, it would not explain why species nesting in humid, forested environments use green plants.
4. *The shading hypothesis.* Green plants are incorporated into nests to cover eggs and thereby serve to reduce insolation. But this hypothesis again fails to account for birds that show the behaviour and nest in shaded environments and cavities, nor does it explain why some species weave fresh material throughout an otherwise dry nest matrix.
5. *The nest insulation hypothesis.* Fresh vegetation incorporated into the nest matrix may retard heat flux from the nest cup. However, an insulative advantage is unlikely because dry grasses or moss retards heat loss better than damp nest material (Mertens 1977).
6. *The nest protection hypothesis.* Early workers believed that birds, like

herbalists, used the intrinsic properties of plants as protection against vermin (Widmann 1922).

More formally, the avian nest protection hypothesis explains the widespread use of fresh plants for nest material as an evolutionary adaptation against ectoparasites and pathogens. Birds are presumably exploiting the chemicals within plants for their antipathogenic or insecticidal properties. If the nest protection hypothesis is a plausible explanation for the use of green plants in nest construction, then several conditions must hold true:

1. Birds must face a selective pressure from pathogens and parasites.
2. Use of plant material should correspond to the probability of being parasitized or encountering pathogens, and to some extent be related to the parasite load encountered.
3. Plants used as nest material should be effective at reducing parasite and pathogen load.
4. Birds should be able to discriminate among plants and select plants which have a high probability of containing biological properties of selective advantage to the bird.

There is no doubt that some species of birds encounter a strong selective pressure from ectoparasites and pathogens. For example, irruptions of ticks in seabird colonies have caused birds to abandon colonies (Feare 1976; Duffy 1983; Chapter 13, this volume). Similar disruptive patterns are seen for colonially breeding passerines such as barn swallows (*Hirundo rustica*) and cliff swallows (*Hirundo pyrrhonota*) (Brown and Brown 1986; Shields and Crook 1987; this volume: Chapman and George, Chapter 4; Loye and Carroll, Chapter 12). Because of the frequency of intraspecific encounters, and the associated higher probability of encountering ectoparasites, colonially breeding birds may face an increased likelihood of contracting pathogens via arthropod vectors (Whiteman and Bickford 1983; see also Scott and Edman, Chapter 10, this volume). However, the effects of parasites and pathogens on birds are not always so dramatic. Parasites and pathogens may not directly affect mortality; rather, they may more subtly affect morbidity. Parasitism by haematophagous parasites can result in poor growth of swallow and thrasher chicks (Moss and Camin 1970; Arendt 1985; Brown and Brown 1986; Shields and Crook 1987). As a consequence the resultant low fledging weight can decrease the probability of first-year survivorship (Kluyver 1966; Garnett 1981). Growth measurement is the most frequently used measure for the well-being of chicks, yet it can fail to detect the effects of parasites and pathogens. Gold and Dahlsten (1983) failed to find a statistically significant effect of *Protocalliphora* on weight gain or mortality of chickadees, despite the obvious energetic load parasitism represented (see also Rogers *et al.*, Chapter 7, this volume). At present, the consequences on lifetime fecundity of supporting the energetic load of parasitism are unknown.

Even under relatively homogeneous conditions, ectoparasites may have variable effects on their hosts. For example, northern fowl mites (*Ornithonyssus sylviarum*) feeding on incubating domestic hens can cause increased agitation, which disrupts the incubation pattern and results in decreased hatching success (DeVaney 1979). When infestations reach 1000–10 000 mites per bird, egg production can decrease by 5–15 per cent, and hens show signs of microcytic, normochromic anaemia (Matthysse *et al.* 1974; DeVaney 1979); it should be noted that earlier studies failed to demonstrate such effects (Loomis *et al.* 1970). Heavy mite infestations on roosters failed to result in anaemia, but did cause decreases in weight, seminal fluid, sperm concentration, and plasma testosterone levels (Matthysse *et al.* 1974). What factors contribute to the variable results in seemingly similar studies is unknown, though a variety of factors may modulate the effect of ectoparasites on the host, e.g. the nutritional status of the host, other factors influencing energy expenditure of the host, whether ectoparasites act as vectors for pathogens, and the nature of the pathogen pool (Whiteman and Bickford 1983; Scott and Edman, Chapter 10, this volume). Thus, ectoparasite load itself may not be the single crucial factor determining morbidity or mortality of parasitized birds, rather ectoparasites may be a correlated, synergistic or exacerbating factor affecting survivorship and fecundity. The time-scale of observations that define reproductive success may be critical to assessing the impact of ectoparasites. For wild populations of birds the effects of ectoparasites may not be seen during the breeding season, rather effects may be delayed until winter or future reproductive activity. For example, if ectoparasite load has no other effect than to increase the daily existence energy cost it is feasible that this could, in the case of females, negatively affect the subsequent year's clutch size (viz. Bryant 1979). Thus the effects of ectoparasites may be subtle, being detectable only when comparative demographic information is available. Current studies are seldom long enough or of sufficient detail to detect such effects. Nonetheless, it can be assumed that ectoparasites probably have some effect on fitness, though the magnitude of the selective pressure is not known at this time.

A second condition necessary for the support of the nest protection hypothesis is that birds facing high ectoparasite loads must also have a high probability of using green plants in nest construction. An implicit assumption here is that large parasite loads are directly related to selection pressure. Unfortunately, a rigorous test of this hypothesis is not possible because a systematic survey of avian parasite load is not currently available. However, a slightly less rigorous test is available if we accept the general pattern that the reuse of nest sites is associated with higher ectoparasite loads relative to nests used once (Rothschild and Clay 1952; Wasylick 1971). In a survey of the nesting phenology of 137 species of passerines breeding in eastern North America, Clark and Mason (1985) estimated that 19.7 per cent of the species reused nests across years, and that 34 per cent of these species incorporated green vegetation into their nests. Clark and Mason found that passerine species reusing nest sites were 5.9 times as likely to use green plants as species that used nest sites only once. Raptorial birds are more likely than passerines to use green vegetation in nest construction (57 per cent, $n = 49$; Wimberger 1984). But, based on Wimberger's data, the odds are about the same (5.96) that raptors reusing nest sites will incorporate green plant material. The similarity in the odds ratios are intriguing. Further data on the relationship between nest site reuse and addition of greenery for other orders of birds are needed before definitive statements can be made about the similarity of the odds ratio. Nonetheless, as a general pattern, the data are consistent with the nest protection hypothesis: species likely to encounter higher parasite loads or experience unsanitary nesting conditions are more likely to use green plants in nest construction.

The comparative data bases on parasitism and the use of green plant material are consistent with the first two conditions necessary for support of the nest protection hypothesis. However, more detailed studies on single species are required for critical tests of the hypothesis. The objective of this chapter is to review the evidence on the above conditions with special reference to the breeding biology of the European starling (*Sturnus vulgaris*). Starlings are ideally suited to test specifics of the nest protection hypothesis because they are cavity nesters, often reusing nest sites, and they can be induced to breed in artificial cavities in numbers where experimental manipulation of the nest environment is possible. In addition, starlings are plagued by a host of ectoparasites (Boyd 1951). These parasite loads can be experimentally manipulated. Addition of pyrethrins will kill arthropods without affecting the hosts and thereby depress ectoparasite load. Alternatively, parasite load can be increased by inoculating nests with ectoparasites or pathogens. Finally, starlings incorporate green plant material in nest construction.

Methods

Study site

The breeding colony of starlings used for this study was located at the Stroud Water Research Center (SWRC) of the Academy of Natural

Sciences of Philadelphia, Avondale, Pennsylvania. Nest boxes at the colony have been in use since the colony's establishment in 1973. The habitat surrounding SWRC is a mix of lawns, old fields, pastures, and secondary growth deciduous forest.

Nesting phenology

In south-eastern Pennsylvania, males begin placing pieces of green plant material into nest boxes in early March. By the beginning of April, nests are one-half to three-quarters completed, composed mostly of dried grass with green plant material interwoven into the dry nest matrix. At this time females help the males finish the nest. Starlings add green vegetation until the eggs are laid and incubation begins. After this time, no additional material is added to the nest (Kessel 1957). Because starlings begin nest construction in the early spring, green vegetation is generally not abundant, though it is available. My own observations, based on birds breeding in south-eastern Pennsylvania, indicate that starlings gather the bulk of their nest material within 20 m of the nest, and never more than 50 m away from it. Thus the type of vegetation used by starlings is strongly influenced by small variations in the local distribution of plants. Within a single colony site, starlings showed a distinct preference ($p < 0.001$) for at least six species of herbs relative to all plants available to them during nest construction: agrimony, *Agrimonia parvaflora*; wild carrot, *Daucus carota*; fleabane, *Erigeron philadelphicus*; rough goldenrod, *Solidago rugosa*; yarrow, *Achillea millefolium*; and red dead nettle, *Lamium purpureum*. Fifty-seven species either were selected in proportion to their availability in the habitat or were actually avoided (Clark and Mason 1985). Observations over the years suggest that the age of the nest box is correlated with the amount of green vegetation used by starlings for nest construction. To control for this possibility only those boxes constructed in 1980 were used for this study.

In 1984 field experiments were designed to manipulate northern fowl mite populations at nest boxes and determine the effect of different infestation densities on nestling starlings. All nest boxes were cleared of debris and scraped clean in January to control for previous infestation history. During nest construction in March and April all fresh green vegetation was removed from the nest box and dry nest matrix as an effort to control for putative antiparasitical effects of fresh vegetation. Beginning on the second day of incubation (−10, day zero being the day of hatching) nest boxes were randomly assigned to one of three treatments:

1. *Nest replacement* (NR). The dry nest matrix was removed and replaced with an equivalent amount of clean dry grass at −10, −5, 1,

7, and 13 days of the breeding cycle. This treatment was designed to maintain small mite populations throughout the nestlings' development.

2. *Plant removal* (PR). These nests received no additional treatment, and were devoid of putative miticidal or repellent effects of green plants. This treatment would presumably allow mite populations to grow unconstrained.
3. *Plant addition* (PA). Approximately 5 g of fresh leaves from wild carrot was interwoven into the dry nest matrix at −10, −5, 1, 7, and 13 days posthatching. Wild carrot was chosen because it was a plant commonly preferred by starlings for nest construction and was readily available.

At 1, 7, 13, and 19 days posthatching three nests from each treatment were destructively sampled for mite populations. Initially, nest matrix was placed into a 1-gallon 'zip-lock' bag. Mites emerging from the matrix moved to the top of the bag and were collected by aspiration. After several days, the nest material was removed and washed with 70 per cent alcohol. Arthropods were collected and sorted.

The northern fowl mite is an ectoparasite which feeds on the blood of birds (Krantz 1978). Adult mites are small, being no more than 2 mm long. Northern fowl mites can complete a five-stage life cycle in 7–14 days (Sikes and Chamberlain 1954), while adult forms are capable of enduring fasts of up to 3–4 months (personal observation). Eggs are laid on birds or in nest material. After 24 h, non-feeding, six-legged larva hatch. After an additional 24 h, larvae metamorphose into eight-legged feeding protonymphs. Protonymphs remain in the nest matrix except when feeding on the host. Normally, protonymphs require at least two separate feeding bouts (36 h) before they are fully engorged (0.025 mg blood) and change into a non-feeding deutonymph. After 24 h, deutonymphs give rise to a continuously feeding adult form. Adult females may engorge with 0.041 mg of blood every 1–5 days before laying an average clutch of 2.27 eggs (Sikes and Chamberlain 1954). This life history allows a small founding population to expand to considerable numbers during the 37–40 days starlings are active at the nest (nest construction through fledging of the chicks).

Results

By the end of the first breeding attempt of birds in May, nests protected with wild carrot contained an average of 3000 mites, whereas nests without plant material contained an average of 80 000 mites (Fig. 11.1). By the end of the second breeding attempt in June, nests with wild carrot

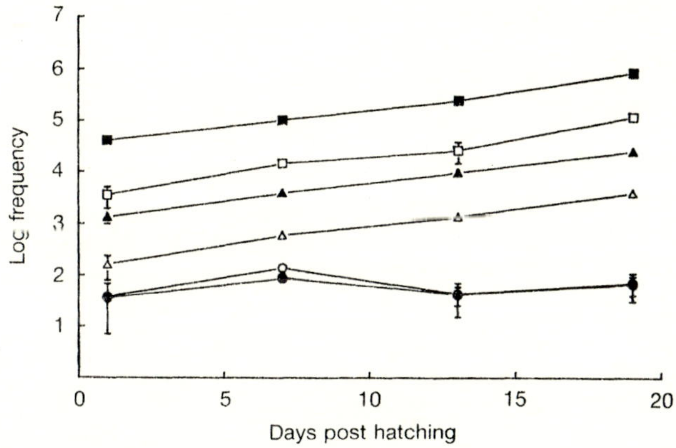

Fig. 11.1. The number of mites collected from starling nests. Open symbols depict nests collected in May. Closed symbols depict nests collected in June. Circles depict nests where the mite population was kept low by periodically replacing nest material every 7 days. Triangles depict nests where leaves from wild carrot were added every 7 days. Squares depict nests from which all green plant material was removed (Clark and Mason 1988).

added contained an average of 11 000 mites, while nests with no green plants contained approximately 500 000 mites. Clearly, wild carrots can effectively depress the population growth of at least one species of ectoparasite. The fact that the numbers of mites seen in 'unprotected' nests are not normally seen in the wild is also suggestive that plants normally found in nests keep mite populations in check.

Discussion

Consequences of ectoparasitism

Using the estimates of population size of mites, age structure of mite populations, volume of blood meals taken per feeding female (Sikes and Chamberlain 1954), and assuming a 1:1 sex ratio, I estimated the daily per capita blood loss for a brood of four starlings raised in nests protected with plant material and reared in nests where plant material was removed (Fig. 11.2). The huge mite populations observed in the second breeding attempt in June for the unprotected nests should have theoretically exsanguinated chicks. Throughout most of this development period, per capita daily blood loss was about 20 per cent of the standing blood volume, and by the time chicks fledged, blood loss was theoretically in the range of 90 per cent of the standing blood volume of each chick. Yet no mortality occurred that could be attributed to exsanguination. The

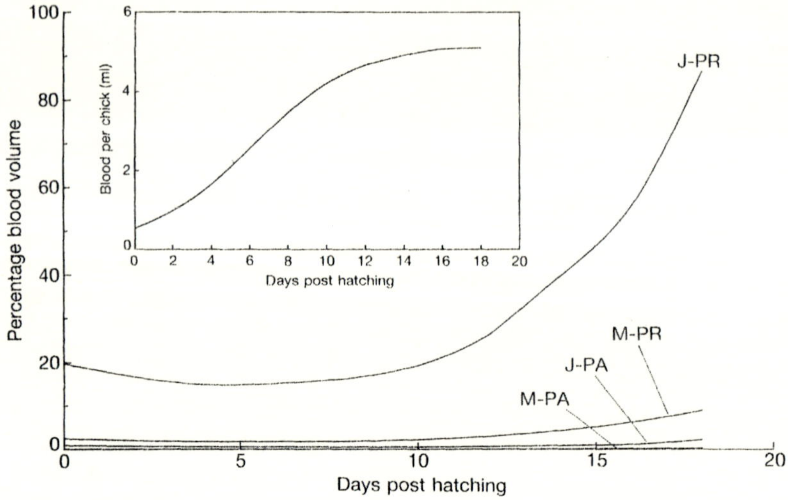

Fig. 11.2. The theoretical per capita blood loss as a percentage of total blood volume of a brood of four caused by the northern fowl mite. The different levels of blood loss are based on the numbers of mites found in nests protected by wild carrot (PA) and for nests not protected by any green plant material (PR). M, May nests; J, June nests. The inset depicts the total blood volume of a nestling starling chick as a function of age.

theoretical loss of blood due to mites during the first reproductive episode in May was less dramatic, but nonetheless substantial. The per capita blood loss for chicks reared in 'unprotected' nests theoretically amounted to 10 per cent of the standing blood volume by the time of fledging.

Many birds can withstand a loss of blood that would prove lethal to mammals (Kovach *et al.* 1969). Adult pigeons can survive at least 7 days after suffering a blood loss of 5–8 per cent of their body weight, providing haemorrhage is protracted (two 10-min bleedings 1 h apart, Kovach and Szasz 1968). Pigeons have a standing blood volume of about 27 ml (Andersen 1966). The average blood loss of the 16 pigeons in the Kovach study surviving at least 7 days was 23.1 ml/bird, representing 85.6 per cent of the standing volume. The reason birds can withstand such dramatic haemorrhaging is because they tend to preserve their blood volume better than mammals by reflexively recovering interstitial fluid (Djojosugito *et al.* 1968; Kovach and Balint 1969; Kovach *et al.* 1969). The ability to preserve blood volume may explain why nestling starlings can endure overwhelming haematophagic parasite load. However, as a consequence of massive blood loss, birds experience high levels of haemodilution, i.e. lower concentrations of whole blood constituents such as haemoglobin (Kovach and Balint 1969).

Starling chicks reared in nests protected by wild carrot may have suffered only moderate to light blood loss due to mites. The maximum per capita loss of standing blood volume was theoretically less than 3 per cent (Fig. 11.2). Because there was no difference in mortality between chicks reared in nests protected by wild carrot and those reared in unprotected nests, the protection afforded by at least some plants may be in terms of preserving blood constituency or limiting the energetic cost of parasite load. Chicks reared in the nests not protected by plants were anaemic relative to those reared in nests with wild carrot (Clark and Mason 1988). Anaemia may negatively affect a chick's fitness. The degree to which fledgelings can sustain flight and repay oxygen debt is directly related to the haemoglobin concentration of the blood. Pigeons subjected to a blood loss of 7 per cent of their body weight (approximately equal to their standing blood volume) experienced a 56 per cent increase in respiration rate during forced flight. After flight, bled birds maintained a 22 per cent increase in respiration rate relative to their prebleeding values, and the time required to eliminate oxygen debt was three times as long (Kovach and Szasz 1968). Thus, anaemic birds must not only support the extra energetic costs of parasitism by replacing blood constituents, but they must also support the extra energetic costs associated with increased ventilation. For birds just learning to forage these energetic costs may mean the difference between postfledging mortality and survival. We must also consider the possibility that endurance for sustained activity is curtailed, and this may have an effect on the probability that a bird can escape predators. Judicious experiments on the energetic costs of anaemia are needed to evaluate the costs associated with ectoparasitism.

Chicks may not die under heavy parasite load because of their ability to withstand heavy blood loss. Alternatively, actual per capita blood loss of chicks may not be as high as theoretically calculated because mites may stop feeding or reduce the size of their blood meals, perhaps due to severe competition amongst themselves when populations are large. The internal physiological state of the chicks may inhibit feeding by mites. Heavy parasitism can cause an induced physiological response resulting in protection from further exsanguination. Chickens infested by fowl mites increase levels of cortical steroids which subsequently inhibit feeding by mites (Hall and Gross 1975).

Nest material as a defence against parasites and pathogens

Plants contain myriads of compounds that can act as biocides and toxicants (Secoy and Smith 1983; Balandrin *et al.* 1985). As a consequence of the natural diversity of secondary plant metabolites starlings might benefit from even a random selection of plant material. However, the general level of biological activity among all plants would have to be high and, more importantly, it would have to be high against the suite of pathogens and parasites that birds face for the behaviour to evolve. Because starlings select plants non-randomly, Clark and Mason (1985) hypothesized that the plants preferred as nest material contained high biological activity towards the starling's parasites and pathogens. Laboratory assays showed that preferred plants had a significantly higher probability of inhibiting bacterial growth of mixed cultures obtained from nestboxes in the field, and of inhibiting the growth of potentially pathogenic pure cultures (e.g. *Staphlococcus aureus*, *Streptococcus epidermis*, *Pseudomonas aerginosa*) when compared to a random set of other available but non-preferred plants. No plant test was effective at inhibiting the growth of *Escherichia coli* cultures (Clark and Mason 1985).

Plants preferred as nest material also were more likely to inhibit hatching of the haematophagous louse *Menacanthus* sp. relative to a random sample of other available plants. However, there was no evidence to indicate that any plant tested was capable of killing adult fowl mites (Clark and Mason 1985). Paradoxically, subsequent field studies, as reported here, suggest that at least one of the preferred plants, wild carrot, effectively reduced mite populations in nestboxes.

To resolve the discrepancy, laboratory tests were conducted to determine the effects of plants on the emergence pattern of feeding forms of mites (Clark and Mason 1988). Nests with green plants were systematically removed during construction, transported to the laboratory, and divided into halves. In one-half of each nest about 5 g of fresh plant material was added to the nest matrix. The other half served as a control. Over a period of 14 days, feeding forms of mites emerged, crawled to the top of the sample bags, and were collected. Emergence of mites from nests where garlic mustard (*Alliaria officinalis*), a non-preferred plant, was added did not differ from 'unprotected' controls. Nests containing wild carrot or fleabane, two preferred plant species, significantly reduced fowl mite emergence. Because geotaxis is one of the principal behaviours associated with feeding mites, one plausible interpretation is that wild carrot and fleabane contain substances that act to disrupt feeding and behaviours associated with feeding. Such compounds may stimulate inhibitory sensory processes which in turn suppress appropriate orientation behaviour (Mustaparta 1984; Stadler 1984). Because blood meals are critical to achieving reproductive status, suppression of feeding would decrease the population growth of mites and thus relax the selective pressure on chicks.

Wild carrot and fleabane may contain compounds which contain arthropod juvenile hormone analogues. Such compounds can delay the transition from one developmental stage to the next (Slama 1979). Again, the net result is a delay of the breeding cycle of mites due to the prolongation of non-adult stages. The data are consistent with this interpretation because, of the two treatments involving preferred plants, the largest difference in emergence of feeding forms occurred for the earliest feeding stage (protonymph) (Clark and Mason 1988).

Plants may also function to lower mite populations by causing a delay in the initial colonization of the nest by mites via repellency, inhibition of oviposition, or disruption of the physiological readiness of females to lay eggs. One possible means by which the starlings achieve this is the use of wild carrot. Wild carrot contains the steroid, *B*-sitosterol (Duke 1985), which is effective as a repellent and oviposition inhibitor for mites (Ambasta 1980). Coincidentally, *B*-sitosterol is also found in the leaves of the margosa tree (*Azadirachta indica*). Margosa leaves are used in nest-building by house sparrows (*Passer domesticus*), much as wild carrot is used by starlings (Sengupta 1981). Margosa leaves are also used as a tick repellent by Indians.

Nests not protected by plants needed approximately 100 mites as colonists in order to yield the populations observed (based on back calculations of the estimates of population growth depicted in Fig. 11.1). Mites colonize nests by leaving the bodies of already infested adults throughout nest-building and incubation (Powlesland 1978). Infestations of 50 to 200 mites are well within the range reported to exist on adults (Boyd 1951; DeVaney *et al.* 1980). Alternatively, those nests protected by wild carrot need have only reduced the colonizing population size to 10–30 mites to yield observed population levels seen in Fig. 11.1.

Regardless of the mechanisms involved, any delay or decrease in the emergence of mites would decrease the per capita daily blood loss of the chicks. A delay in the onset of population growth of mites would presumably be effective until the chemical agents in plants decomposed or volatilized. More work is needed to identify biologically active agents in plants and the effect these agents have on the starling's suite of parasites and pathogens. Once isolated such agents would prove to be useful probes in determining what sensory cues birds use to make discriminations among plants.

The selection of green plant material

Starlings breeding in England and Ohio also seem to show preferences for some plants, but the species involved are not necessarily the same as those preferred in Pennsylvania (P. W. Greg-Smith and R. A. Dolbeer, personal communications). Starlings probably use chemical, in addition to visual, cues to differentiate among plants that act as pesticides. Visual cues, such as leaf shape or colour, may correlate less well with the biologically active properties of plants than do the emitted volatiles. While there are characteristic chemical differences that do correlate with morphological traits, there is still sufficient chemical variation between individuals to suggest that visual discrimination is an inadequate basis for the selection of biologically active leaves (Parks 1974). Clark and Mason (1985) were unable to detect any visual morphological similarity among preferred plants.

It is unlikely that starlings discriminate among plants based upon specific volatile cues that are matched to specific biological activity. This capacity would require birds to assess the nature of parasite or pathogen load prior to laying eggs, and make a specific association for the type of plant to be used. These events are highly improbable given the temporal and spatial variation of infestation and infection. This scenario is even more unlikely given the fact that starlings do not live for many reproductive seasons; thus the time needed to form the experience to make the numerous plant–pathogen/parasite associations probably does not exist in this system. Rather, plants that have a high chemical diversity may have a greater likelihood of containing biologically active material; hence starlings using general volatility and chemical complexity as a cue for selecting plants may improve their chances of encountering biologically active plant material which is toxic to parasites and pathogens. Consistent with the latter interpretation are the observations of Clark and Mason (1985) where gas chromatographic analysis showed that volatiles from plants preferred by starlings contained more compounds at higher concentrations than a random subset of plants drawn from the same environment.

Starlings may use olfaction to discriminate among the volatile cues emitted by plants (Clark and Mason 1987). Despite earlier claims to the contrary (Welty 1972), it is now apparent that avian olfactory abilities rival those of some mammals. Seabirds of the order Procellariiformes and cathartid vultures possess highly developed peripheral and central olfactory anatomy (Bang and Cobb 1968), and are known to use olfactory cues to locate food during foraging (Stager 1964; Grubb 1972, 1974; Hutchison *et al.* 1984; Houston 1987). Even among species that have average relative olfactory bulb sizes for birds (e.g. quail, pigeons, and domestic fowl), detection thresholds lie in the 0.11 to 30.0 p.p.m. range (Henton *et al.* 1966; Henton 1969; Stattleman *et al.* 1975; Walker *et al.* 1986), and are similar to those of rats and rabbits (Davis 1973), species usually considered to have a good sense of smell. However, the passeriiforms, with their disproportionately small and simply organized olfactory anatomy (Bang and Wenzel 1986), are often perceived as lacking a well developed sense of smell. Nonetheless, at least two passerines, the brown-headed cowbird (*Molothrus ater*) (Clark and Mason 1989) and the European starling (Clark and Smeraski 1990), have an olfactory sensitivity that is well within the range reported for non-passerines with relative bulb sizes two times the size of these species. Clark and Smeraski (1990) showed that, during the breeding season (March–June), the starling's threshold to a standard reagent grade stimulus, cyclohexanone (CH), was 0.3 per cent vapour saturation (2.5 p.p.m.). Despite displaying detection thresholds comparable to those of non-passerines, the same starlings tested in November were barely able to detect CH at much higher concentrations (10 per cent vapour saturation or 80 p.p.m.). Similar results were obtained for starlings tested with odours derived from plant material. These findings suggest that olfactory responsiveness is variable, and that this variability could reflect seasonal changes in either sensitivity or performance. More detailed analysis of olfactory responsiveness in starlings revealed that olfactory responsiveness varies circannually (Clark and Smeraski 1990). The timing of peak olfactory responding was concordant with reproductive activity (i.e. nest-building) in the wild, while the timing of ebb responding was concordant with gonadal regression and photorefractoriness (light insensitivity) of wild birds (Dawson 1983; Dawson and Goldsmith 1983; Temple 1974).

The question remains as to why there is a seasonal expression of olfactory responding. In the past, investigators have concluded that olfaction in birds is a primitive trait which has declined as species evolved. Recent forms, such as passerines, are presumed to rely primarily on visual and auditory cues. Hence, the diminution of the olfactory system is presumed to have occurred as other centres of the brain have enlarged (Bang and Wenzel 1986). The seasonally expressed olfactory ability of starlings might be the result of an evolutionary response to selective pressures favouring olfaction in a taxon not generally characterized by acute olfactory sensibility. Because the clear annual pattern of olfactory response is highly concordant with peak nest-building activity in the wild, I speculate that olfactory capacity in starlings is retained (at least seasonally) to maximize the likelihood of selecting plants with high protective potential by using the most relevant cue to effect this end, i.e. volatile chemical cues. Further tests are needed to determine whether starlings use olfaction in the field as a basis in selecting plants for nest construction. The foundations exist which indicate that selection pressures from parasites and pathogens may have brought about an intricate adaptive response in nest-building behaviour by a passerine which may involve a link between its endocrine status and olfactory behaviour.

Acknowledgements

I thank my two collaborators, J. R. Mason and C. A. Smeraski, for their contributions to this work. D. Clark and K. Heinzel were most helpful in assisting in laboratory and field work. This research was broadly supported by the US Fish and Wildlife Service, the National Geographic Society, NSF grant CHE 8509557, and NIH grant 2 RO1 NS19424–13.

References

Ambasta, S.P. (1980). No title. *Science Reports*, **17**, 210.

Andersen, H.T. (1966). Physiological adaptations in diving vertebrates. *Physiological Review*, **46**, 212–43.

Arendt, W.J. (1985). *Philornis* ectoparasitism of pearly-eyed thrashers. 1. Impact on growth and development of nestlings. *Auk*, **102**, 270–80.

Balandrin, M.F., Kloche, J.A., Wurtele, E.S., and Bollinger, W.H. (1985). Natural plant chemicals: sources of industrial and medicinal materials. *Science*, **228**, 1154–60.

Bang, B.G. and Cobb, S. (1968). The size of the olfactory bulb in 108 species of birds. *Auk*, **85**, 55–61.

Bang, B. and Wenzel, B.M. (1986). Nasal cavity and olfactory system. In *Form and function in birds* (ed. A.S. King and J. McLelland), pp. 195–225, vol. 3. Academic Press, London.

Boyd, E.M. (1951). A survey of the parasitism of the starling *Sturnus vulgaris* L. North America. *Journal of Parasitology*, **37**, 56–84.

Brown, C.R. and Brown, M.B. (1986). Ectoparasitism as a cost of coloniality in cliff swallows *Hirundo pyrrhonota*). *Ecology*, **67**, 1206–18.

Bryant, D.M. (1979). Reproductive costs in the house martin (*Delichon urbica*). *Journal of Animal Ecology*, **48**, 655–75.

Clark, L. and Mason, J.R. (1985). Use of nest material as insecticidal and anti-pathogenic agents by the European starling. *Oecologia, Berlin*, **67**, 169–76.

Clark, L. and Mason, J.R. (1987). Olfactory discrimination of plant volatiles by the European starling. *Animal Behaviour*, **35**, 227–35.

Clark, L. and Mason, J.R. (1988). Effect of biologically active plants used as nest material and the derived benefit to starling nestings. *Oecologia, Berlin*, **77**, 174–80.

Clark, L. and Mason, J.R. (1989). Odor discrimination by brown-headed cowbirds. *Condor*, **91**, 944–56.

Clark, L. and Smeraski, C.A. (1990). Seasonal shifts in odor responding by starlings are related to breeding behavior. *Journal of Experimental Zoology*. **255**, 22–9.

Collias, N.E. and Collias, E.C. (1984). *Nest building and bird behavior*. Princeton University Press, Princeton, New Jersey.

Davis, R.G. (1973). Olfactory psychophysical parameters in man, rat, dog, and pigeon. *Journal of Comparative and Physiological Psychology*, **85**, 221–32.

Dawson, A. (1983). Plasma gonadal steroid levels in wild starlings (*Sturnus vulgaris*) during the annual cycle and in relation to the stages of breeding. *General Comparative Endocrinology*, **49**; 286–94.

Dawson, A. and Goldsmith, A.R. (1983). Plasma prolactin and gonadotrophins during gonadal development on the onset of photorefraetoriness in male and female starlings (*Sturnus vulgaris*) on artificial photoperiods. *Journal of Endocrinology*, **97**, 253–60.

DeVaney, J.A. (1979). The effects of the northern fowl mite, *Ornithonyssus sylivarum* (sic) on egg production and body weight of caged white leghorn hens. *Poultry Science*, **58**, 191–4.

DeVaney, J.A., Quisenberry, J.H., Doran, B.H., and Bradley, J.W. (1980). Dispersal of the northern fowl mite, *Ornithonyssus sylviarum* (Canestrini and Ganzago), and the chicken body louse, *Menacanthus stramineus* (Nitzsch), among thirty strains of egg-type hens in a caged laying house. *Poultry Science*, **59**, 1745–9.

Djojosugito, A.M., Folkow, B., and Kovach, A.G.B. (1968). The mechanism behind the rapid blood volume restoration after hemorrhage in birds. *Acta Physiologica Scandinavica*, **74**, 114–22.

Duffy, D.C. (1983). The ecology of tick parasitism on densely nesting Peruvian seabirds. *Ecology*, **64**, 110–19.

Duke, J.A. (1985). *Handbook of medicinal herbs*. CRC Press, Boca Raton, Florida.

Feare, C.J. (1976). Desertion and abnormal development in a colony of sooty terns *Sterna fuscata* infested by virus-infected ticks. *Ibis*, **118**, 112–15.

Garnett, M.C. (1981). Body size, its heritability and influence on juvenile survival among great tits *Parus major*. *Ibis*, **123**, 31–41.

Gibb, J. (1950). The breeding biology of the great and blue titmice. *Ibis*, **92**, 507–39.

Gold, C.S. and Dahlsten, D.L. (1983). Effects of parasitic flies (*Protocalliphora* spp) on nesting of mountain and chestnut-backed chickadees. *Wilson Bulletin*, **95**, 560–72.

Grubb, T.C. Jr (1972). Smell and foraging in shearwaters and petrels. *Nature*, **237**, 404–5.

Grubb, T.C. Jr (1974). Olfactory navigation to the nesting burrow in Leach's Petrel (*Oceanodroma leucorrhoa*). *Animal Behaviour*, **22**, 192–202.

Hall, R.D. and Gross, W.B. (1975). Effects of social stress and inherited plasma corticosterone levels in chickens on populations of the northern fowl mite, *Ornithonyssus sylviarum*. *Journal of Parasitology*, **61**, 1096–100.

Henton, W.W. (1969). Conditional suppression to odorous stimuli in pigeons. *Journal of the Experimental Analysis of Behavior*, **12**, 178–85.

Henton, W.W., Smith, J.C. and Tucker, D. (1966). Odor discrimination in pigeons. *Science*, **153**, 1138–9.

Houston, D.C. (1987). Scavenging efficiency of turkey vultures in tropical forests. *Condor*, **88**, 318–23.

Hutchison, L.V., Wenzel, B.M., Stager, K.E., and Tedford, B.L. (1984). Further evidence for olfactory foraging by sooty shearwaters and northern fulmars. In *Marine birds: their feeding ecology and commercial fisheries relationships*, Special Publication Canadian Wildlife Service (ed. G.A. Sanger and P.F. Springer). pp. 72–77. Canadian Wildlife Service, Ottawa.

Jacobs, C.H., Collias, N.E. and Fujimoto, J.T. (1978). Nest colour as a factor in nest selection by female village weaverbirds. *Animal Behaviour*, **26**, 463–9.

Kessel, B. (1957). A study of the breeding biology of the European Starling (*Sturnus vulgaris* L) in North America. *American Midland Naturalist*, **58**, 257–331.

Kluyver, H.N. (1966). Regulation of a bird population. *Ostrich Supplement*, **6**, 389–96.

Kovach, A.G.B. and Balint, T. (1969). Comparative study on Haemodilution after haemorrhage in the pigeon and the rat. *Acta Physiologica Academiae Scientiarum Hungaricae*, **35**, 231–43.

Kovach, A.G.B. and Szasz, (1968). Survival of pigeon after graded haemorrhage. *Acta Physiologica Academiae Scientiarum Hungaricae*, **34**, 301–9.

Kovach, A.G.B., Szasz, E., and Pilmayer, N. (1969). Mortality of various avian and mammalian species following blood loss. *Acta Physiologica Academiae Scientiarum Hungaricae*, **35**, 109–16.

Krantz, G.W. (1978). *A manual of acarology*. Oregon State University Press, Corvallis.

Loomis, E.C., Bramhall, E.L., Allen, J.A., Ernst, R.A., and Dunning, L.L. (1970). Effects of the northern fowl mite on white leghorn chickens. *Journal of Economic Entomology*, **63**, 1885–9.

Matthysse, J.G., Jones, C.J., and Purnasirl, A. (1974). Development of northern fowl mite populations on chickens, effects on the host, and immunology. *Searches in Agriculture*, **4**, 39.

Mertens, J.A.L. (1977). Thermal conditions for successful breeding in great tits (Parus major major L) II. Thermal properties of nest and nestboxes and their implication for the range of temperature tolerance of great tit broods. *Oecologia, Berlin*, **28**, 31–56.

Moss, W.W. and Camin, J.H. (1970). Nest parasitism productivity and clutch size in purple martins. *Science*, **168**, 1000–3.

Mustaparta, H. (1984). Olfaction. In *Chemical ecology of insects* (ed. W.J. Bell and R.T. Carde), pp. 37–69. Chapman and Hall, London.

Parks, C.R. (1974). Chemical evidence. In *Vascular plant systematics* (ed. A.E. Radford, W.C. Dickison, J.R. Massey, and C.R. Bell), pp. 285–305. Harper and Row, New York.

Powlesland, R.G. (1978). Behavior of the haematophagous mite *Ornithonyssus bursa* in starling nest boxes in New Zealand. *New Zealand Journal of Zoology*, **5**, 395–400.

Rothschild, M. and Clay, T. (1952). *Fleas, flukes, and cuckoos*. Collins, London.

Secoy, D.M. and Smith, A.E. (1983). Use of plants in control of agricultural and domestic pests. *Economic Botany*, **37**, 28–57.

Sengupta, S. (1968). Studies in the life history of the common myna. *Proceedings of the Zoological Society of Calcutta*, **21**, 1–27.

Sengupta, S. (1981). Adaptive significance of the use of margosa leaves in nests of the house sparrow *Passer domesticus*. *Emu*, **81**, 114–15.

Shields, W.M. and Crook, J.R. (1987). Barn swallow coloniality: a net cost for group breeding in the Adirondacks? *Ecology*, **68**, 1373–86.

Skutch, A.F. (1976). *Parent birds and their young*. University of Texas Press, Austin.

Slama, K. (1979). Insect hormones and antihormones in plants. In *Herbivores. Their interaction with secondary plant metabolites*, (ed. G.A. Rosenthal and D.H. Janzen), pp. 683–706. Academic Press, New York.

Stadler, E. (1984). Contact chemoreception. In *Chemical ecology of insects* (ed. W.J. Bell and R.T. Carde), pp. 3–36. Chapman and Hall, London.

Stager, K.E. (1964). The role of olfaction in food location by the turkey vulture (*Cathartes aura*). *Contributions to Science of the Los Angeles County Museum*, **81**, 1063.

Stattelman, A.J., Talbot, R.B. and Coulter, D.B. (1975). Olfactory thresholds of pigeons (*Columba livia*), quail (*Colinus virginianus*) and chickens (*Gallus gallus*). *Comparative Biochemistry and Physiology*, **50A**: 807–9.

Sikes, R.K. and Chamberlain, R.W. (1954). Laboratory observations on three species of bird mites. *Journal of Parasitology*, **40**, 491–7.

Taverner, P.A. (1933). Purple martins gathering leaves. *Auk*, **50**, 110–11.

Temple, S.A. (1974). Plasma testosterone titres during the annual reproductive cycle of starlings (*Sturnus vulgaris*). *General and Comparative Endocrinology*, **22**, 470–9.

Walker, J.C., Walker, D.B., Tambiah, C.R. and Gilmore, K.S. (1986). Olfactory and nonolfactory odor detection in pigeons: elucidation by a cardiac acceleration paradigm. *Physiology and Behavior*, **38**, 575–80.

Wasylik, A. (1971). Nest types and abundance of mites. *Ekologia Polska*, **19**, 689–99.

Welty, J.C. (1972). *The life of birds*. Alfred A. Knopf, New York.

Whiteman, E.E. and Bickford, A.A. (1983). *Avian disease manual* (2nd edn). American Association of Avian Pathology, University of Pennsylvania Press, Kennet Square.

Widmann, O. (1922). Extracts from the diary of Otto Widmann. *Transactions of the Academy of Science of St. Louis*, **24**, 1–77.

Wimberger, P.H. (1984). The use of green plant material in bird nests to avoid ectoparasites. *Auk*, **101**, 615–18.

12 *Nest ectoparasite abundance and cliff swallow colony site selection, nestling development, and departure time*

JENELLA E. LOYE and SCOTT P. CARROLL

Introduction

Although theory suggests that parasitism will have a measurable effect on evolution in host populations (e.g., Anderson and May 1979; Hamilton and Zuk 1982), the influence of ectoparasitism on nesting and migration behaviour in birds has not been quantified. One can well imagine that a bird's nest full of bugs, ticks, and fleas could be more than just unpleasant for the host (Rothschild and Clay 1952). Because blood loss may reduce growth rates and survivorship of nestlings (Moss and Camin 1970; Chapman 1973; Brown and Brown 1986; Møller 1987; this volume: Chapman and George, Chapter 4; Møller, Chapter 17) and threaten adults as well, the presence of ectoparasites could make some sites less attractive for nesting, and infested nests might be used for a shorter period of time than uninfested nests due to a buildup of ectoparasites (Møller, Chapter 17, this volume). In addition, parasite buildup during the nesting season might force or favour mass synchronous departure from the breeding area.

The measurement of parasite effects, as well as speculation on the potential selective nature of parasitism, is of particular importance in colonial animals, in which the probability of parasite transmission among hosts is increased (Alexander 1974; Hoogland and Sherman 1976; Price 1980; Duffy 1983, Chapter 13, this volume; Brown 1985). Several studies have examined the effects of ectoparasites on the breeding biology of colonial birds. Ectoparasitism as a factor in colony site selection was discussed as early as 1930 by Grinnell *et al.* and demonstrated in fumigation experiments by J. T. Emlen in the 1950s (Emlen 1986). In both seabirds and swallows, colony abandonment has been attributed to ectoparasites (Feare 1976; Duffy 1983; Emlen 1986). In addition, positive correlations have been found between swallow colony size and ectoparasite population size (Hoogland and Sherman 1976; Brown and Brown 1986). Such results strongly suggest that ectoparasites have important fitness consequences for colonial birds including swallows, and that they should be important selective factors in the evolution of colony use patterns.

In the colonial cliff swallow, *Hirundo pyrrhonata*, the choice of the colony site, the length of brooding period, and the timing of departure are three colony use patterns that vary among years, colony sites, and geographic regions (Mayhew 1958). Emlen and Demong (1975) suggested that the primary factor determining the synchrony of brooding and of departure date in colonial bank swallows is the benefit of social foraging (information about food sources). Colonial swallows are often associated with aquatic habitats, where the emergence of the aquatic insects on which they feed is both seasonal and associated with weather patterns, so that the availability of foraging information could be important. Findings consistent with this hypothesis in cliff swallows include observations of synchronous nesting and departure occurring within, but not among, colonies in the same area (Emlen 1952; Myres 1957; Mayhew 1958; Brown and Brown 1987; this study). Snapp (1973) reported similar patterns in the barn swallow (*Hirundo rustica*). But, while the nesting period of swallows does coincide with periods of high availability of aquatic insects, they also forage on a variety of terrestrial insects that remain abundant into the autumn (Bent 1942; Johnston 1965; Bryant 1975; J. Loye, personal observation) and may be less patchily distributed in space. Since the publication of Alexander's seminal paper (1974) parasitism has been widely recognized as an important potential cost of colonial living, but, to date, quantifications of its effect have been wanting for many host traits. Hence, major colonial nesting patterns, such as the temporal dispersion of individual nesting events, may be determined by factors other than foraging efficiency, such as costs related to ectoparasitic attack.

In this study, we examine the effects of ectoparasites on nesting behaviour over a 6-year period, and report that ectoparasite density is correlated with cliff swallow colony site selection, shortened nesting period, and early migration. We take advantage of the natural experiment provided by contrasting numbers of ectoparasites in colonies on cliff faces versus those beneath concrete bridge overhangs.

Methods

From 1978 to 1983 we sampled cliff swallow nests in 11 natural cliff colonies and two colonies under concrete bridge overhangs (Fig. 12.1). All colonies were located near water, in the mesic canyons of Caddo

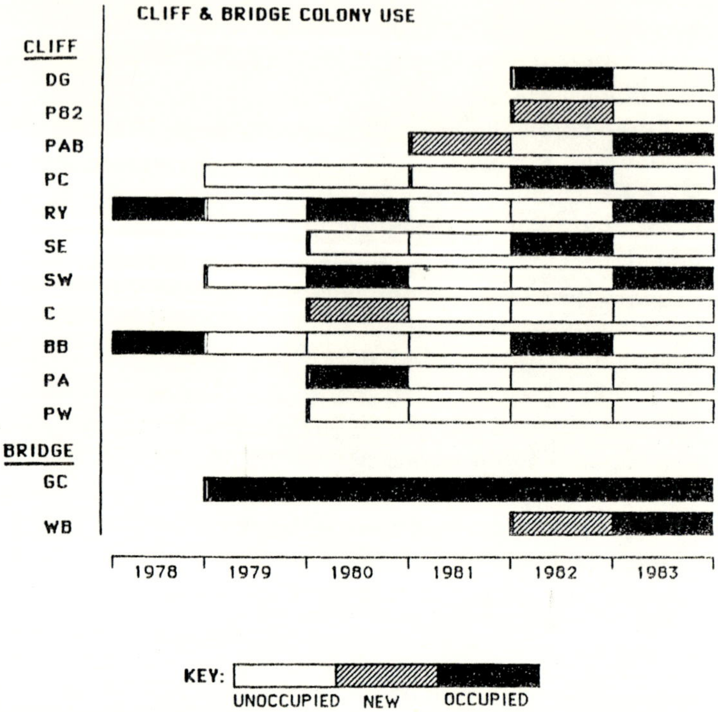

Fig. 12.1. Swallow usage patterns of 11 cliff and two bridge colonies, 1978–83, Caddo County, Oklahoma, $n = 45$ observations. Double line, first observation.

county and adjacent Grady county in western Oklahoma, USA (the environment is described by Little 1939). Colony size ranged from 125 to 1300 nests in use during the breeding period (Table 12.1). Large colonies were designated as those of 550 nests or more; data from these colonies were used for statistical comparison of ectoparasite numbers. Groups of four nests were removed from randomized quadrats in the colonies each month (Loye 1985*a*).

Cliff swallow nests contained four species of ectoparasites: the flea, *Ceratophyllus celsus* (Siphonaptera: Ceratophyllidae); ticks of two families, *Ixodes baergi* (Acari: Ixodidae) and *Ornithodoros concannensis* (Acari: Argasidae); and the cliff swallow bug, *Oeciacus vicarius* (Hemiptera: Cimcidae). Swallow bugs were by far the most numerous and consistently present of the ectoparasites. Fleas feed only as adults early in the spring on returning swallows and to a lesser extent on nestlings; the tick species were rare in comparison with bugs in the colonies sampled. Bugs feed on adult and nestling swallows as long as they are nesting, and their population increase is rapid (Loye 1985*a*). Bug abundance was used as the measure of ectoparasite infestation.

Table 12.1. Cliff swallow colony size

Colony†	Nests in use	Total nests in colony
1. P82	125*	125
2. SW	250–300	300
3. PAB	390*–1300	1300
4. RY	350–850	800–1000
5. PC	800	800+
6. SE	870	900+
7. DG	570	570+
8. C	150*	150
9. BB	150	1000
10. PH	150	150
11. PW	50	50
12. WB	550*, 100	550
13. GC	1000	1500

* Newly built.
† Colonies 1–6 used for ANOVA and regression analysis.

Differences in ectoparasite abundance were compared between nests on cliffs and bridges. Cliffs had extensive cracks behind nests which offered overwintering harbourage for large numbers of ectoparasites, while bridges provided no such overwintering sites. Each randomly collected nest was considered independently of colony and year due to an inordinate variation in individual infestation rates (Loye 1984). Initial analysis indicated no significant variation in infestation rates between colonies or years within substrate type, so that data were pooled and tested for effects of cliff versus bridge substrates.

To determine the effect of ectoparasites on nesting decisions, three nesting patterns were contrasted between colonies: (1) annual pattern of colony use or abandonment; (2) nesting period; and (3) swallow departure time. Nestling abandonment was evaluated by checking nests soon after fledgling departure. Non-parametric statistical tests were used to examine the differences (Siegel 1956).

Because swallows do not nest at random within a colony (earliest nesters tend to occupy colony centres; Brown 1985; Emlen 1986; this study), the distribution of bugs within colonies was examined by sampling them in nests in the centre, intermediate, and colony edge throughout the breeding season. This type of sampling was limited to larger colonies which had clearly identifiable centre versus edge regions.

Within-colony effects on nestling development were examined in 1982 and 1983. Nestlings were removed for banding and weighing (via a small hole pierced in the nest wall behind the entrance) with fingers or long forceps with padded tips. No subsequent mortality was related to the

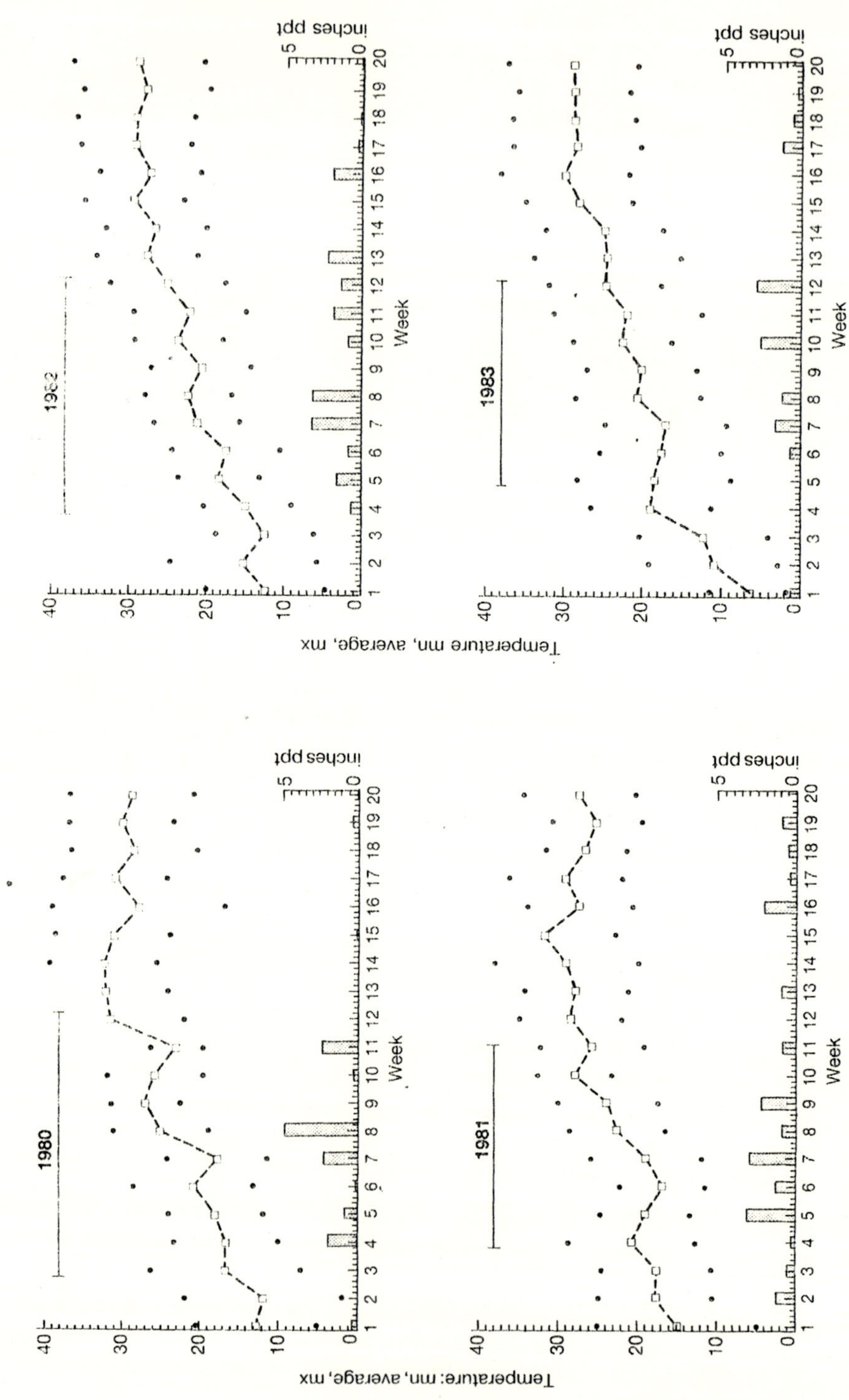

Fig. 12.2. Temperature and rainfall levels and average nesting period of cliff swallows using cliff colonies in Caddo County, Oklahoma, 1980–83.

handling and the hole was patched within 48 h by the parents. Nestlings were weighed from a transect of 60 nests in each of six colonies that began with 20 nests from the dense centre portion of each colony and ended with 20 nests from the sparsely-nested edge area in four colonies in 1982. In 1982, nestlings in three cliff colonies and one bridge colony were banded and weighed 10 to 12 days after the first eggs were laid (as by Stoner 1945; Brown and Brown 1986). In 1983, 20 nests each from the centre and edge of one cliff and one bridge colony were similarly treated. The average number of bugs per nest in all transects was measured by collecting the nests 2–5 days after fledging. Regression analysis of colony size versus nestling weight and colony size versus average number of bugs per nest was used to examine the data from the six colonies (SE, PC, WB, P82, SW, and PAB) used in transect samples (Sokal and Rohlf 1981).

Results

Arrival in the breeding area

As temperature and rainfall increased in late April and early May, cliff swallows returned to the Caddo Canyons (Fig. 12.2). Food was plentiful at this time and adult swallows were frequently mist-netted with boluses of insects in their mouths, including some aquatic dipterans (Chironomidae and Tipulidae), but primarily terrestrial insects (Dipterans: Sarcophagidae, Syrphidae, Tachinidae; Homopterans: Cicidellidae, Membracidae; small Lepidopterans, and flying ants). For a week or more the newly arrived swallows explored the old nests with a 'flighting' behaviour which consisted of vertical hovering before, and occasional entrance into, a succession of intact nests from previous years (cf. Jolley and Storer 1945).

Ectoparasite population dynamics and behaviour in relation to host biology and colony substrate

Annual pattern of colony occupation

The clay nests remained intact at colony sites for up to several years even in perennially unoccupied sites. The 13 colony sites provided a total of 52 potential nesting areas over the four year course of the study. Of these, 22 (42 per cent) were occupied, so that nesting cliffs were unavailable as hosts to ectoparasites in more years than not (Fig. 12.1). Disuse of colony sites was especially prevalent on cliff substrates (see below).

Pre-nesting interactions

Swallows encountered great numbers of ectoparasites clustered in the nest entrances when they returned and evaluated nests that had been occupied in the previous year. For example, in cliff colony nests sampled in May 1982, eight neck-like entrances of such nests contained a total of 668 bugs. (These ectoparasites cling to the exploring birds and disperse to the colony that is ultimately used for nesting (Foster and Olkowski 1968; Loye 1985*a*).) In contrast, in colonies not occupied in the previous year, no ectoparasites clustered in the nest entrances. Whereas clustering of ectoparasites at nest entrances occurred for only 1 year after colony occupation, ectoparasite populations persisted at a low level of 1–5 bugs per nest in collections for 4 years in the absence of swallow breeding (Fig. 12.1, colonies PC, BB, PA, PW; Loye 1985*a*).

Cliff versus bridge effects

Occupied cliff colonies supported larger bug populations than did occupied bridge colonies. After one swallow breeding season, nests in cliff colonies contained an average (± 1 SD) of 184.8 ± 527.2 bugs per nest ($n = 402$ nests), while bridge colonies averaged 96.7 ± 107.7 bugs per nest ($n = 99$ nests). This difference in infestation level is significant ($t = 2.68$, $p < 0.008$, data normalized by log-transformation). Of the other three ectoparasite species only *O. concanensis* were more numerous in certain cliff colonies; these colonies were not used for analysis of ectoparasite abundance.

Perennially discontinuous occupation of cliff colony sites resulted in diminished bug population sizes. In August 1980, at the end of the swallow breeding season, the average number of bugs per nest in RY colony was 1212.96 ± 1880.76 ($n = 12$ nests). By May 1981, the average number per nest had dropped to 137 ± 114.73 ($n = 12$ nests). The colony was not occupied in the summer of 1981, and bugs were absent from eight nests sampled that August.

Three differences in colony use by cliff swallows were found between cliff and bridge colonies. First, no cliff nesting site was used in 2 successive years, suggesting an effect of previous occupation on colony site choice ($\chi^2 = 7.51$, $p = 0.006$; Table 12.2). Cliff colonies were abandoned for 2 or more years in 14 of 45 observations (Fig. 12.1). In contrast, swallows occupied old nests at bridge sites in each year (Fig. 12.1).

Second, the nesting period of colonies on cliffs was significantly shorter than that of those on bridges (64.3 ± 9.6 days and 75 ± 8.1 days respectively; $U = 10.0$, $p = 0.041$; Table 12.3). Time of departure in the bridge colonies was a week later than in cliff colonies in 1980 and 1981 (Table 12.3). This difference occurred although nestling development

Table 12.2. Chi-square analysis of cliff swallow nesting site preference in 11 cliff colonies, 1978–83*

	Used year $x+1$	Not use year $x+1$
Used year x	0	12
Not used year x	11	14
Chi square = 7.51, $p = 0.006$, $df = 1$, $n = 37$		

* This analyses the frequency of occupation in the previous year (x), for sites occupied in a given year ($x+1$), compared to sites unoccupied in $x+1$ and includes the first year of occupancy.

rates were similar in all colonies except P82, a new colony (weights shown in Table 12.4).

Third, migration occurred synchronously (over 1 to 2 days) in cliff colonies ($n = 8$), while birds departed over a period of 2 weeks in bridge

Table 12.3. Cliff swallow arrival and departure times, 1980–83, Caddo County, Oklahoma

Year	Colony	Arrival date	Departure date	Nesting period (days)
1980	Cliff			
	RY	20 April	2 July	73
	SW	20 April	27 June	68
	Bridge			
	GC	25 April	11 July	77
1981	Cliff			
	PAB	24 April	20 June	57
	Bridge			
	GC	25 April	9 July	75
1982	Cliff			
	P82	29 May	15 July	65
	SE	29 April	15 July	77
	Bridge			
	GC	23 April	1 July	69
	WB	23 April	30 June	68
1983	Cliff			
	PAB	17 May	6 July	50
	RY	5 May	14 July	70
	SW	13 May	6 July	54
	Bridge			
	GC	25 April	24 July	90
	WB	3 May	13 July	71

Table 12.4. Effect of location in colony (c, centre; i, intermediate; e, edge) on nestling weight in transect samples taken 10–15 days after the first egg laid ($n = 827$ nestlings)

Colony	Section	Nestling weight (g + SD)	n	F-value ANOVA
1982				
Cliff				
SE	c	25.2 ± 2.0	62	26.9, $p < 0.0001$
	i	20.8 ± 6.3	67	
	e	18.7 ± 5.5	66	NS, 2.0, $p < 0.1$
PC	c	22.0 ± 3.9	60	
	i	20.0 ± 5.3	61	
	e	20.8 ± 5.9	54	6.3, $p < 0.01$
P82*	c	13.5 ± 5.8	24	
	i	10.6 ± 3.4	7	
	e	8.7 ± 2.6	20	
Bridge				
WB	c	20.6 ± 4.5	52	
	i	19.3 ± 5.6	41	
	e	20.7 ± 2.7		
1983				
Cliff				
RY	c	19.9 ± 2.79	68	136.7, $p < 0.0001$
	e	13.8 ± 3.35	68	
Bridge				
WB	c	20.9 ± 3.14	65	2.33, $p < 0.126$
	e	9.8 ± 4.58	58	

* Newly built colony; with significant difference in weights from the other three colonies: ANOVA, $F = 57.01$, $p < 0.0001$, SNK, $p < 0.05$.

colonies ($n = 6$). Immatures were abandoned when synchronous departure occurred at the cliff colonies (Table 12.5). In contrast, immatures

Table 12.5. Nestling abandonment in five Oklahoma cliff swallow colonies in 1982

Type	Colony	Total nests	Percentage of nests with dead nestlings	(n)
Cliff*	PC	75	30.6	(23)
	P82	45	85	(38)
Bridge	DG	45	31	(14)
	WB	85	5	(4)
	GC	240	5.8	(14)

* Percentage of nests with dead nestlings in cliff nests: $U = 0$, $p = 0.1$. Mann–Whitney U-test.

Table 12.6. Effect of location in colony (c, centre; i, intermediate; e, edge) on total bugs in transect samples (30 nests per colony taken after fledging July 1982)

	Colony	Section	Total bugs	Variation (*G* value)
Cliff				
	SE	c	430	511.0*
		i	1048	
		e	299	
	PC	c	1	339.6*
		i	0	
		e	159	
	P82	c	304	30.0*
		i	184	
		e	245	
	DG	c	3810	2516.0*
	($n=15$)	i	3729	
		e	345	
Bridge				
	GC			
	WB	c	1130	430.8*
		e	170	
		c	710	20.3*
		i	552	
		e	606	
Not used in 1982				
	SW	c	138	166.2*
		i	449	
		e	579	
	PAB	c	68	36.0*
		i	108	
		e	20	

* $p \leqslant 0.001$, *G*-test.

were never abandoned in bridge colonies, with subgroups leaving asynchronously as the young fledged.

Within-colony variation and effects of ectoparasite infestation

Effects of nest placement in the colony

Bugs were non-randomly distributed within cliff colonies sampled in transects of central, intermediate, and edge nests (Table 12.6). The mean number of bugs per nest was significantly greater in the central region in the colonies WB, P82, DG, GC, and in SE (where central and intermediate samples were pooled; Table 12.6). However, at PC the bugs were clustered in edge nests. In comparison, in nests from two unoccupied colonies sampled at the same time, bugs were not clustered in central nests. Additional within-colony, central-location effects were observed in the RY cliff colony which had a significantly greater infestation rate in the central nests both in 1980 ($G=917.6$, $p<0.01$, $n=62$) when it was a breeding colony, and in 1981 ($G=71.0$, $p<0.01$, $n=29$) when swallows were absent.

Nestling weight

Although bugs were generally more numerous in the central nests, nestlings were heavier in central nests in four of six colonies ($F=84.84$, $p<0.001$, $n=6$ colonies, $n=827$ nestlings, Table 12.4). Nestlings from the two colonies that had no difference in weight associated with location were both colonies with densely packed, intact nets at the edge as well as in the centre (WB and PC, 1982). In 1983, nestlings (8 to 10 days old) were heavier in the central nests at WB in its second year of use and at RY cliff colony.

Nestling weight was positively associated with colony size (the number of nests) ($r^2=0.26$, $p<0.001$, $n=827$ weights, from six colonies, 60 nests per colony), and ectoparasites were significantly more numerous in large colonies ($r=0.74$, $p<0.001$, $n=30$ nests/colony from six colonies). Because the mean number of bugs per nest in the transect samples (at the population peak density in July) was also positively correlated with nestling weight ($r=0.74$, $p<0.001$, $n=180$ nests in six colonies), the largest nestlings may also be those with the most bugs in their nests.

Discussion

Ectoparasite effect on swallow breeding behaviour

Cliff swallow nesting biology differed significantly between bridge and cliff substrates in a variety of major characters. The most striking difference was in the frequency of colony occupation: bridge colonies were reused every year, but cliff colonies were perennially disused for at least 1 year after occupation. In addition, annual nesting periods were briefer in cliff colonies in spite of similar growth rates of nestlings in both colony types. Briefer nesting on cliffs was related to mass synchronous departure from breeding colonies on this substrate, often resulting in the abandonment of large numbers of immature nestlings. Such behaviour was not observed in bridge colonies; parents remained with nestlings until they could fledge. This is comparable to the differences in colonies with and without ectoparasites observed by Chapman and George (Chapter 4, this volume).

What differences between cliff and bridge nesting substrates might be responsible for evoking such vivid differences in cliff swallow behaviour? Presuming that the explanation is in some way related to energy balance or risk, several factors in addition to ectoparasitism may be considered, including foraging success, predation on adults or nestlings, and other aspects of the suitability of the substrate for rearing nestlings, such as the thermal environment or proximity of mud for nest building and repair.

Similarity in nestling growth rates among most of the bridge and cliff colonies measured argues against differences in food availability or thermal aspects of the nesting environment. Moreover, all nesting colonies were situated near continuous water supplies, and clear flyways over cultivated fields. Predation, by a rat snake (*Elaphe obsoletus*), was observed only once.

Interestingly, the observed similarity in nestling growth rates among cliff and bridge colonies may also suggest that the greater ectoparasite populations in cliff nests had little detrimental effect on the birds (also observed by Brown and Brown 1986; see also Rogers *et al.*, Chapter 8, this volume). This perspective is supported by the observation that the largest nestlings within a colony were usually located near the colony centre, precisely where bug numbers tended to be highest. Yet perennial disuse of cliff colonies is in marked contrast to the pattern on bridges, and is consistent with the hypothesis that the birds only use substrates in which the ectoparasite population has dropped below some threshold size at the beginning of the nesting season. The average number of bugs per nest in cliff colonies abandoned for a year did not differ significantly from that in the annually occupied bridge nests. Ectoparasite populations may be directly evaluated by the birds during the pre-nesting 'flighting' period. Spring clustering of ectoparasites near the entrances of nests, and flighting, has also been observed in the nests of other species of swallows (Rothschild and Clay 1952; Humphries 1969; Hoogland and Sherman 1976).

At least three ectoparasite-induced conditions may present greater energetic costs or risks in cliff colonies and thus alter breeding behaviour. First, even with perennial disuse, ectoparasite populations in cliff colonies grew at twice the rate of those in bridge colonies within nesting seasons. Therefore, while the amount of blood removed from adults may be similar between substrates at the onset of nesting, more blood is removed from adults and nestlings in cliff colonies as the season progresses. This condition must require that cliff-nesting adults catch more insects to offset their own energetic needs and those of their offspring than must bridge-nesting adults (cf. Brown, 1988; see Clark, Chapter 11, this volume). Second, synchronous departure from cliff colonies may be favoured because ectoparasites would (or do) move quickly to the nests of remnant families in which young are not ready to fledge. The cliff swallow bug is mobile within colonies, and may often be seen moving across cliffs, between nests, at night. Furthermore, the concentration of bugs at colony centres, where the earliest breeding swallows nest, implies that bugs move toward birds. The abandonment of nesting attendant to synchronous departure may therefore be related to the rapid growth of an impossible ectoparasite load on straggling families. A third, related condition may be a dramatic increase in the probability of contracting ectoparasite-transmitted disease in straggling families.

Assuming that such affects of ectoparasites do influence cliff swallow nesting decisions, the pattern of annual versus non-annual occupation of bridge versus cliff substrates could result from swallows deciding to nest first each year at sites where ectoparasites are least dense (often bridges), while later nesters are forced to occupy sites with higher ectoparasite densities. Nesting initiation dates were often, but not always, earlier at bridge sites than at cliff sites (Table 12.3).

Other factors affecting ectoparasite populations

The annual colony use occupation patterns of swallows had extreme effects on bug population sizes. However, substrate differences in bug population growth rates within nesting periods, and over winter, indicate that other mortality factors were also important. In contrast to bridges, the sandstone cliffs provide numerous moist cracks and crevices into which bugs may retreat to escape desiccation during the summer and winter. In addition, cliff nests offered a winter harbourage for the bugs several degrees above the ambient temperature, while bridge nests were slightly below ambient temperature (C. E. Hopla, unpublished data). Another mortality factor, predation by pseudoscorpions, was found only on bridges. A similar substrate difference between ectoparasite number in cliff swallow nests was noted by Brown and Brown (1986) for Nebraska culverts compared to bridges.

Ectoparasite density was positively correlated with colony size. This probably results from larger absolute population sizes, and therefore greater populations concentrated in nests at the beginning of the next breeding season.

Ectoparasite distribution within colonies

Several factors could result in the proportionally greater abundance of bugs near colony centres, including within-colony movement, within-colony variation in the duration of bird occupation, and within-colony variation in bug overwintering survivorship. First, bugs that have over-wintered may be attracted from throughout the colony site to the activity of the early, central nesters, and remain to breed in this region throughout the season. Second, bugs that happen to overwinter in the centre of the colony (or those that have been attracted to the centre) may have a longer breeding season, and therefore a longer period of population growth, than bugs on later nesting, peripheral birds (migration of birds from cliff colonies is synchronous). The central abundance of bugs is apparent only when birds are present or immediately after fledging. Ectoparasites migrate to nestlings at the edge after the central nests fledge.

Nestling weight

Nestling weights were usually greater in colony centres, probably because the dominant birds are the earliest breeders and favour central nests, which are normally protected by an overhang in cliff colonies (Emlen 1952). Such protected central nests are less likely to require repair than peripheral nests (the energy saved by reusing a complete nest, with its 2000 clay balls, must be substantial), so central nesters can immediately lay eggs and provision the young up to several days ahead of peripheral nesters. This would result in slightly older, heavier nestlings. Early nesting will also mean that the young are developing when the rapidly reproducing bugs are at relatively lower numbers, although this effect could be offset by the movement to or accumulation of bugs in central nests. Earlier nesters may also be more experienced birds that provision their young more effectively. Nestling weights were greater in the largest colonies. The weights of nestlings in bridge colonies varied less from centre to edge than the weights of those on cliffs (Table 12.4). If the daily synchronized departures of swallows from relatively large colonies is indicative of the enhanced foraging success of individuals (proposed by Brown and Brown 1986), this behaviour could make such sites preferred in spite of the fact that ectoparasite numbers are also higher.

Conclusions and relations to other studies

Several studies have documented detrimental effects of blood-feeding insects on nestling fitness (Moss and Camin 1970; Chapman 1973; Brown and Brown 1986, 1987; Møller 1987; this volume: Chapman and

George, Chapter 4; Duffy, Chapter 13; but see Rogers *et al.*, Chapter 7). The avoidance of the most heavily bug-infested colonies by cliff swallows in Oklahoma is consistent with this pattern, but the complex relationships discovered between swallow breeding behaviour, nesting substrate, and colony size indicate that the birds are evaluating multiple factors simultaneously. In particular, while birds that choose to nest in large colonies are predictably subjected to greater parasitic attack, they also may benefit from group foraging information (Emlen and Demong 1975; Brown and Brown 1986; Brown 1988), or perhaps from enhanced predator detection (e.g. Hoogland and Sherman 1976).

The argument of Emlen and Demong (1975) that synchronous nesting is favoured because birds with the modal breeding period have greatest access to group foraging information is not supported by the contrast between bridge and cliff colonies in synchrony. Birds from colonies of similar size on each substrate type forage in the same habitats in the Caddo Canyons study region, so that the potential benefits of information transfer should not differ between the colony types. The principal environmental difference between cliff and bridge colonies was instead the size of the ectoparasite populations.

The disadvantage of late nesting, due to ectoparasite concentration, may be so great for the parents using peripheral nests that abandonment of the doomed nestlings is adaptive. The fragile balance on which hangs the survival of such nestlings may be the presence of abundant insects early in the breeding season, allowing rapid growth. However, even when the nesting period was approximately the same on both substrates (1982, Table 12.3), nestling abandonment still occurred in cliff colonies (Table 12.4).

Purple martins (*Progne subis*), barn swallows, and cliff swallows infested with ectoparasites have shown an increased tendency to desert nests, and the weight and number of offspring fledged is lower than in uninfested nests (Moss and Camin 1970; Brown and Brown 1986; Møller 1987; Chapman and George, Chapter 4, this volume). Abandonment of nestlings and eggs has been observed in California cliff swallows (Foster 1968) and ectoparasite infestation has been regarded as a cause of nest abandonment in seabird colonies as well (Feare 1976; King *et al.* 1978; Duffy 1983; Chapter 13, this volume).

The presence of the abundant, specifically adapted ectoparasites in cliff swallow nests suggests that there has been a long evolutionary history of association between these animals (Myers 1928; Eads 1956; Usinger 1966; Loye and Hopla 1983; George 1987). The results presented here provide evidence that a blood-sucking, nest-dwelling arthropod has strongly influenced the evolution of nesting behaviour in the colonial cliff swallow. Its impact on the energetics and survival of nestlings and adults suggests that parasitism, in addition to predation and competition (e.g., Lack 1954), should be regarded as a potentially important agent in the regulation of host populations.

Future work

While the study of parasitic impact on natural bird populations in evolutionary and contemporary time is still in its infancy, we now have several documentations that parasites do affect host reproductive success and undoubtedly have fitness consequences. Birds may adapt to ectoparasites on a number of levels, from the evolution of immune response and other physiological defences to behavioural responses ranging from preening behaviour (this volume: Clayton, Chapter 14; Saumier *et al.* Chapter 15) and nest sanitation (Clark, Chapter 11, this volume) to the choice of a nesting area (Duffy, Chapter 13, this volume). In relatively long-lived organisms such as birds, behavioural plasticity should afford especially valuable adaptive opportunities because of the highly variable distribution of parasites likely to be encountered in the course of an individual's lifetime.

Behavioural plasticity can be monitored at several levels. While we still need to measure directly the energetic costs and responses of nestlings to ectoparasites, it is equally important to measure the degree to which parents enhance feeding to compensate for nestling blood loss, and the limits to which they can or choose to extend themselves to provision heavily parasitized nestlings (Rogers *et al.*, Chapter 7, this volume). In addition, spatial and temporal avoidance of ectoparasites probably entails a variety of costs. In a bird such as the cliff swallow, prime candidate costs are delayed nesting, exposure to multiple ectoparasites and pathogens while examining potential nesting sites, and the energy required to build new nests or repair old nests at sites unoccupied for extended periods. Quantifications of the fitness consequences of these costs should give insight into the magnitude of the evolutionary significance of ectoparasite damage (Chapman and George, Chapter 4, this volume).

Predicting the general form of behavioural adaptations to parasitism requires knowledge of the life-cycle organization and feeding strategies of the parasites. Are there particular phases of ectoparasite life cycles that are particularly vulnerable to response from avian hosts? Two obvious critical periods for cliff swallow parasites are mortality during attempted migration to new colonies, and mortality from starvation or other stress in unoccupied sites. Variation in swallow behaviour during the pre-nesting colony-site exploration phase, and cyclical abandonment of nesting sites are both factors likely to impinge upon parasite life-cycle evolution and population dynamics.

One difficulty in using knowledge of ectoparasite life-cycle organization to predict host behavioural response is the fact that a guild of disparate parasites may converge upon a single host. While bird blood is a relatively easily quantified resource base for nidicolous ectoparasites, allowing the ready generation of life history predictions about the parasites, the presence of a phylogenetically diverse community of parasites, with strategies ranging from '*r*-selected' (e.g., short-lived, readily-dispersing fleas) to 'adversity selected' (e.g., seemingly immortal soft ticks that become part of the geological matrix of a cliff, personal observation), may render less likely or impossible the fine-tuning of the host evolutionary response and coevolutionary relationship to any one ectoparasite species.

What ultimately limits the extent of habitat use for nesting in a region? In the cliff swallow, nesting on shear vertical substrates probably evolved as a deterrent to nest predators, so habitat use may be limited by the availability of physically suitable substrate. In our study region, however, many apparently suitable cliff sites were never occupied. Limited habitat use may result from the costs of building nests at completely new sites, dependence on social nesting that favours later nesters joining early nesters that have already occupied all of the intact nests at a traditional site, or the availability of substrates that will not foster rapidly growing and persistent populations of ectoparasites. To what extent do abiotic and biotic limitations in colony site availability, or interactions among them, regulate bird population size?

Several methodological tactics can greatly enhance the collection of data to address these ecological and evolutionary issues. First, one of the great difficulties in quantifying ectoparasite abundance in colonial birds is the distribution of ectoparasites within the nesting site (e.g., all four species of ectoparasites found in cliff swallow colonies cluster, and thus samples on hosts or in nests fit a negative binomial distribution, making it difficult to estimate parasite population parameters). This sampling problem can be alleviated to some extent by extensive stratified sampling (Southwood 1978), although colony disturbance is a risk of increasing the sample size. Further, collecting and observing ectoparasites at a fixed time of day, stage of host development, and season is necessary for making valid ecological comparisons among sites. (The best times for sampling most bird ectoparasites are during their crepuscular activity periods and when the hosts are fledging.)

Little is known of the population biology of ectoparasites (Price 1980), and refinement of study procedures will help to alleviate this deficiency. Manipulative approaches, including the use of artificial nests (Moss and

Camin 1970), fumigation of nests (Chapman 1973) or antibiotic treatments (Boyce and Manly, Chapter 21, this volume); and laboratory observation of hosts and parasites (Loye 1985*b*; this volume: Scott and Edman, Chapter 10; Boyce and Manly, Chapter 22), including marked populations of ectoparasites as well as simple observations aimed at critical aspects of parasite behaviour (Duffy and Daturi 1987), are all methods that should be utilized in future experimental and observational research. Data on parasite demography and behaviour, including distributions in nature, movement patterns from host to host and among sites, host preferences, and main sources of mortality, will be of great value both for answering general biological questions and for applied purposes such as avian conservation.

Acknowledgements

Our thanks to Ken Gage and Rex Thomas for help in the field. Brian Chapman, John Endler, John George, and Cliff Hopla encouraged and advised the project. Sigma Xi provided two grants.

References

Alexander, R.D. (1974). The evolution of social behavior. *Annual Review of Ecology and Systematics*, **5**, 325–83.

Anderson, R.M. and May, R.M. (1979). Population biology of infectious diseases. *Nature*, **280**, 361–7, 455–61.

Bent, A.C. (1942). Life histories of North American flycatchers, larks, swallows, and their allies. Smithsonian Institution US National Museum Bulletin, **179**, 1–555.

Brown, C.R. (1984). Laying eggs in a neighbor's nest: benefit and cost of colonial nesting in swallows. *Science*, **224**, 518–19.

Brown, C.R. (1985). The costs and benefits of coloniality in the cliff swallow. Unpublished Ph.D. thesis, Princeton University, Princeton, New Jersey.

Brown, C.R. (1988). Enhanced foraging efficiency through information centers: a benefit of coloniality in cliff swallows. *Ecology*, **69**(3), 602–13.

Brown, C.R. and Brown, M.B. (1986). Ectoparasitism as a cost of coloniality in cliff swallows (*Hirundo pyrrhonota*). *Ecology*, **67**(5), 1206–18.

Brown, C.R. and Brown, M.B. (1987). Group living as an advantage in avoiding predators. *Behavioural Ecology and Sociobiology*, **21**, 97–107.

Bryant, D.M. (1975). Breeding biology of house martins *Delichon urbica* in relation to aerial insect abundance. *Ibis*, **117**, 180–216.

Chapman, B.R. (1973). The effects of nest ectoparasitism on cliff swallow populations. Unpublished Ph.D. thesis, Texas Technical University, Lubbock.

Duffy, D.C. (1983). The ecology of tick parasitism on densely nesting Peruvian seabirds. *Ecology*, **64**(1), 110–19.

Duffy, D.C. and Daturi, A. (1987). Diel rhythms of tick parasitism on incubating African penguins. *Medical and Veterinary Entomology*, **1**, 103–6.

Eads, R.B. (1956). Ectoparasites from swallow nests with the description of a new ceratophyllid flea. *Journal of Parasitology*, **42**, 73–5.

Emlen, J.T. Jr (1952). Social behavior in nesting cliff swallows. *Condor*, **54**(4), 177–99.

Emlen, J.T. (1986). Responses of breeding cliff swallow to nidicolous parasite infestations. *Condor*, **88**, 110–11.

Emlen, S.T. and Demong, N.J. (1975). Adaptive significance of synchronized breeding in a colonial bird: a new hypothesis. *Science*, **188**, 1029–31.

Feare, C.J. (1976). Desertion and abnormal development in a colony of sooty terns *Sturna fuscata* infested by virus-infected ticks. *Ibis*, **118**, 112–15.

Foster, W.A. (1968). Total brood mortality in late-nesting cliff swallows. *Condor*, **70**, 275.

Foster, W.A. and Olkowski, W. (1968). The natural invasion of artificial cliff swallow nests by *Oeciacus vicarius* (Hemiptera: Cimicidae) and *Ceratophyllus petrochelidoni* Siponaptera: Ceratophyllidae). *Journal of Medical Entomology*, **5**(4), 488–91.

George, J.E. (1987). Field observations on the life cycle of *Ixodes baergi* and some seasonal and daily activity cycles of *Oeciacus vicarius* (Hemiptera: Cimicidae), *Argas cooleyi* (Acari: Argasidae), and *Ixodes baergi* (Acari: Ixodidae). *Journal of Medical Entomology*, **24**(6), 683–8.

Grinnell, J., Dixon, J.S., and Linsdale, J.M. (1930). Vertebrate natural history of a section of northern California through the Lassen Peak region. *University of California Publications in Zoology*, **35**, 1–594.

Hamilton, W.D. and Zuk, M. (1982). Heritable true fitness and bright birds: a role for parasites? *Science*, **218**, 384–7.

Hoogland, J.L. and Sherman, P.W. (1976). Advantages and disadvantages of bank swallow (*Riparia riparia*) coloniality. *Ecological Monographs*, **46**, 33–58.

Humphries, D.A. (1969). Behavioural aspects of the ecology of the sand martin flea *Ceratophyllus styx jordani* Smit (Siphonaptera). *Parasitology*, **59**, 311–34.

Johnston, R.F. (1965). Seasonal variation in the food of the purple martin *Progne subis* in Kansas. *Ibis*, **109**, 8–13.

Jolley, A.E. and Storer, R. (1945). Report on the Sand Martin 1944. In: Proceedings of the Leicester Literary and Philosophical Society, Ornithological section for Leicestershire and Rutland, pp. 10–21.

King, K.A., Keith, J.O., Mitchell, C.A., and Keirans, J.E. (1978). Ticks as a factor in nest desertion of California brown pelicans. *Condor*, **79**(4), 507–9.

Lack, D. (1954). *The natural regulation of animal numbers*. Clarendon Press, Oxford.

Little, E.L. Jr (1939). The vegetation of the Caddo County canyons, Oklahoma. *Ecology*, **20**(1), 1–10.

Loye, J.E. (1984). The cliff swallow bug *Oeciacus vicarius* (Cimicidae: Hemiptera), natural history and population ecology. Unpublished Ph.D. thesis, University of Oklahoma, Norman.

Loye, J.E. (1985*a*). The life history and ecology of the cliff swallow bug, *Oeciacus vicarius* (Hemiptera: Cimicidae). *Cahiers Office de la Recherche Scientifique et Technique Outre Mer, Serie Entomologie Medicale et Parasitologie*, **23**, 133–9.

Loye, J.E. (1985*b*). Host effects on feeding and survival of the polyphagous cliff swallow bug, *Oeciacus vicarius* (Hemiptera: Cimicidae). *Bulletin of the Society of Vector Ecology*, **10**(1), 7–13.

Loye, J.E. and Hopla, C.E. (1983). Ectoparasites and microorganisms associated with the cliff swallow in west-central Oklahoma. II. Life history patterns. *Bulletin of the Society of Vector Ecology*, **8**, 79–84.

Mayhew, W.W. (1958). The biology of the cliff swallow in California. *Condor*, **60**, 7–37.

Møller, A.P. (1987). Advantages and disadvantages of coloniality in the swallow, *Hirundo rustica*. *Animal Behaviour*, **35**, 819–32.

Moss, W.W. and Camin, J.H. (1970). Nest parasitism, productivity, and clutch size in purple martins. *Science*, **168**, 1000–3.

Myers, L.E. (1928). The swallow bug, *Oeciacus vicarius* Horvath (Hemiptera, Cimicidae). *Parasitology*, **20**, 159–72.

Myres, M.T. (1957). Clutch size and laying dates in cliff swallow colonies. *Condor*, **59**, 311–16.

Price, P.W. (1980). *Evolutionary biology of parasites*. Princeton University Press, Princeton, New Jersey.

Rothschild, M. and Clay, T. (1952). *Fleas, flukes and cuckoos*. Collins, London.

Siegel, S. (1956). *Nonparametric statistics for the behavioral sciences*. McGraw-Hill, New York.

Snapp, B.D. (1973). The occurrence of colonial breeding in the barn swallow (*Hirundo rustica*) and its adaptive significance. Unpublished Ph.D. thesis, Cornell University, Ithaca, New York.

Sokal, R.R. and Rohlf, F.J. (1981). *Biometry: the principles and practice of statistics in biological research*. W.H. Freeman and Co, San Francisco.

Southwood, T.R.E. (1978). *Ecological methods*. Chapman and Hall, London.

Stoner, D. (1945). Temperature and growth studies of the northern cliff swallow. *Auk*, **62**, 207–16.

Usinger, R.L. (1966). *Monograph of Cimicidae (Hemiptera-Heteroptera)*, Thomas Say Foundation, no. 6. Philadelphia, PA.

13 *Ants, ticks, and nesting seabirds: dynamic interactions?*

DAVID C. DUFFY

Introduction

Colonial nesting or use of traditional sites must be advantageous to birds or they would not be widely used nesting strategies (Hoogland and Sherman 1976; Wittenberger and Hunt 1985). However, such sites are also beneficial to the birds' parasites, providing them with a renewable, easily-locatable source of hosts, removing starvation and host-location as major sources of parasite mortality (Rothschild and Clay 1952). Large concentrations of ticks and other parasites are frequently observed at colonies of densely-nesting birds and at individual avian nest sites which are reused year after year (Rothschild and Clay 1952; Marshall 1981). Heavy infestations of parasites can lead to nest desertion by adult birds or to mortality of their young (Duffy 1983*a*, and references therein).

Avian hosts can avoid excessive parasite loads during breeding by a variety of means. For example, birds can switch nest or colony sites (King *et al.* 1977*a*; in this volume: Chapman and George, Chapter 4; Loye and Carroll, Chapter 12). Many passerines build new nests for successive clutches (Skutch 1976). Other bird species may use plants with insecticidal properties as nest material (Wimberger 1984; Clark and Mason 1985; Clark, Chapter 11, this volume). Seabird guano, ammoniacal and cement-like, may also discourage parasites (personal observation). Many birds nest in areas where physical forces such as freezing or flooding reduce survival of parasites between nesting seasons (Baerg 1944; Faaborg, in Duffy 1983*a*).

Where suitable nest sites are limited, so are options for breeding birds to protect their young from parasitism. Birds can forgo breeding until parasite populations decrease to acceptable limits because of starvation (Duffy 1983*a*; Brown and Brown 1986), but this may take more than a decade for parasites such as ticks (Anastos 1957). Such intermittent breeding is unlikely to be a successful reproductive strategy for even the most long-lived seabird. Alternatively, birds could nest early in the spring in order to fledge young before parasite populations build up to levels dangerous to birds (see Burtt *et al.*, Chapter 6, this volume).

Food limitation and predation on nestlings have usually been cited as explanations of synchronous nesting in birds (Emlen and Demong 1975; Wittenberger and Hunt 1985), but parasitism needs to be investigated as another potential cause of synchrony (this volume: Chapman and George, Chapter 4; Loye and Carroll, Chapter 12). Young birds still present in the colony after the majority have fledged might attract a disproportionate number of parasites. Nidifugous behaviour, where young leave their nest before full-grown, may be another antiparasite adaptation, although such behaviour has usually been explained in terms of reducing nestling predation (Lack 1967). Finally, adult birds may develop immunological defences against parasites (e.g. Trager 1939). Such defences may not have time to develop in nestlings, and an adult defence would not directly reduce nestling mortality.

Because seabirds are primarily colonial (Lack 1967) and frequently reuse the same colony and nest sites for generations, if not centuries (Nelson 1978), they may be correspondingly vulnerable to disease and parasites (e.g. Harris 1965; Clifford 1979; Chastel 1980). One of the most conspicuous groups of parasites at colonies of seabirds are ticks (Ixodoidea) which have been shown to cause nest desertion, and subsequent nestling mortality, in a range of species throughout the world (Duffy 1983*a*). Ticks also harbour appoximately 50 arboviruses which may cause illness and death in seabirds (Nutall 1984; Morgan *et al.* 1985).

Only two of the three families of ticks parasitize seabirds: the Argasidae, or soft ticks, and the Ixodidae, or hard ticks. As seabird parasites, argasid ticks tend to be more tropical and ixodids more polar, although there is extensive overlap in distribution (Clifford 1979). The two families differ greatly in ecology (details from Hoogstraal 1978, 1985). The argasids lack a hard dorsal scutum and are rubbery rather than hard. They spend little time on their hosts and complete their meals in a matter of hours, compared to days spent by ixodids. Argasids prefer to shelter in crevices or nest litter where they may attain great densities. Argasids appear to be specialists, occurring where a concentrated food supply such as a human dwelling, animal colony, or domestic animal enclosure ensures adequate and reliable access to food, even if interrupted by periods of host absence. Most argasids are adapted to a narrow choice of suitable hosts (Hoogstraal and Aeschlimann 1982). In contrast, the ixodids have a hard dorsal scutum, a more dispersed distribution, and tend to find their prey by chance encounter. Once a host is found, ixodids may stay on it for days or weeks, because finding another host is extremely unlikely.

There has been relatively little work on population regulation of ticks in seabird colonies. Predators, such as spiders, scorpions, and lizards, may be important in regulating these parasites (Murphy 1921; LaValle

1923; Vogt 1939), as may unfavourable conditions in the resting substrate (Vogt 1939; Duffy 1983*a*).

In relatively permanent or stable habitats, tick parasitism on nesting seabirds may be reduced or controlled by the predatory behaviour of one of the world's most ubiquitous organisms, ants. Ants occur over a wide range of habitats and latitudes and are usually important components of ecosystems when they occur (Pectal 1978; Pisarski 1978). Although diet specializations occur, many ants are generalized predators, taking anything that cannot escape them (Sudd 1967; Carroll and Janzen 1973; Stradling 1978). Ants have been shown to reduce populations of ticks (Burns and Melancon 1977 and references therein; Bull 1986) and other arthropods (Kajak *et al.* 1972; Pectal 1978; Risch and Carroll 1982). Landbirds also rid themselves of ectoparasites by 'anting' or lying on an anthill and letting ants swarm over their bodies (Simmons 1985).

This chapter presents two hypotheses: (1) ants are present in all but the most hostile terrains and densest colonies of seabirds; and (2) when ants are present, their predatory activity normally prevents massive tick infestations, allowing seabirds to reproduce at permanent nests and colonies. Evidence is presented from several field sites in initial support of these hypotheses, predictions are made about where tick infestations should occur (and where ants should be absent), and ways are suggested to test more rigorously the hypothesis that ants control tick populations.

Methods

During field work on ticks and seabirds in Ecuador, Peru, Namibia, South Africa, and the United States, I made casual observations in seabird colonies of ant and tick interactions (Duffy 1983*a*; Duffy and Duffy 1986; Duffy and Daturi 1987).

Methods of estimating tick numbers varied with substrate and seabird species. In guano and sand, 500 cm^3 scoop samples were filtered for ticks (e.g. Duffy 1983*a*). At nests on rock and in trees, ticks often occurred in deep crevices, so that population estimates were highly dependent on observer skill and effort. In such cases, daylight counts of ticks on adult or nestling birds allowed greater standardization, but were dependent on time of day (e.g. Duffy and Daturi 1987). I avoided counting ticks in early morning or early evening when ticks might have been more active than during midday hours.

I have divided population levels into three subjective classes: high, low, and absent (Table 13.1). High tick densities produced swarms of nymphal ticks on birds, large numbers of adult ticks in resting places, and frequent bites on human intruders. Low densities typically produced

0–2 ticks per bird, with occasional higher numbers; adult ticks could be found in resting places after considerable searching; and ticks rarely, if ever, occurred on investigators. Ticks were considered to be absent when they were not found after repeated examination of dozens of nestling or adult birds.

Those wishing a more quantitative approach might consider timing how long they can sit still in the midst of a seabird colony and how many ticks they remove from their bodies afterwards (cf. Møller 1987). Given the impressive array of arboviruses carried by ticks (Clifford 1979; Nuttall 1984) and the discomfort caused by tick bites (Hoogstraal and Gallagher 1982; personal observation), the resulting turnover in field staff may limit replication.

In each colony visited, I noted the presence or absence of individual ants within 0.5 m of seabird nests. I did not collect ants for identification.

To provide more case studies, I have used data from the literature and information generously provided by colleagues. Because my requests for data often came years after field-work had concluded and because their studies usually focused on the birds rather than their parasites or other arthropods, data are often incomplete.

Field examples

Table 13.1 presents a summary of the results from 10 island ecosystems where sufficient data could be assembled to examine tick abundances in relation to vegetation, ant, and seabird densities.

Guano islands, Peru

The guano islands off the coast of Peru are among the driest ecosystems on earth, with almost no vegetation except where caretakers manually irrigate small gardens. Although breeding populations of guanay cormorant (*Phalacrocorax bougainvilli*), Peruvian booby (*Sula variegata*), and Peruvian brown pelican (*Pelecanus* (*occidentalis*) *thagus*) are now greatly reduced from historical levels, they still cover substantial portions of many islands at densities of 1.9–3.0 nests/m^2 (Duffy 1983*b*). Nest structures included bare guano deposited by boobies, beds of feathers, guano, and seaweed used by pelicans, and cylindrical nests, with a mean height of 17 cm, constructed by cormorants (Duffy 1983*b*). Argasid tick (*Ornithodoros amblus and O. spheniscus*) densities are frequently very high and tick-induced desertions by breeding seabirds are common, occurring in about 30 per cent of breeding seasons (Duffy 1983*a*;

Table 13.1 Tick densities, vascular plant abundance, ant presence, and the species and densities of nesting seabirds

LOCATION	Seabird species	Seabird density	Tick density	Vascular plant density	Ant presence	Effect on seabirds
Peru	Cormorant, booby, pelican	High	High	None	No	Desertion
Southern Africa						
Marcus Island	Penguin	Low	Low	High	Yes	None
Ichaboe Islands	Cormorant, penguin, Gannet	High	High	None	No	Not observed
Platforms	Cormorant	High	High	None	No	Not observed
Seychelles	Tern	High	High	Sparse	No?	Desertion
Texas	Pelican	Medium	High	Sparse	No?	Desertion
Great Gull Island	Terns	Medium	None	High	Yes	None
Bird Island	Terns	Medium	None	High	Yes	None
Australia	Noddy	Medium	High	None	No	Mortality?
Saint Paul's Rock	Booby	High	High	None	No	?

Hoogstraal *et al.* 1985). Numerous workers have reported large numbers of ticks and tick bites during visits to colonies (e.g. LaValle 1923; Clifford *et al.* 1980; Hoogstraal *et al.* 1985).

Ants are very rare on the islands. Murphy (1921) did not report any from the islands and I found only a single, unidentified species in a caretaker's garden on one island (Duffy 1983*a*). Murphy (1921) and LaValle (1923) suggested that a spider (*Dysdera murphyi*) was an important predator of ticks. Murphy indicated, however, that such predation would be less effective within seabird colonies, because guano would inundate spider burrows and the spiders would risk being crushed. Vogt (1939) and Duffy (1983*a*) found numerous ticks in the stomach of a lizard (*Tropidurus* spp.) on several guano islands, but Vogt (1939) noted that lizards rarely occurred in the colonies themselves, because the colonies lacked suitable shelters for lizards and because nesting birds attacked lizards. Lizards were also absent from many islands (Vogt 1939).

Guano islands: southern Africa

Three main seabird species nest on islands off the coasts of Namibia and South Africa: the African penguin (*Spheniscus demersus*), the Cape cormorant (*Phalacrocorax capensis*), and the Cape gannet (*Morus *capensis*). The penguin nests both on the surface and in burrows, frequently with several meters between nests; in contrast, the gannet and cormorant nest in densely-packed surface colonies. Gannets make minimal scrapes in the guano substrate, whereas cormorants construct elaborate nests, with feathers, guano, and seaweed (Rand 1960; Nelson 1978).

Cape Islands, South Africa

Marcus Island, Cape Province, South Africa, is in a Mediterranean climate, with most vegetation growth in winter, which is the peak nesting period of the African penguin (La Cock *et al.* 1987). Nests are dispersed at sufficiently low densities to allow herb and grass cover within the colonies. Nest sites, especially burrows, are used repeatedly.

It was difficult to find ticks in penguin nest burrows and mean tick load per adult penguin did not rise above 1.0 ± 1.64 SD (Duffy and Daturi 1987). For nestlings, the median was 2.0, range 1–50 (Duffy *et al.*, unpublished).

Nest desertion by adult penguins is common, but seems associated with food shortages during El Niño/Southern Oscillation events (La Cock 1986; Duffy *et al.* 1984) and has not been linked to tick infestations.

Ants of unknown species were common. Ants were not observed feeding on ticks, but during casual experiments, placing ticks on ant runways within the penguin colony, larval ticks were seized by ants, while nymphal and adult ticks were investigated, but usually left alone.

Namibian islands

In contrast to Marcus Island, on Mercury and Ichaboe islands off the desert coast of Namibia, vegetation was absent, and nesting and roosting Cape cormorants, bank cormorants (*P. neglectus*), Cape gannets, and African penguins covered both islands, preventing plant growth. Argasid ticks (*Ornithodoros capensis*) were inconspicuous during the day, but careful searching under guano slabs and rocks, or in nests in cormorant and gannets nesting areas uncovered large numbers. Desertions by nesting seabirds have been frequent (Rand 1960), but normally have been ascribed to the unavailability of food (e.g. La Cock 1986). No tick-induced desertions have been reported for the three species, but this may reflect a past lack of awareness of this cause of nest desertion. Ants were absent.

On artificial wooden nesting-platforms used by densely-nesting Cape cormorants off Namibia, 'vast numbers' of ticks occurred, although no effects on nesting were noted (H. H. Berry, *in litt.*). Ants were again absent (H. H. Berry, *in litt.*)

Sooty tern colonies: Seychelles

The Seychelles are tropical and humid, lying just south of the Equator. They have a dry season from April to November during the south-east trade winds, followed by a wet season when the trade winds cease (Stoddart and Fosberg 1981).

Bird Island, 70 ha, is the northernmost of the Seychelles archipelago and one of two sand islands in the otherwise granitic group (Feare 1979). Sooty terns (*Sterna fuscata*), pantropical seabirds, nest on treeless regions of Bird Island, using shallow ground scrapes. Low vegetation occurs in the tern colony but is burned off in dry years by guano (Feare 1979).

During 1973, argasid ticks (*Ornithodoros capensis*), elsewhere rare on the island, were abundant in a nesting area deserted by 5000 pairs of sooty terns at a colony of 400 000 dense ground nests (Converse *et al.* 1975; Feare 1976). Hoogstraal and Feare (1984) reported that 'innumerable ticks crawling on the ground quickly covered the observers' legs and attached to the pubic area and on the back'. Ticks remained sufficiently abundant in subsequent years to prevent successful nesting through 1976 (Feare and Feare 1984). Ants were 'incredibly numerous' in the Seychelles, including the island where the desertions occurred, but were 'certainly not a problem for observers sitting in the colony' (Feare *in litt.*), suggesting that ants were rare or absent in the deserted area.

Pelican colonies

False Live Oak Point, now eroded, was one of three islands of dredge spoil, located on the subtropical Texas coast in Arkansas County (Texas Colonial Waterbird Society 1982). The substrate of the islands was 'unvegetated shell and sand, sparse halophytic vegetation to 0.3 m' (Texas Colonial Waterbird Society 1982).

Brown pelicans (*Pelecanus occidentalis*) nested on the ground on one of the islands, 'a near-barren oyster shell reef that has since eroded away' (K. A. King *in litt.*). Argasid ticks (*Ornithodoros capensis*) were 'unusually abundant', with three collected nests yielding 2389 adult and nymphal ticks (King *et al.* 1977*b*). King *et al.* (1977*b*) reported nest desertion by pelicans. 'Because it contained little vegetation and was subjected to repeated flooding, there may not have been ants there' (K. A. King *in litt.*). In contrast, ticks and fire ants co-occur in other bird colonies along the Texas coast (K. A. King *in litt.*). More details of the latter situation are needed, as they appear contrary to one of the hypotheses presented in this chapter.

Common tern colonies, north-eastern United States

Common tern colonies in the temperate eastern United States are occupied only in summer. Colony sites range from low sand-bars, flooded during winter and hurricanes and possessing little vegetation, to larger, higher islands with extensive vegetation (Nisbet 1973; Duffy 1977). Most common terns nest in open areas near vegetation with distances of 0.5 to several meters between sites (personal observation). Nests are usually simple scrapes in the ground, with pebbles, sticks, and other debris added as a rim.

In contrast to the high levels of ticks at other seabird colonies, parasitism at colonies of common terns (*Sterna hirundo*) in the north-eastern United States was rare or absent. Dispersed nest sites, changes in sites between years, and nidifugous young terns may also serve to restrict tick populations.

At Great Gull Island, New York, ticks have not been seen on nestling terns in 20 years of field-work (H. Hays personal communication). Several species of ant occur on the island, and they have occasionally killed nestlings, especially those struggling to emerge from the egg. Similarly, ticks have not been observed on nestlings at common tern colonies in Massachusetts during more than a decade of field-work (I. C. T. Nisbet *in litt.*). Ants (*Lasius neoniger*) caused mortality of nestling terns at six of seven Massachusetts study colonies (Nisbet and Welton 1984; Nisbet *in litt.*).

Coral islands, Australia

Low coral islands or cays in the Great Barrier Reef and Coral Sea off Australia have a subtropical maritime climate with an air temperature range of approximately 10–33.5°C (Heatwole *et al.* 1981). Rainfall occurs throughout the year, with two peaks (Heatwole *et al.* 1981). Strong southerly winds are present, pruning vegetation and spraying salt over islands (Heatwole *et al.* 1981).

Heatwole (1971) studied 12 bare cays with simple terrestrial ecosystems characterized by the absence of vascular plants. Seabirds provided the nutrients in the form of guano, feathers, and carcasses for the islands' heterotroph communites (Heatwole 1971). Six of 12 islands examined and five of seven islands with nesting seabirds had tick (the argasid tick: *Ornithodoros capensis* and the ixodid ticks: *Amblyomma loculosum*, and *Ixodes* cf. *pterodromae*) populations, several of them at high densities. At one island, on two separate visits, Heatwole (1971) found brown noddy (*Anous stolidus*) dead young, abandoned eggs, and dead and moribund adults, but did not mention tick numbers. Ants were not present (Heatwole 1971).

Saint Paul's Rocks, equatorial Atlantic

Saint Paul's Rocks are the most isolated oceanic islands, situated just off the Equator and about 1000 km from Brazil. The islands are swept by strong south-east trade winds and westward-setting surface currents. Only three of the islands are above the spray zone (Edwards and Lubbock 1983). Algae appears to be the only terrestrial vegetation; vascular plants are absent (Hemsley 1884).

Only three seabird species nest on the rocks: brown noddy, white-capped noddy (*A. minutus*), and the brown booby (*Sula leucogaster*). The booby colony appears to be unusually dense compared to other colonies of this species and to have remained stable for at least a quarter-century (Edwards *et al.* 1981).

Argasid ticks (*O. capensis*) were first collected from Saint Paul's Rocks in 1832 (Darwin 1852) and were common in nests of brown boobies (Edwards and Lubbock 1983). Sixty-three were found on a dead booby and a 'few' on a dead brown noddy (Edwards and Lubbock 1983), but no effects of the ticks were noted on nestlings. Ants have not been reported from Saint Paul's (Edwards and Lubbock 1983).

Analysis of the island ecosystems

To examine the relationships between ants, plants, ticks, and seabirds, plant densities were grouped into absent and high/sparse; tick densities into high or low/absent; and seabird densities into low or medium/high (Table 13.1). Ants were scored as present or rare/absent, the latter including cases where ants were present on the island, but not necessarily in the colony.

Analysis of the 10 island ecosystems showed that at three islands ants were present and ticks were low/absent; at seven islands ants were rare/absent and tick densities were high. These data indicate a significant inverse relationship between ants and tick densities (Fisher exact probability test: $p = 0.008$).

While this supports the hypothesis that ants control tick populations, an alternative hypothesis is that the two groups have different habitat requirements. To test this, I examined plant density in relation to ant presence and tick density. When plant densities were sparse/high, ants were present in three cases and rare/absent in two cases. In five cases, ants and plants were both absent, but the difference in distributions was not significant (Fisher exact probability test: $p=0.08$). In contrast, tick densities were high at two islands and rare/absent at three others where plant densities were high/sparse. Ticks were present in five cases where plants were absent, but again the difference is not significant (Fisher exact probability test: $p=0.08$).

Seabird density may also be important in determining ant presence, or plant and tick densities. In seven cases, seabird densities were medium/high and tick densities were high; in one case, seabird and tick densities were both low. In two other cases, low tick densities occurred with medium/high seabird densities; however, the difference was not significant (Fisher exact probability test: $p=0.30$).

Ants were present at only two localities (Great Gull and Bird Islands) where seabird densities were medium/high and one site (Marcus Island) where seabird density was low. In contrast, ants were rare/absent in seven cases where seabird densities were medium/high, but not significantly so (Fisher exact probability test: $p=0.30$). Finally, plant densities were sparse/high in four cases where seabird densities were medium/high and absent in five other cases. Plant density was sparse/high in the single case where seabird density was low, but the differences were not significant (Fisher exact probability test: $p=0.50$).

Discussion

Unfortunately, it is difficult to prove absence of ticks or ants, especially on islands such as Mercury and Ichaboe off southern Africa or St. Paul's in the Atlantic that are rarely visited by biologists. In addition, tick-induced seabird mortality or nest desertion may be occasional, requiring relatively intensive and continuous observation to detect.

The data presented here are preliminary and limited, and the analysis, exploratory; nevertheless, they suggest that ants tend to occur on vegetated islands and to be absent on islands or parts of islands where vascular plant life does not occur. Ecosystems without higher plant life may not be stable or productive enough to support ant colonies. In addition, ants seemed to be excluded from parts of islands when local nesting densities of seabirds were high. Densely nesting seabirds may exclude ants because ant bodies are fragile and cannot survive being trampled or defecated on by seabirds walking in the colony.

In contrast, ticks are squat, thick-bodied, and tolerate being squashed (personal observation), as would occur in areas of densely-nesting seabirds. Large infestations of ticks tended to occur when ants were rare or absent, which appears to support the hypothesis that ants restrict tick populations.

Ants may reduce tick parasitism on nesting seabirds, but they are not necessarily beneficial to birds. Most ant species are opportunistic predators (Sudd 1967). Perhaps initially attracted to a nest by the presence of parasites or faeces, ants may proceed to attack and kill nestlings. This has been reported for common terns (Nisbet and Welton 1984 and references therein), swallows (*Hirundo rustica* and *H. pyrrhonota*) (Kroll *et al.* 1973; Sikes and Arnold 1986), and quail (*Colinus virginianus* (J. T. Emlen *in litt.*).

Seabird nesting on small offshore islands has often been ascribed to avoiding terrestrial predation, especially by mammals (Lack 1967). Ants may be sufficiently important predators in their own right to force ground-nesting seabirds to nest in habitats inhospitable to ant colonies. This could explain dense nesting behaviour by gannets and other species, even where sufficient space would appear to be present to allow more dispersed nesting (Lack 1967; Nelson 1978). Such dense nesting may protect against aerial predation, through a 'phalanx' effect of jabbing bills (Duffy 1983*b*); it may also provide an impenetrable environment for ants because of guano and avian perambulation.

Ticks, ants, and seabirds may be in a dynamic balance with certain habitat types favouring the predator, others the parasite. For example, as succession occurs on islands such as dredge-spoil deposits, coral keys, or sand-bars, insufficient vegetation may exist initially to support ants. If hosts are available, and if the substrate is not subject to flooding, ticks should be present. With time, as vegetation arrives, ants should colonize, suppressing the tick population and perhaps also inhibiting nesting by some host species. A similar pattern may be generated across dune systems, although the large foraging areas of some ant species may make this more difficult to observe (Sudd 1967).

Conclusions and future work

Much of seabird nesting ecology has been explained in terms of mammalian predators and food availability (e.g. Lack 1967), but many of the same arguments apply equally well to parasites. Greater understanding may await us through thinking small—about parasites, disease, and arthropod predators usually ignored by biologists.

On a wider scale, host resistance or acquiring a suitable host are often assumed to limit parasite populations (May 1977; Randolph 1979; Crofton 1981). There has been relatively little attention paid to the role of predation in regulating parasite populations (e.g. LaValle 1923; Ault and Ault 1979).

Predation on parasites may be most obvious and easiest to study in hosts such as seabirds where dense and permanent nesting colonies virtually eliminate the problem of finding a host for parasites.

If ants and other arthropods prove to be important tick regulators, this finding may provide new approaches to medical and veterinary problems. For example, control of introduced fire ants could actually increase the incidence of arbovirus transmission to humans by allowing an increase in tick populations. Use of pesticides in newly-cleared rain forest might have the same result.

Further tests of these ideas require directed work. An ideal study would include identification of the tick species and quantification of their density. Determination of effects of ticks on their hosts should be made at the disease, parasite, and behavioural levels. The ant species should be identified and their trophic roles determined. Field examination of ant predation (e.g. Finnegan 1969) might be combined with feeding experiments in the laboratory.

Several aspects of the biologies of the two groups suggest caution in the design of field studies. For example, ant and tick species may not be active at the same times, reducing tick vulnerability to predation. Similarly, field-work during daylight hours may be uninformative if most tick activity occurs at night (e.g. Belozerov 1982; Duffy and Daturi 1987). Finally, ticks are often overdispersed, so that, at lower densities, many potential hosts need to be sampled to document the presence of ticks (cf. Southwood 1978).

Experimental removal of ticks might be a profitable research approach, but it may be difficult to find acaricides or repellants that are effective in the harsh environments used by nesting seabirds. Wind, spray, intense insolation and guano accretion made DDT and pyrethrins ineffective in small trials at Peruvian guano islands (personal observation). Burning the substrate was similarly ineffective; ticks immigrated from unburned areas. Kerosene is an effective barrier and has traditionally been used to exclude ticks from human habitation on Peruvian guano islands (personal observation), but it may result in feather contamination if used in nesting colonies.

Interesting host species for research might include the brown pelican which nests at varying densities in trees and on the ground, the common tern which nests in a variety of substrates and densities, and boobies which nest in a wide range of habitats and densities (Nelson 1978).

Acknowledgements

The ideas presented here have fermented for over a decade and have benefited from discussions and correspondence with A. Berruti, H. H.

Berry, B. J. Cole, H. A. Denmark, M. J. Duffy, J. T. Emlen, C. J. Feare, J. Frazier, M. Gochfeld, H. Hays, H. Hoogstraal, H. Horn, K. A. King, R. M. May, I. C. T. Nisbet, E. Ortiz, R. T. Paul, W. B. Robertson, Jr., B. Tomkins, and M. P. and R. P. Wilson. I thank R. Carroll, J. Frazier, G. L. Hunt, Jr., H. Korn, and J. Loye for comments on the manuscript. My field work has been supported by the Benguela Ecology Programme, the National Geographic Society, the National Science Foundation, the Great Gull Island Project of the American Museum of Natural History, and the World Wildlife Fund.

References

Anastos, G. (1957). *The ticks or Ixodides of the U.S.S.R., a review of the literature.* US Public Health Service Publication, no. 548. United States Public Health Service, Washington, DC.

Ault, S.K. and Ault, K.O. (1979). Spider predation (Araneae, Salticidae) on *Ornithodoros coriaceus* (Acarina, Argasidae), with a survey of the predators of the genus *Ornithodoros. Journal of Medical Entomology*, **15**, 570–1.

Baerg, W.J. (1944). Ticks and other parasites attacking Northern cliff swallows. *Auk*, **61**, 413–14.

Belozerov, V.N. (1982). Diapause and biological rhythms in ticks. In *Physiology of ticks*, (ed. F.D. Obenchain and R. Galun), pp. 469–500. Pergamon Press, Oxford.

Brown, C.R. and Brown, M.B. (1986). Ectoparasites as a cost of coloniality in cliff swallows (*Hirundo pyrrhonota*). *Ecology*, **67**, 1206–18.

Bull, C.M. (1986). Ticks on reptiles, a difference in scale. In *Morphology, physiology, and behavioural biology of ticks* (ed. J.R. Sauer and J.A. Hair), pp. 391–405. Wiley and Sons, New York.

Burns, E.C. and Melancon, D.G. (1977). Effect of imported fire ant (Hymenoptera: Formicidae) invasion on lone star tick (Acarina: Ixodidae) populations. *Journal of Medical Entomology*, **14**, 247–9.

Carroll, C.R. and Janzen, D.H. (1973). Ecology of foraging ants. *Annual Review of Ecology and Systematics*, **4**, 231–57.

Chastel, C. (1980). Arbovirus transmis par des tiques et associés à des oiseaux de mer; une revue generale. *Médecine Tropicale*, **40**, 535–48.

Clark, L. and Mason, J.R. (1985). Use of nest material as insecticidal and anti-pathogenic agents by the European starling. *Oecologia*, **67**, 169–76.

Clifford, C.M. (1979). Tick-borne viruses of seabirds. In *Arctic and tropical arboviruses* (ed. E. Kurstack), pp. 83–100. Academic Press, New York.

Clifford, C.M., Hoogstraal, H., Radovsky, F.J. Stiller, D., and Keirans, J.E. (1980). *Ornithodoros* (*Alectorobius*) *amblus* (Acarina: Ixoidea: Argasidae): identity, marine bird and human hosts, virus infections and distribution in Peru. *Journal of Parasitology*, **66**, 312–23.

Converse, J.D., Hoogstraal, H., Moussa, M.I., Feare, C.J., and Kaiser, M.N. (1975). Soldado virus (Hughes group) from *Ornithodoros* (*Alectorobius*) *capensis* (Ixodoidea: Argasidae) infesting sooty tern colonies in the Seychelles, Indian Ocean. *American Journal of Tropical Medicine and Hygiene*, **24**, 1010–18.

Crofton, H.D. (1981). A model of host–parasite relationships. *Parasitology*, **63**, 343–64.

Darwin, C. (1852). *Journal of researches into the natural history and geology of the countries visited during the voyage of H.M.S. Beagle around the world, under the command of Capt. Fitz Roy, R.N.* (2nd edn). John Murray, London.

Duffy, D.C. (1977). Breeding populations of terns and skimmers on Long Island Sound and eastern Long Island: 1972–1975. *Proceedings of the Linnaean Society of New York*, **73**, 1–48.

Duffy, D.C. (1983*a*). The ecology of tick parasitism on densely nesting Peruvian seabirds. *Ecology*, **64**, 110–19.

Duffy, D.C. (1983*b*). Competition for nesting space among Peruvian guano birds. *Auk*, **100**, 680–8.

Duffy, D.C. and Campos de Duffy, M.J. (1986). Tick parasitism at nesting colonies of blue-footed boobies in Peru and Galapagos. *Condor*, **88**, 242–4.

Duffy, D.C. and Daturi, A. (1987). Diel rhythms of tick parasitism on incubating African penguins. *Medical and Veterinary Entomology*, **1**, 103–6.

Duffy, D.C., Berruti, A., Randall, R.M., and Cooper, J. (1984). Effects of the 1982–1983 warm water event in the breeding of South African seabirds. *South African Journal of Science*, **80**, 65–9.

Edwards, A. and Lubbock, R. (1983). The ecology of Saint Paul's Rocks (Equatorial Atlantic). *Journal of Zoology (London)*, **200**, 51–69.

Edwards, A.J., Wilson, K., and Lubbock, H.R. (1981). The sea-bird populations of Saint Paul's Rocks. *Ibis*, **123**, 233–8.

Emlen, S.T. and Demong, N.J. (1975). Adaptive significance of sychronized breeding in a colonial bird: a new hypothesis. *Science*, **188**, 1029–31.

Feare, C.J. (1976). Desertion and abnormal development in a colony of sooty terns infested by virus-infected ticks. *Ibis*, **118**, 112–15.

Feare, C.J. (1979). Ecology of Bird Island, Seychelles. *Atoll Research Bulletin*, **226**, 1–7.

Feare, C.J. and Feare, D.C. (1984). Seabird ecology and tick distribution in the western Indian Ocean. *National Geographic Research Report, 1976 Projects*, pp. 341–58.

Finnegan, R.J. (1969). Assessing predation by ants on insects. *Insectes Sociaux*, **16**, 61–5.

Harris, M.P. (1965). Puffinosis among Manx Shearwaters on Skokholm. *British Birds*, **58**: 426–33.

Heatwole, H. (1971). Marine-dependent terrestrial biotic communities on some cays in the Coral Sea. *Ecology*, **52**, 363–6.

Heatwole, H., Done, T., and Cameron, E. (1981). *Community ecology of a coral cay*. Dr. W. Junk, The Hague.

Hemsley, W.B. (1884). Report on the botany of the Bermudas and various other islands of the Atlantic and Southern Oceans. Second part. *Reports of the Scientific Results of the Exploratory Voyage of the Challenger (Botany)* **1**(3), 1–229.

Hoogland, J.L. and Sherman, P.W. (1976). Advantages and disadvantages of bank swallow (*Riparia riparia*) coloniality. *Ecological Monographs*, **46**, 33–58.

Hoogstraal, H. (1978). Biology of ticks. In *Tick-borne diseases and their vectors* (ed. J.K.H. Wilde), pp. 3–14. Proceedings of International Congress, Edinburgh, Scotland, September–October 1976. University of Edinburgh, Centre for Tropical Veterinary Medicine, Edinburgh.

Hoogstraal, H. (1985). Argasid and Nutalliellid ticks as parasites and vectors. *Advances in Parasitology*, **24**, 135–238.

Hoogstraal, H. and Aeschlimann, A. (1982). Tick-host specificity. *Mitteilungen der Schweizerischen Entomologischen Gesellschaft*, **55**, 5–32.

Hoogstraal, H. and Feare, C.J. (1984). Ticks and tickborne viruses. In *Biogeography and ecology of the Seychelles Islands* (ed. D.R. Stoddart), pp. 267–80. Dr. W. Junk, The Hague.

Hoogstraal, H. and Gallagher, M.D. (1982). Blisters, pruritis and fever after bites by the Arabian tick *Ornithodoros (Alectorobius) muesebecki*. *Lancet* **ii**, 288–9.

Hoogstraal, H., Wassef, H.Y., Hays, C., and Keirans, J.E. (1985). *Ornithodoros (Alectorobius) spheniscus* n. sp. (Acarina: Ixoidea: Argasidae: *Ornithodoros (Alectorobius) capensis* group), a tick parasite of the Humboldt Penguin in Peru. *Journal of Parasitology*, **71**, 635–44.

Kajak, A., Breymeyer, A., Pectal, J., and Olechowicz, E. (1972). The influence of ants on the meadow vertebrates. *Ekologia Polska*, **20**, 163–71.

King, K.A., Keith, J.O., Mitchell, C.A., and Keirans, J.E. (1977*a*). Ticks as a factor in nest desertions of California brown pelicans. *Condor*, **79**, 507–9.

King, K.A., Blankinship, D.R., Paul, R.T., and Rice, R.C.A. (1977*b*). Ticks as a factor in the 1975 nesting failure of Texas brown pelicans. *Wilson Bulletin*, **89**, 157–8.

Kroll, J.C., Arnold, K.A., and Gotie, R.F. (1973). An observation of predation by native fire ants on nestling barn swallows. *Wilson Bulletin*, **85**, 478–9.

Lack, D. (1967). Interrelationships in breeding adaptations as shown by marine birds. In *Proceedings of the XIV International Ornithological Congress*, pp. 3–42. Blackwell, Oxford.

La Cock, G.D. (1986). The Southern Oscillation, environmental anomalies, and mortality of two southern African seabirds. *Climatic Change*, **8**, 173–84.

La Cock, G.D., Duffy, D.C., and Cooper, J. (1987). Population dynamics of the African penguin *Spheniscus demersus* at Marcus Island in the Benguela upwelling ecosystem: 1979–85. *Biological Conservation*, **40**, 117–26.

LaValle, J.A. (1923). La destrucion de las garrapatas en las islas guaneras por sus enimigos naturales. *Memorias de la Compania Administradora del Guano*, **14**, 165–70.

Marshall, A.G. (1981). *The ecology of ectoparasitic insects*. Academic Press, New York.

May, R.M. (1977). Dynamic aspects of host–parasite associations: Crofton's model revisited. *Parasitology*, **75**, 259–76.

Møller, A.P. (1987). Nest lining in relation to the nesting cycle in the swallow *Hirundo rustica*. *Ornis Scandinavica*, **18**, 148–9.

Morgan, I.R., Westbury, H.A., and Campbell, J. (1985). Viral infections of little blue penguins (*Eudyptula minor*) along the south coast of Australia. *Journal of Wildlife Diseases*, **21**, 193–8.

Murphy, R.C. (1921). Los invertebrados terrestres de las islas guaneras del Peru. *Memorias de la Compania Administradora del Guano*, **12**, 95–116.

Nelson, J.B. (1987). *The Sulidae: gannets and boobies*. Oxford University Press, Oxford.

Nisbet, I.C.T. (1973). Terns in Massachusetts: present numbers and historical changes. *Bird-Banding*, **44**, 27–55.

Nisbet, I.C.T. and Welton, M.J. (1984). Seasonal variations in breeding success of common terns: consequences of predation. *Condor*, **86**, 53–60.

Nutall, P.A. (1984). Tick-borne viruses in seabird colonies. *Seabird*, **7**, 31–41.

Pectal, J. (1978). The role of ants in ecosystems. In *Production ecology of ants and termites* (ed. M.V. Brian), pp. 293–325. Cambridge University Press, Cambridge.

Pisarski, B. (1978). Comparison of various biomes. In *Production ecology of ants and termites* (ed. M.V. Brian), pp. 326–40. Cambridge University Press, Cambridge.

Rand, R.W. (1960). The biology of guano-producing seabirds: 3. The distribution, abundance, and feeding habits of the cormorants Phalacrocoracidae off the south west coast of the Cape Province. *Investigational Report of the Division of Fisheries of the Union of South Africa*, **42**, 1–32.

Randolph, S.E. (1979). Population regulation in ticks: the role of acquired resistance in natural and unnatural hosts. *Parasitology*, **79**, 141–56.

Risch, S.J. and Carroll, C.R. (1982). Effect of a keystone predaceous ant, *Solenopsis geminata*, on arthropods in a tropical agroecosystem. *Ecology*, **63**, 1979–83.

Rothschild, M. and Clay, T. (1952). *Fleas, flukes, and cuckoos*. Collins, London.

Sikes, P.J. and Arnold, K.A. (1986). Red imported fire ant (*Solenopsis invicta*) predation on cliff swallows (*Hirundo pyrrhonota*) nestlings in east-central Texas. *Southwestern Naturalist*, **31**, 105–6.

Simmons, K.E.L. (1985). Anting. In *A dictionary of birds* (ed. E. Lack), p. 19. Buteo Books, Vermillion, South Dakota.

Skutch, A.F. (1976). *Parent birds and their young*. University of Texas, Austin.

Southwood, T.R.E. (1978). *Ecological methods*. Chapman and Hall, London.

Stoddart, D.R. and Fosberg, F.R. (1981). Bird and Denis Islands, Seychelles. *Atoll Research Bulletin*, **252**, 1–54.

Stradling, D.J. (1978). Food and feeding habits of ants. In *Production ecology of ants and termites* (ed. M.V. Brian), pp. 81–106. Cambridge University Press, Cambridge.

Sudd, J.H. (1967). *An introduction to the behaviour of ants*. Edward Arnold, London.

Texas Colonial Waterbird Society (1982). An atlas and census of Texas waterbird colonies, 1973–1989. Kleberg Wildlife Research Institute, Kingsville.

Trager, W. (1939). Acquired immunity to ticks. *Journal of Parasitology*, **25**, 57–81.

Vogt, W. (1939). Las lagartijas y las aves guaneras. *Boletín de la Companía Administradora del Guano*, **15**, 346–8.

Wimberger, P.H. (1984). The use of green plant material in bird nests to avoid ectoparasites. *Auk*, **101**, 615–18.

Wittenberger, J.F. and Hunt, G.L. Jr (1985). The adaptive significance of coloniality in birds. In *Avian biology*, Vol. VIII, (ed. D.S. Farner, J.R. King, and K.C. Parkes), pp. 1–78. Academic Press, New York.

14 *Coevolution of avian grooming and ectoparasite avoidance*

DALE H. CLAYTON

Introduction

Ectoparasites severely reduce avian fitness when they occur in large numbers or when they serve as intermediate hosts for avian pathogens. In most cases, however, ectoparasites occur in small populations with little or no effect on the host. These small populations may be the result of host–parasite coevolution. If a parasite exerts selection on its host and the host exerts reciprocal selection on the parasite, repeated bouts of coevolutionary change can lead to coadaptation of host defence and parasite resistance. Although host defence can prevent the buildup of large parasite loads, energetic constraints will probably prohibit the evolution of perfect defence against parasites with coadaptations for resistance. Persistence of small loads on birds may therefore reflect a coevolutionary balance between defence and resistance.

Coevolution is not the only explanation for the persistence of small loads, however. If small loads have no effect on the host, they may persist simply because a bird has nothing to gain by removing them. Coevolution cannot be assumed to occur in a given system, but must be empirically inferred, which entails a three-step process. First, the adaptive function of presumably coevolved traits must be documented. Second, reciprocal selection on those traits must be measured. Third, the opportunity for coevolutionary response to reciprocal selection must be demonstrated. In this chapter I report the results of three experiments designed to address the first step—documentation of the adaptive functions of avian grooming and ectoparasite avoidance.

Coevolution of birds and ectoparasites: background

> Adaptation. Process of evolutionary modification which results in improved survival and reproductive efficiency.
>
> (Lincoln *et al.* 1982)

Coevolution is evolutionary change in two or more species resulting from reciprocal selection exerted by those species on one another (Janzen

1980; Thompson 1982, 1989; Futuyma and Slatkin 1983; Nitecki 1983). Persistant bouts of coevolution can lead to coadaptation of species living in close ecological association. Patterns of apparent coadaptation are widespread among competitors, mutualists, predators and prey, and hosts and parasites (*sensu* Price 1980). The reciprocity of coevolution makes it a unique process that can lead to complex evolutionary outcomes, such as the elaborate patterns of chemical coadaptation observed among plants and their herbivores (Spencer 1988).

Host–parasite interactions, which often show circumstantial evidence of coadaptation, are ideal testing grounds for coevolution because they include some of the most intimate associations known (Price 1980; Rollinson and Anderson 1985). In a comprehensive overview of the evolution of vertebrate–insect associations, Waage (1979) hypothesized 'extensive reciprocal coevolution' between vertebrates and their ectoparasites. Waage cited examples of presumed coadaptation between birds and ectoparasites, noting that the coevolution of these and other vertebrate–insect systems was unexplored relative to the expanding literature on plant–herbivore coevolution. Today, more than a decade after Waage's review, the situation remains the same; almost no research has been conducted on the coevolution of birds and their ectoparasites.

Birds are host to a variety of ectoparasites, including mites and ticks (Acarina), flies (Diptera), bugs (Hemiptera), chewing lice (Mallophaga), and fleas (Siphonaptera). Members of all five orders have the potential to severely reduce avian fitness when present in large numbers (Boyd 1951*a*; Moss and Camin 1970; Duffy 1983; Brown and Brown 1986; Shields and Crook 1987; Clayton 1989). However, most individual birds are host to small, relatively innocuous loads of ectoparasites (Rothschild and Clay 1952; Ash 1960; Baum 1968; Marshall 1981). These small loads appear to be kept in check by the birds themselves through a variety of defence mechanisms ranging from behavioural responses (Murray 1990) to immune reactions (Matthysse *et al.* 1974).

The most obvious defence against ectoparasites is grooming behaviour (Nelson *et al.* 1977; Waage 1979; Murray 1990). The predominant form of avian grooming is preening, which may be operationally defined as manipulation of the plumage with the bill. The other major form of grooming is foot scratching, which appears to control ectoparasites on the head and other regions inaccessible to preening (Brooke 1985). The precise grooming behaviour of birds is partly a function of morphology. For example, species with long, unwieldy bills scratch more than species with short bills, as illustrated by a significant correlation between the ratio of bill to body size, and the ratio of scratching to preening among families of neotropical birds (Clayton, in preparation).

Ectoparasites show apparent morphological and behavioural adaptations for resistance to host grooming. These include: (1) small size and flattened shape; (2) heavily sclerotized integuments; (3) spines, ctenidia, and numerous setae; (4) strong mouthparts and grasping claws; and (5) avoidance behaviour (Marshall 1981). For the purposes of this chapter, 'avoidance' refers to any morphological or behavioural trait that facilitates resistance to grooming. The microhabitat distributions of some ectoparasites also appear to reflect grooming-imposed selection (Waage 1979). This is best documented for species of 'chewing lice' (Insecta: Mallophaga), which tend to be more or less restricted to the wings, head, or abdomen of their hosts (Dubinin 1947; Clay 1949, 1957). Species on the wings are often elongate, compressed forms capable of flattening themselves against the surface of feathers, or inserting themselves between the barbs of feathers. Species on the head and neck, where they are protected from preening, are typically round-bodied, sluggish forms with no apparent adaptations for avoidance. Species on the abdomen tend to be intermediate in form and behaviour (Marshall 1981).

Observational data such as these suggest that avian grooming and ectoparasite avoidance are coadapted traits, but this is not necessarily the case. The elongate morphology and insertion behaviour of wing lice, for example, may not be adapted for the avoidance of preening, but to prevent lice from being blown off the host's feathers during flight (Stenram 1956). Moreover, traits may not be adaptive at all, but neutral with respect to fitness. As stated by Futuyma (1986, p. 283), 'Adaptation is an onerous concept, and the adaptive value of a trait should be demonstrated rather than assumed, for numerous factors other than adaptation can influence the evolution of a trait.' In short, if a trait's effect on fitness has not been measured, adaptation can be addressed only in a speculative fashion (Williams 1966; Gould and Lewontin 1979).

This chapter presents the results of a series of three experiments designed to explore the adaptive functions of host preening and ectoparasite avoidance. The goal of the first two experiments was to measure the relationship of preening behaviour to host fitness by determining: (1) the extent to which preening controls ectoparasite populations on the host (experiment I); and (2) the impact of ectoparasite populations on a probable component of host fitness (experiment II). The goal of the third experiment was to measure the relationship of avoidance behaviour to ectoparasite fitness by comparing the impact of preening on species of ectoparasites with different avoidance responses. The experimental system chosen for study was the rock dove, *Columba livia*, or feral pigeon, and two species of chewing lice: *Columbicola columbae* (Linnaeus 1758), an elongate, agile species found primarily on the wings, and

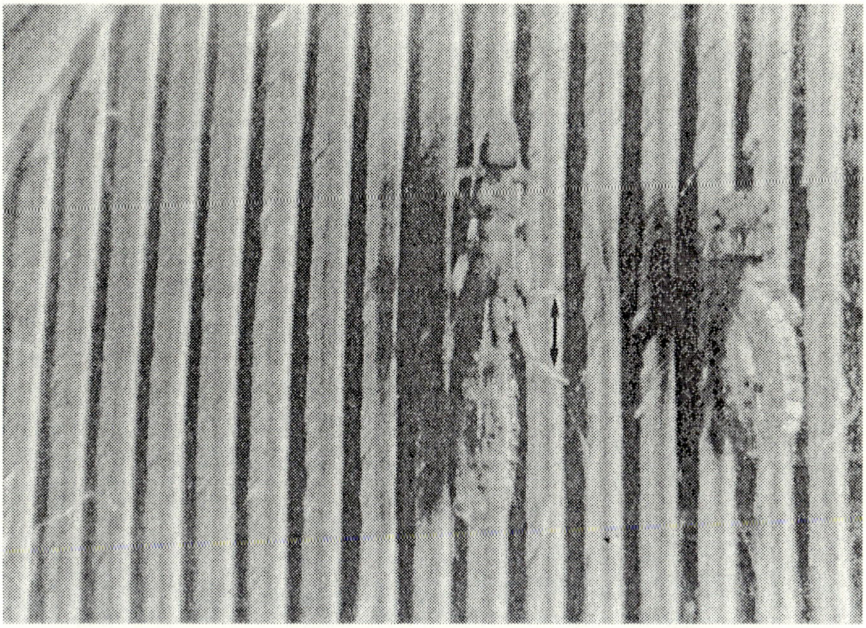

Fig. 14.1. *Columbicola columbae* (left, 2.5 mm long) and *Campanulotes bidentatus* (1.5 mm) on the vane of a flight feather. *C. columbae* uses its large posterior legs (arrows) to orient and run across the ventral surfaces of flight feathers. *C. bidentatus*, which cannot manoeuvre on flight feathers because of its small legs (not visible), is normally restricted to abdominal contour feathers. (Photo: J. Barabe.)

Campanulotes bidentatus (Burmeister 1838), a slow moving species found primarily on the abdomen (Fig. 14.1; Nelson and Murray 1971).

Rock doves and chewing lice: natural history

The rock dove was chosen for this research because it is plentiful, accessible, and adjusts well to captivity and the techniques described below. Furthermore, the composition of its parasite fauna is well known (Levi 1957; Rand 1959; Brown 1971). *C. livia* was domesticated around 4500 BC (Zeuner 1963), and introduced to North America in 1606 (Schorger 1952). North American rock doves retain considerable morphological and behavioural similarity to ancestral populations of *C. livia* in Morocco and other regions (Goodwin 1983). Although urban pigeons habituate to people, rural pigeons are as timid as other wild species of birds. The rock doves in this study were rural-trapped individuals or their offspring.

Columbicola columbae and *Campanulotes bidentatus* are members of the largest Mallophagan suborder, Ischnocera, which contains 2954 species, 2619 (88.7 per cent) of which parasitize birds; the rest are parasites of mammals (Price, Hellenthal, and Emerson, in preparation). Most species of Ischnocera are restricted to a single species or genus of host (e.g. Price and Clayton 1983; Clayton 1990*a*). This extreme level of host specificity makes Ischnocera an attractive group for coevolutionary analysis because specificity implies long-term association, which reduces the likelihood that current patterns of coadaptation are the result of transient past associations. *Columbicola* and *Campanulotes* are genera found only on members of Columbiformes (Keler 1939; Tendeiro 1965). *C. columbae* and *C. bidentatus* are specific to the rock dove, and are the only species of Ischnocera typically found on this host in North America (Emerson 1972). The Ischnoceran *Physconelloides zenaidurae* '... has occasionally been collected from the domestic pigeon', but its true host is the mourning dove, *Zenaidura macroura* (Emerson 1960).

Avian Ischnocera are obligate parasites that complete their entire life cycle on the host (Marshall 1981). They are dependent for reproduction on the warmth of the bird's body which they do not leave except to pass from one individual to another during periods of direct contact, e.g. between parents and offspring (except in cases of phoresy on Hippoboscid flies or other insects—see Keirans 1975). The life cycle, which requires 3 to 4 weeks, includes a nit (egg), three nymphal instars, and the adult (Martin 1934). Nits are glued to the host's feathers with a glandular cement, often in positions more or less protected from grooming, such as the gulum (throat). *C. columbae* and *C. bidentatus* are so highly specialized for locomotion on feathers that they rarely if ever venture on to the host's skin (Stenram 1956; Clayton 1990*b*). They also feed exclusively on the barbules of feathers (Fig. 14.2; Nelson and Murray 1971), which they metabolize with the aid of symbiotic bacteria (Eichler *et al.* 1972; Marshall 1981).

No published test exists of the direct impact of Ischnocera on host fitness. Derylo (1974*a,b*) reported a reduction in the egg production of leghorn chickens parasitized by Ischnocera, but the birds in Derylo's study were also host to Amblycera, a smaller suborder of Mallophaga that lives and feeds on the skin and blood of the host in addition to feathers. Amblycera are known to promote dermatitis and scratching, and are responsible for up to a 46 per cent reduction in the egg production of poultry (DeVaney 1976; Nelson *et al.* 1977). Amblycera are also known to vector avian endoparasites and pathogens (reviewed by Clayton 1990*b*). To date, Ischnocera are not demonstrated vectors of avian endoparasites or pathogens.

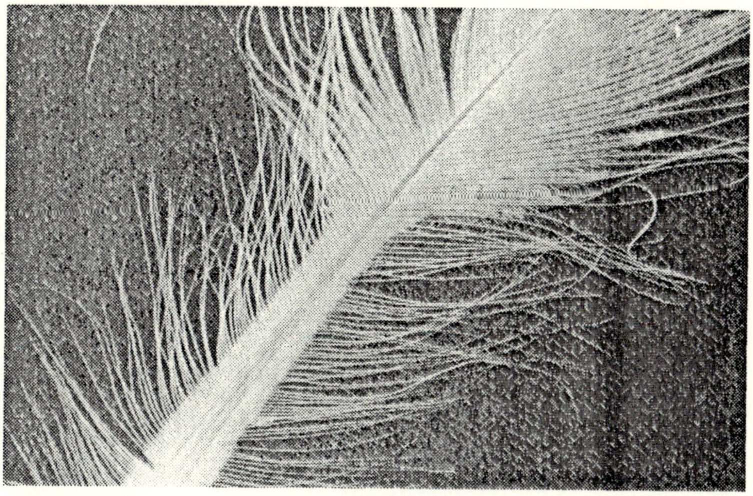

Fig. 14.2. Rock dove feather showing damage from lice. The barbules of the basal and medial regions have been consumed. The barbs and shaft are never eaten, possibly because they are too large to ingest (Clay 1950). *C. columbae* and *C. bidentatus* feed only on abdominal contour feathers. (Photo: B. Edinger.)

Experiment I: role of host preening in louse control

Observational data and previous experiments

Birds with deformed beaks or feet often have large ectoparasite loads, apparently because of their inability to groom properly (Boyd 1951*b*; Rothschild and Clay 1952; Ilienko 1959; Ash 1960; Pomeroy 1962; Baum 1968; Ledger 1969; Marshall 1981). For example, I obtained a female house sparrow, *Passer domesticus*, that was missing the upper mandible of its bill, and which was host to over 1200 lice (*Brüellia subtilis*) and over 400 mites (mostly *Proctophyllodes truncatus*). By comparison, 10 normal house sparrows (five males, five females) collected in the same county 1 year later (Ramsey County, Minnesota; May 1982) were host to a mean of 20 lice (range, 0–56) and 38 mites (range, 0–128). In a year-round study of 300 house sparrows in north-western Indiana, McGroarty and Dobson (1974) reported mean loads of 19 *B. subtilis* (range, 0–146) and 62 *P. truncatus* (range, 1–1204). Thus, the louse and mite loads of the deformed sparrow exceeded those of normal birds by more than an order of magnitude!

Birds with deformed or missing feet sometimes have elevated populations of ectoparasites that are restricted to the head and upper body, i.e. those regions that must be groomed with the feet. For example, two one-legged sanderlings, *Calidris alba*, at Bodega Bay, California had elevated numbers of nits on their heads and upper breasts, compared to 78 normal sanderlings handled at the same site (Myers, personal communication). A one-legged ruddy turnstone, *Arenaria interpres*, in Jamaica also had an unusually large number of lice and nits on the feathers of its head and neck (Tordoff, personal communication). These and other observations suggest that grooming plays an important role in ectoparasite resistance, but they do not control for the possibility that increased loads are due to the poorer general condition of deformed hosts. Proper nutrition and health are critical for ectoparasite resistance (Nelson *et al.* 1975; Nelson 1984).

Ectoparasite increases also have been documented on birds with experimentally impaired preening (Kartman 1949; Nelson and Murray 1971; Brown 1972; DeVaney 1976). In all cases, however, birds were impaired by 'debeaking', viz. removing the distal end of the upper mandible of the bill, which results in a dramatic increase in louse population. Although such studies convincingly demonstrate that preening controls lice, they do not factor out the impact of debeaking-induced stress on louse loads. Stress is known to reduce general resistance to parasites (Esch *et al.* 1975), including ectoparasites (Nelson 1962; Lodmell *et al.* 1970). The goal of experiment I was to measure louse increases on hosts impaired with a less stressful technique, both under natural conditions and captive conditions that maximize the transmission and reproductive potential of lice.

Methods

I impaired the preening ability of pigeons with metal 'bits': small (<0.8 g), C-shaped pieces of plated steel inserted between the mandibles (Fig. 14.3). The bits were crimped slightly in the nostrils to prevent dislodging, but not so far as to pierce the tissue. Bits create a 1.0–3.0 mm gap between the mandibles, preventing the full occlusion of the bill necessary for efficient preening. Birds typically shake their heads for several minutes immediately after receiving bits, but ignore them thereafter. Bits have no detectable effect on the general condition or reproductive performance of rock doves. During a pilot study of captive, louse-free pigeons, the weights and reproductive success of bitted adult pairs did not differ significantly from those of non-bitted pairs (Clayton 1989). Because rock doves feed their young by regurgitation rather than by the forceps-action of the bill typical of most birds, bits do not interfere with the rearing of offspring.

During March–May 1981 I captured 35 wild rock doves from a population of several hundred individuals residing in an abandoned grain elevator in St. Paul, Minnesota. The birds were banded and

Fig. 14.3. Mandibular displacement of a bitted host.

weighed, then placed in a 2.5 m × 1.5 m × 2.5 m outdoor enclosure constructed of wood and wire mesh, where they were provided with a limited daily supply of cracked corn and *ad libitum* grit and vitamin-supplemented water. The enclosure exposed birds to ambient temperature and photoperiod, but protected them from wind and precipitation. In May, 28 birds were randomly fitted with bits, and seven were designated non-bitted controls. One of the 28 bitted individuals served as a fortuitous control because it was fitted with a bit too thin to create a mandibular gap. In August, after approximately 15 weeks, nine of the captive birds were weighed, killed, and frozen for later censusing of their total louse loads. The nine birds consisted of six bitted and two non-bitted birds selected at random, and the 'control-bitted' bird selected intentionally.

During May–June 1981, I captured, banded and weighed another 41 wild rock doves from the same population. To estimate natural loads, five birds were selected at random, killed, and frozen for louse censuses. The 36 remaining birds were released at the capture site after randomly fitting 23 with bits, and leaving 13 as controls. In September, after approximately 15 weeks, eight birds from the wild population were shot and frozen for louse censuses. Of the eight, two were bitted birds and six were previously uncaptured birds. The mandibles of one of the latter were crossed, a common congenital deformity (Pomeroy 1962).

To isolate the louse populations of frozen birds for censusing total load, I used a modified version of the KOH dissolution technique (Cook 1954; Kim 1972). After thawing, the entire plumage of each bird was plucked carefully into a beaker, then dissolved in 15 per cent (by weight) KOH solution heated over a Bunsen burner for 15 min. After dissolution, the hot solution was filtered through an 80 mesh stainless steel screen. Undissolved solids were then incubated at 37°C for 24 h in a solution of 3 per cent trypsin (4 × U.S.P. pancreatin), buffered to pH 7.8–8.3 with 0.2 M Na_2 PO_4. This procedure dissolved the host's feathers and attached tissue, but not the chitinous exoskeletons of lice, which were filtered with the screen, stained in acid fuchsin, and counted under a dissecting microscope. Mean recovery of the technique was 95 per cent (range, 91–100 per cent), based on four blind tests with predetermined numbers of adult and nymphal lice (range, 50–520) added to clean feathers. Mean recovery was only 82 per cent (76–93 per cent) in four additional tests using only nymphs (37–89) on clean feathers. Thus, reliability of the technique was less for nymphs than adults, presumably because some nymphs penetrated the screen. I found no indication that nymphal exoskeletons are dissolved by the technique, however, as suggested by Ash (1960).

For statistical analyses louse census data were log transformed to meet the assumptions (homoscedasticity) of multivariate (MANOVA) and univariate (ANOVA) analyses of variance, which were performed using the general linear model procedure (GLM) of the Statistical Analysis Software package (SAS Institute 1985). June and September natural loads were compared non-parametrically (Siegel 1956) because of significantly different variances, despite transformation. Host weight data were analysed non-parametrically for the same reason.

Results and discussion

The *Campanulotes bidentatus* loads of wild, non-impaired rock doves did not change substantially over the summer (Fig. 14.4). The *C. bidentatus* loads of June birds were not significantly different from those of September birds (Mann–Whitney $U=9$, $p=0.55$, two-tailed; Table 14.1). *Columbicola columbae* loads, however, showed a highly significant increase on the same birds over the course of the summer ($U=0$, $p=0.009$). This seasonal increase was similar to those reported for other species of avian lice (reviewed in Marshall 1981).

Loads of both *C. bidentatus* and *C. columbae* increased dramatically on the impaired, wild rock doves (one deformed, two bitted) (Fig. 14.4), compared to the non-impaired, wild birds. Both species of lice also increased on the impaired, captive birds (six bitted), compared to the non-impaired captives (two non-bitted, one control-bitted). The effect of host preening ability on each species of louse, for both wild and captive

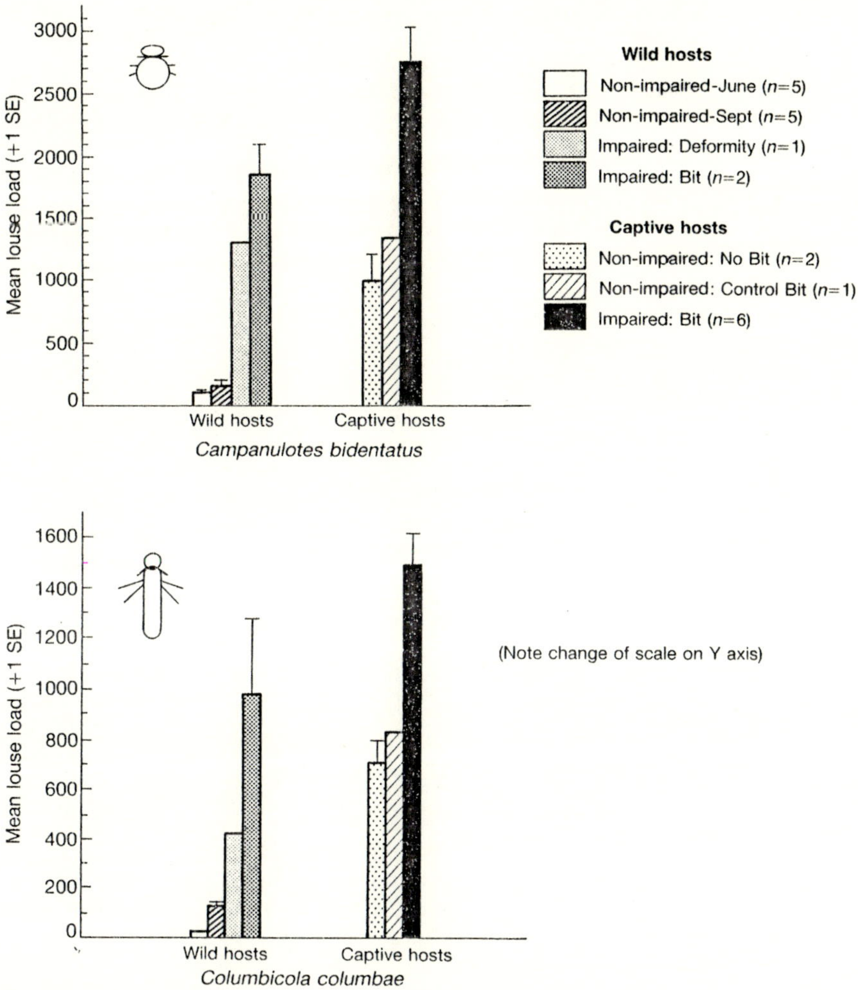

Fig. 14.4. Louse loads of experiment I rock doves.

hosts, was highly significant (ANOVA, $p=0.0001$; Tables 14.2 and 14.3).

Likewise, a highly significant overall effect of preening on louse load occurred (MANOVA, Wilks' Criterion $F=49.7$, $p=0.0001$, $df=2,12$). The preening-related increase in the *C. columbae* loads of wild birds was nearly an order of magnitude greater than the seasonal increase in *C. columbae*.

Wild pigeons had fewer lice than captive pigeons (Fig. 14.4): impaired wild birds were host to smaller loads than impaired captives, and non-impaired wild birds were host to much smaller loads than non-impaired

Table 14.1. Mean louse loads and weights of experiment I rock doves (ranges)

Host location & preening ability (*N*)	*Campanulotes bidentatus*	*Columbicola columbae*	Host weight (to nearest 5 g) (g)	
			May/June	Aug/Sept
Wild: non-impaired				
June (5)	94 (34–130)	21 (5–39)	336 (280–450)	—
September (5)	148 (44–287)	133 (91–168)	—	430 (390–490)
Wild: impaired				
Deformed (1)	1307	475	—	360
Bitted (2)	1847 (1595–2100)	975 (671–1280)	355 (320–390)	400 (350–450)
Captive: non-impaired				
Non-bitted (2)	995 (792–1198)	711 (624–799)	405 (400–410)	360 (300–420)
Control-bitted (1)	1336	832	410	370
Captive: impaired				
Bitted (6)	2749 (1619–3666)	1488 (963–1810)	383 (360–400)	345 (330–370)

captives. The effect of host location on both species of lice was highly significant (ANOVA, $p = 0.0001$; Tables 14.2 and 14.3), and a highly significant overall effect of location on louse load occurred (MANOVA, $F = 41.8$, $p = 0.0001$, $df = 2,12$).

Table 14.2. Analysis of variance, testing the relationship of *Campanulotes bidentatus* load to preening ability (impaired versus non-impaired) and host location (wild versus captive)*

Source of variation	Degrees of freedom	Sums of squares†	Mean square	*F*
Preening ability	1	2.24068	2.24068	54.90‡
Host location	1	1.30846	1.30846	32.06‡
Preening by location interaction	1	0.52765	0.52765	12.93¶
Model	3	5.33031	1.77677	43.54‡
Error	13	0.53054	0.04081	
Total	16	5.86085		

* Log transformed data.
† Type IV sum of squares, reflecting adjustment for other factors in the model.
‡ $p < 0.0001$.
¶ $p < 0.01$.

Table 14.3. Analysis of variance, testing the relationship of *Columbicola columbae* load to preening ability (impaired versus non-impaired) and host location (wild versus captive)*

Source of variation	Degrees of freedom	Sums of squares†	Mean square	*F*
Preening ability	1	1.02985	1.02985	59.02‡
Host location	1	1.10651	1.10651	63.41‡
Preening by location interaction	1	0.19650	0.19650	11.26¶
Model	3	3.13229	1.04410	59.84‡
Error	13	0.22684	0.01745	
Total	16	3.35913		

* Log transformed data.
† Type IV sum of squares, reflecting adjustment for other factors in the model.
‡ $p<0.0001$.
¶ $p<0.01$.

The larger loads of captive birds probably resulted from several factors. First, the small size and design of the enclosure forced the captive birds to roost side by side in direct contact, probably facilitating louse increases by increasing transmission rates. As a rule, louse populations are larger on birds kept in close quarters (Blagoveshchenskii 1959). Captivity also reduced the efficacy of preening, as indicated by the significant interaction of location and preening ability in the ANOVA for each species of louse ($p<0.01$; Tables 14.2 and 14.3), as well as the significant overall interaction between location and preening demonstrated by the MANOVA ($F=10.5$, $p=0.0023$). This reduction in preening efficiency may have resulted from the poorer general condition of captive birds, whose weights decreased significantly over the course of the experiment (Wilcoxon $T=1$, $p<0.01$, two-tailed, $N=6$ impaired, 3 non-impaired; Table 14.1). This decrease, attributable to the intentionally restricted diet (see methods), did not differ significantly between the impaired and non-impaired captives ($U=8.5$, $p>0.44$, one-tailed). Thus, reduced condition was not responsible for the differential increases of lice on impaired versus non-impaired captives.

Reduced general condition was not responsible for the differential increases of lice on the wild rock doves. Weights of the two bitted birds increased over the summer (Table 14.1), and September weights of the three impaired birds did not differ significantly from those of the five non-impaired birds ($U=3$, $p=0.18$, two-tailed). The frequency of recapture for the bitted birds (2 of 23) and non-bitted birds (0 of 13) did not differ significantly ($\chi^2=1.2$, $p=0.27$; in this case the null hypothesis was not rejected despite the increased probability of a Type I error due to expected frequencies <5; Zar 1984). Subsequent field experiments with bitted birds in Illinois verify that bits do not cause a reduction in general condition during the summer (Clayton 1989).

The louse loads of bitted wild birds did not exceed the maximum loads of wild rock doves in a field study of effects of lice on host reproduction and survival (Clayton 1989). Of 190 wild pigeons (without bits) censused in rural Illinois, six individuals (3.2 per cent) supported loads similar to those on the bitted wild birds in this experiment. The upper mandibles of four of these six wild birds were overgrown by 5–10 mm, and a fifth bird was missing the tip of its lower mandible. The remaining bird showed no obvious deformity. Thus, bit-induced increases mimic naturally occurring increases resulting from bill deformities, which are not uncommon (Rothschild and Clay 1952; Ash 1960; Pomeroy 1962; Ledger 1969).

This experiment demonstrates the important role of preening for louse control, both under natural conditions, and artificial conditions that maximize louse fitness. The effectiveness of preening depends on the conditions of a host's environment, so it is not the only factor influencing louse populations. Preening is clearly an important means of control, however, and may represent the first line of defence against lice and other ectoparasites.

Experiment II: effect and distribution of lice on the host

Defence against parasites is adaptive only if parasites have a negative effect on host fitness. The adaptive function of parasite defence cannot be assumed because many parasites act as commensals much of the time (Noble and Noble 1982). The chief goal of this experiment was to measure the effect of lice on the host, viz. the relationship of louse load to plumage quality. Two additional objectives of the experiment were (1) to determine the relative distributions of *C. bidentatus* and *C. columbae* on the host, and (2) to establish a statistical model for predicting the louse loads of live rock doves from visual counts of specific body regions. These objectives were prerequisites for experiment III (see below).

Methods

I captured rock doves under bridges spanning streams in northern Illinois, May–November 1986. The birds were caged individually or in small groups in the loft of a barn at the University of Illinois, Urbana–Champaign. They were provided with commercial grain (Kaytee Breeder and Conditioner Pigeon Feed #6™), grit, and water, and were exposed to

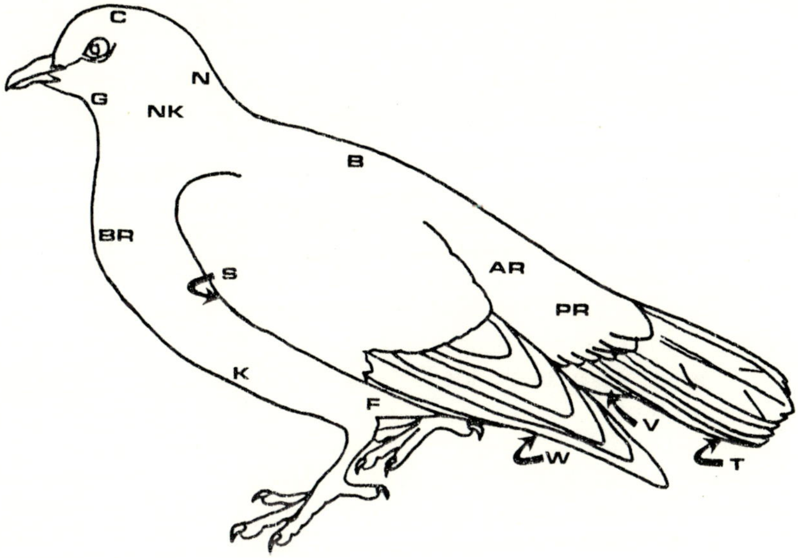

Fig. 14.5. Regions counted for lice and/or sampled for feathers. B, back; BR, breast; C, crown; F, flank*; G. gulum; K, adjacent to keel*; N, nape; NK, side of neck*; AR, anterior rump; PR, posterior rump; S, side* (under wing); T, tail (ventral surface of rectrices); V, ventral caudal tract; W, wing* (ventral surface of remiges). * Indicates counts restricted to one side of the host.

ambient temperature and photoperiod. At least several weeks prior to sampling (see below), all birds were isolated in small cages to prevent them from contacting other birds and to prevent them from flying.

I selected 26 individuals of varying plumage colour and pattern (13 males, 13 females) and bitted 18 of these at various intervals between July and December to create a range of louse loads. The birds were bitted for the following approximate durations: three individuals for 12 weeks, three for 14 weeks, two for 18 weeks, eight for 24 weeks, and two for 34 weeks. Eight individuals were not bitted. In December, 15 of the birds were counted for lice, sampled for feathers, and immediately killed and frozen for censuses of total load (see below). These procedures were repeated in February 1987 for the remaining 11 birds.

I estimated the louse loads of live birds by counting lice on each of the 14 regions illustrated in Fig. 14.5. Counting was performed within 1 m of a bright light, as follows. Each bird was partially immobilized with a Velcro™ strip wrapped once around its tarsi (lower legs). Its right wing was then extended, and lice on the ventral surface of the remiges were counted. The basal region of each remex was observed by first deflecting the underwing covert with forceps. Next, the tail was fanned open, and lice on the ventral surface of each rectrix were counted after deflecting the under-tail covert. A timed (30 s) count of each of the remaining regions was made in the following order: vent, keel, breast, gulum, neck, flank, side, crown, nape, back, anterior rump, and posterior rump (Fig. 14.5). The time restriction was necessary because, in contrast to flight feathers (remiges and retrices), it was impossible to examine thoroughly all of the feathers in the regions listed. An attempt was made to observe feathers from throughout each region during its allotted 30-s interval.

I estimated plumage damage by weighing samples of feathers pulled from seven regions: breast, flank, side, nape, back, anterior rump, and posterior rump. Ten feathers were selected haphazardly from each region, and weighed as a group to the nearest 0.001 g. The weights of all seven regions were then added to create a single index of feather weight for each bird.

To isolate the louse populations of frozen birds for censusing total load, I modified the procedure described in experiment I as follows. Birds were skinned in a 1 gallon plastic bucket and the skin with attached plumage was incubated for a mean of 40 h (range, 15–87 hours) in 1–3 per cent buffered trypsin solution. The mixture was then dissolved in 2.5–5 per cent KOH solution heated on a hot plate for up to 4 h. If undissolved feathers or tissue persisted, the procedure was repeated. Exoskeletons were filtered and rinsed with xylene, then 95 per cent ethanol, to eliminate fat adhering to the material (Lemke 1982). The exoskeletons were then stained in acetocarmine and counted. This protocol required one filtration, instead of two, cf. experiment I, and may have reduced the proportion of nymphs passing through the stainless steel screen. The total loads of 23 of the 26 birds were censused successfully using this procedure (the tissue of three birds did not dissolve satisfactorily).

During the regional counts, I observed another species of louse, *Hohorstiella lata* (suborder Amblycera), in large numbers (100s–1000s) on four of the 23 censused birds. Because *H. lata* exoskeletons could not be distinguished reliably from those of *Campanulotes bidentatus*, I omitted these four birds from further consideration, leaving a sample of 19 censused hosts.

Effect of lice on plumage

The relationship between louse load and feather damage was determined by regressing feather weight on total load, after square root transforming the census data to meet the assumptions (homoscedasticity) of least squares regression. Although feather damage accumulates until moult, louse loads fluctuate temporally (Clayton, unpublished data) owing to the depletion of limited food resources and other factors (see experiment III). For this reason, the analysis of load versus damage was restricted to a large group of birds incorporated into the experiment on the same day (nine bitted and four non-bitted individuals; 4 July 1986).

Distribution of lice

Regional distributions of *C. bidentatus* and *C. columbae* were compared using regional count data from the 17 hosts on which at least small numbers of both species of lice were observed.

Prediction of louse load

Regional counts were also used to generate two regression models for predicting the total loads of live birds. First, multiple regression of total census on regional counts was performed for each species of louse (all variables square root transformed; $N = 19$). Regions used as independent variables in the initial *C. bidentatus* regression were vent, keel, breast, neck, flank, side, nape, back, and rump (anterior and posterior rump combined). The wing, tail, gulum, and crown were not used because *C. bidentatus* was never observed on these regions.

All regional counts, except the tail, were included as independent variables in the initial *C. columbae* regression. Although numerous *C. columbae* were observed on the tail early in the experiment (autumn 1986), few were observed during winter counts (December 1986 and February 1987), presumably because *C. columbae* dispersed from the exposed tail (ambient temp = 0–9°C) to the more protected undersurface of the wings. Because of the extreme seasonal variation in the number of *C. columbae* on the tail the count for this region was subtracted from the total census before using the latter as the dependent variable in the regression. Ultimately, to estimate the total load of a live host, the tail count was added to the prediction of the *C. columbae* regression model (see below). This procedure hopefully reduced the model's sensitivity to changes in ambient temperature.

The final model for each species of louse was generated by subjecting its initial regression to a backward elimination procedure that omitted non-significant variables ($p > 0.01$; Neter *et al.* 1983). Because this procedure, like other stepwise approaches, does not test all possible combinations of independent variables, it does not necessarily generate the best model. The procedure is sufficient, however, for generating a comparative model to estimate the loads of birds subjected to different experimental treatments (see experiment III).

Results and discussion

The staggered bitting schedule produced the desired range of loads (Table 14.4). *C. bidentatus* and *C. columbae* loads did not differ significantly among the 19 censused hosts ($T = 90$, $p = 0.86$, two-tailed).

Table 14.4. Louse loads and feather weights of experiment II rock doves

Host trait measured (*N*)	*Campanulotes bidentatus* Mean	SD	Range	*Columbicola columbae* Mean	SD	Range	Combined species Mean	SD	Range
Total louse load (19)	441	570	5–1966	402	456	5–1565	843	800	21–2399
Feather weight (13)	277	256	7–773	394	413	5–1257	671	642	21–2030

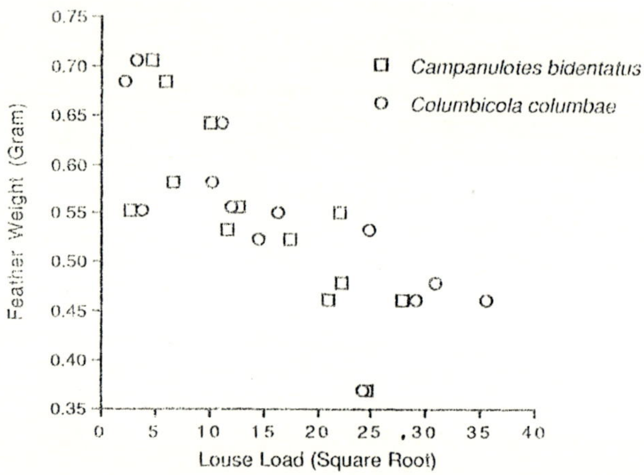

Fig. 14.6. Relationship of feather weight to the number of *Campanulotes bidentatus* and *Columbicola columbae* on 13 rock doves.

Effect of lice on plumage

Louse load was inversely related to feather weight (Fig. 14.6). Combined load was a significant predictor of overall feather weight ($r^2 = 0.64$, $p < 0.001$, $df = 12$; adjusted r^2s reported throughout). Combined load also was a significant predictor of regional weight for six of the seven regions sampled ($df = 12$ for all regions): back ($r^2 = 0.53$, $p = 0.003$); breast ($r^2 = 0.50$, $p = 0.004$); flank ($r^2 = 0.56$, $p = 0.002$); nape ($r^2 = 0.27$, $p = 0.04$); side ($r^2 = 0.39$, $p = 0.01$); anterior rump ($r^2 = 0.54$, $p = 0.003$); posterior rump ($r^2 = 0.19$, $p = 0.08$, non-significant). Because *C. bidentatus* and *C. columbae* loads were highly correlated ($r = 0.87$, $p < 0.001$, $df = 12$, untransformed data), the separate effect of each species on the plumage could not be determined. Nelson and Murray (1971) showed experimentally that both species depend on the abdominal contour feathers for food. Therefore, both species undoubtedly contributed to the extensive feather damage of birds with high loads.

Clayton (1960*b*) reported a 23 per cent mean reduction in overall feather weight for a sample of 12 'lousy' rock doves (mean load, 1923;

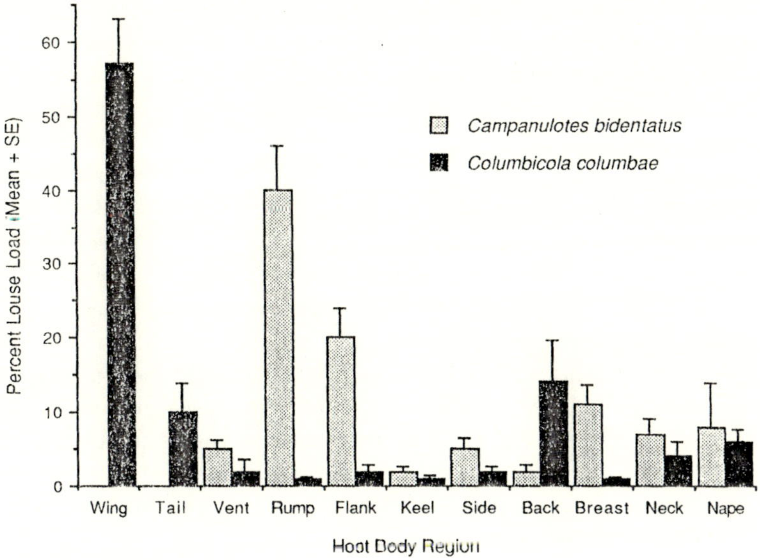

Fig. 14.7. Comparative distribution of lice on 17 rock doves. Data for anterior and posterior rump are combined. Crown and gulum are not displayed because lice were seldom observed in these regions. The data for wing, flank, keel, side, and neck represent one side of the host only. (Percentages calculated within species.)

range, 387–2960), compared to 12 'clean' controls. In the current experiment, mean reduction in feather weight was 28 per cent, comparing weights of the three highest-load birds to weights of the three lowest-load birds. The greater mean damage of the current study probably reflects the fact that the clean birds of the previous study were subjected to some damage prior to the experiment (see Clayton 1990*b*).

Feather weight, which varies seasonally due to feather wear and replacement, is a useful index of plumage insulation (Calder and King 1974). Barnett (1970) measured a 70 per cent autumn increase in the plumage weights of freshly moulted house sparrows, coinciding with a 12°C increase in their cold tolerance. Conversely, a gradual decrease in feather weight during spring coincided with a reduction in cold tolerance. Reduction in feather weight owing to lice presumably has a similar impact on thermoregulation, and may be responsible for reduced mating success and winter survival of rock doves with large louse loads (Clayton 1989, 1990*b*).

Distribution of lice

C. bidentatus and *C. columbae* were observed largely on different regions of the host (Fig. 14.7). These results are in qualitative agreement with those of Nelson and Murray (1971), who used a different division of host regions to document the distributions of these species on several rock doves. Approximately 70 per cent of *C. bidentatus* in the current study were observed on the rump, flank, and breast, while approximately 80 per cent of *C. columbae* were observed on the back, tail, and wing. The proportion of *C. columbae* on the remiges and rectrices probably approached 75 per cent, given that 57 per cent were observed on the tail and only one wing. *C. bidentatus*, in contrast, was never observed on the wing or tail. Thus, *C. columbae* is predominantly a wing louse, compared to *C. bidentatus*, which is essentially a body louse. *C. columbae* must spend some time on the body, however, because it depends on abdominal feathers for food. *C. columbae* cannot survive when supplied solely with wing feathers (Nelson and Murray 1971). Although these data are useful for comparative purposes, they do not accurately quantify the microhabitat distributions of the two species, which can only be accomplished with more tedious methods (e.g. Choe and Kim 1988).

Prediction of louse load

Two regression models were generated for predicting the louse loads of live hosts from regional counts. The model for *C. bidentatus* is

$$\hat{y} = 4.11x_1 + 2.14x_2 + 5.54x_3 + 2.77$$

where x_1, x_2, and x_3 are the square roots of the number of lice counted, respectively, on the host's back, rump, and adjacent to the keel ($r^2_{adj} = 0.91$, $p = 0.0001$, $df = 18$). The model for *C. columbae* is

$$\hat{y} = 3.21x + 4.91$$

where x is the square root of the number of lice counted on one wing ($r^2_{adj} = 0.82$, $p = 0.0001$, $df = 18$).

The first model predicts the host's total *C. bidentatus* load. The second model predicts the number of *C. columbae* on the host, excluding those on the tail. The number of lice on the tail must be added to the square of the number predicted by the model to estimate a host's total *C. columbae* load (see Methods). Because lice on the tail are easy to observe and count, it is unlikely that this procedure reduces the accuracy of the model appreciably.

Experiment III: comparative impact of preening on dissimilar lice

When disturbed by light, streaming air, or simulated preening, *C. columbae* and *C. bidentatus* exhibit different avoidance responses. On a

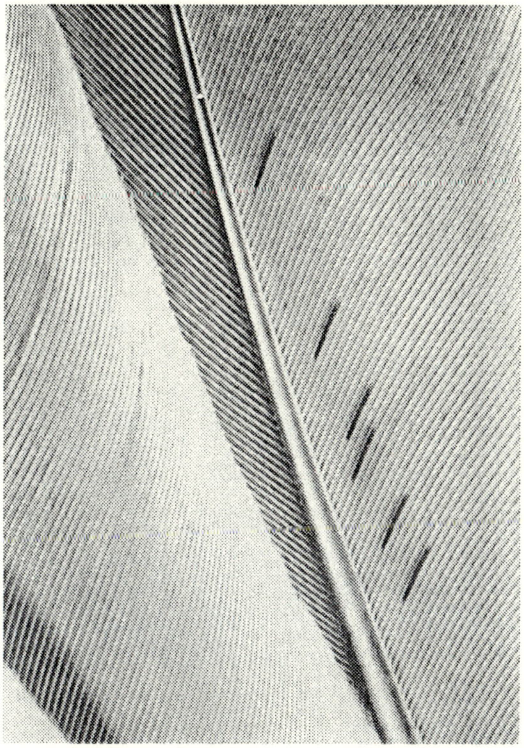

Fig. 14.8. *Columbicola columbae* in typical avoidance posture on a flight feather of the wing. Each louse is inserted between the barbs and oriented toward the shaft, with its mandibles firmly clenched around a barbule. In this position *C. columbae* removes itself from the surface of the feather. (Photo: B. Edinger.)

flight feather *C. columbae* responds to disturbance in at least one of the following ways: (1) it runs rapidly to the base of the feather, where it is hidden by the underwing coverts; (2) it immediately stops moving and flattens itself against the vane; or (3) it inserts itself between two barbs, displacing them with lateral movements of the abdomen, then grasps a barbule firmly between its mandibles. In this position the louse removes itself from the surface of the feather (Fig. 14.8). In an elegant series of experiments, Stenram (1956) demonstrated that *C. columbae* orients to the base of a feather by detecting the directionality of the feather's barbules. Individual lice almost always remain parallel to the barbs of the feather with their heads pointed toward its shaft. If a section of the feather's vane is cut, rotated 180°, and re-attached, individuals placed on the rotated vane orient away from the shaft.

C. bidentatus responds to disturbance on an abdominal contour feather in at least one of two ways: (1) it drops from one feather to

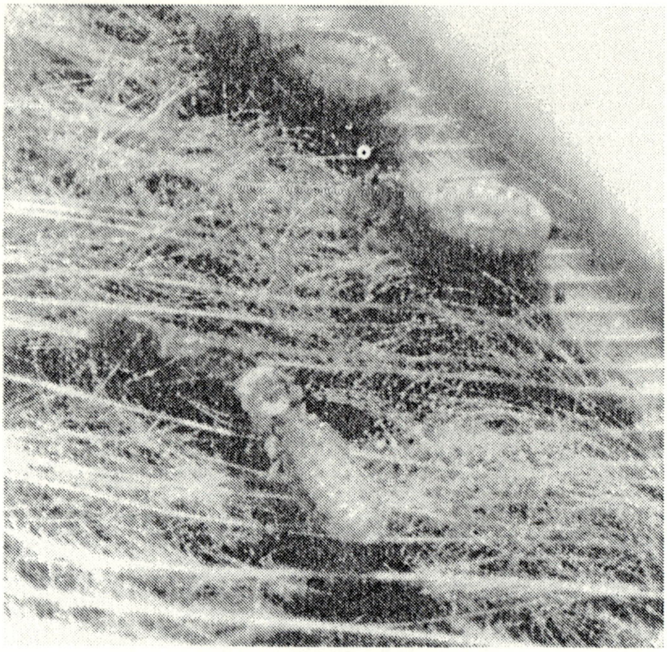

Fig. 14.9. *Campanulotes bidentatus* in avoidance posture on the downy portion of an abdominal contour feather. This species creeps through the downy region at the base of the feather, but remains relatively exposed. (Photo: J. Barabe.)

another; or (2) it creeps to the base of the feather and into the downy region, where it remains relatively exposed (Fig. 14.9). The orientation behaviour of *C. bidentatus* has not been carefully studied, but it is probably less complicated than that of *C. columbae. C. bidentatus* seldom leaves the immediate vicinity of a feather's shaft when present on its exposed vane.

Observations such as these suggest that *C. columbae* is less vulnerable to preening than *C. bidentatus*. In response to preening *C. columbae* can run and hide, whereas *C. bidentatus* can merely run—and more slowly, at that. The avoidance response of *C. columbae* is not necessarily adaptive for the avoidance of preening, however. Stenram (1956, p. 187) contended that the function of this response is to prevent *C. columbae* from being blown off the host's feathers during flight, 'When exposed to a stream of air the specimens of *C. columbae* immediately stop moving and press up to the vane. If the airstream becomes more powerful the specimens force themselves down between the barbs so that only one lateral margin is visible in the vane. On a pigeon the specimens on the distal part of the remiges are mostly found in this position. Here they are exposed to airstreams every time the pigeon flies or flaps its wings. This reaction of the insects has also been released by fastening a remex to a rotating fan.' The goal of the following experiment was to determine whether *C. columbae*, with its complex avoidance response, is more resistant to preening than *C. bidentatus*. An additional objective was to compare the selective effect of preening on the morphology of the two species; the results pertaining to this second objective will not be presented here, however.

Methods

During 1986–87, rock doves were captively bred from wild-caught parents in an aviary located in the barn described above (and see Clayton 1990*b*). Birds several months old were transferred periodically from the aviary to a walk-in enclosure (5.5 × 3.8 × 1.8 m). In May 1987 all birds in the enclosure were bitted, and in late July all birds added to the enclosure since May were bitted, bringing the size of the bitted flock to 175 individuals. Within several weeks of bitting, the louse loads of most birds in the flock were high, owing to the combined effects of the bits and close quarters (4.6 birds/m^3, versus 3.7 birds/m^3 in experiment I). In late August 18 birds with large louse populations were selected for the experiment, including 8 birds bitted in May and 10 birds bitted in July. The birds were divided randomly into experimental and control groups, each with four May-bitted birds, and five July-bitted birds. They were then placed in 18 cages arranged in pairs, with one experimental bird and one control bird to a pair.

During 2–5 September the louse loads of the 18 birds were estimated from regional counts as described in experiment II. Because one region was inadvertently omitted during the counts (Keel, Fig. 14.5), it was necessary to adopt a modified model for predicting *C. bidentatus* loads

$$\hat{y} = 5.83x_1 + 2.25x_2 + 4.27$$

where x_1 and x_2 are the square roots of the number of lice counted, respectively, on the host's back and rump ($r^2_{adj} = 0.81$, $p = 0.0001$, $df = 18$). This model was generated from the multiple regression of total load on the back and rump counts of the experiment II birds. The range of the count data in experiment III exceeded the range of data used in experiment II to generate the predictive models; therefore, the load estimates for birds in experiment III are valid only for comparisons among those birds.

During 6–10 September, approximately 100 *C. bidentatus* and 100 *C. columbae* were sampled (removed with forceps) from each of the

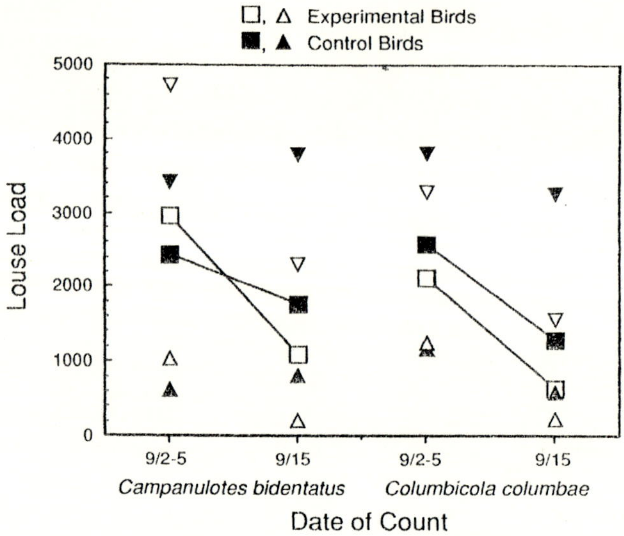

Fig. 14.10. Comparative impact of restored host preening on *Campanulotes bidentatus* and *Columbicola columbae.* Squares denote median loads, triangles denote ranges in load, and line segments denote changes in median loads over the course of the experiment. The bits of experimental birds ($n = 9$) were removed 6–7 September; the bits of control birds ($n = 9$) were not removed. The effect of restored preening was significantly greater on *C. bidentatus* than *C. columbae,* suggesting that the complex avoidance response of the latter facilitates escape from host preening.

experimental and control birds for a future study of the selective effect of preening on louse morphology. After sampling, the preening ability of each experimental bird was restored by removing its bit. The preening ability of control birds was not restored; their bits were left in place. On 15 September, the louse loads of all experimental and control birds were again estimated from new counts. The impact of restored preening on each species of louse was determined by comparing changes in the loads of experimental birds to changes in the loads of their paired controls. The relative impact of preening on the two species of lice was determined by comparing their vulnerability to preening on the experimental birds (see below). All comparisons were non-parametric because of significant changes in the variances of loads over time.

Results and discussion

The average louse loads of both groups of birds decreased over the course of the experiment (Fig. 14.10, Table 14.5). This overall decrease, regardless of host treatment, was caused by the sampling of lice and unwanted side-effects of sampling. The sampling procedure required

Table 14.5. Louse loads of experiment III rock doves. Bits were removed from birds in the experimental group, 6–7 September, 1987

Louse species and date, 1987	Experimentals: bits removed ($N=9$)				Controls: bits not removed ($N=9$)			
	Mean	SD	Median	Range	Mean	SD	Median	Range
Campanulotes bidentatus								
2–5 September	2646	1165	2957	994–4756	2224	944	2431	596–3439
15 September	1046	594	1094	183–2348	1855	957	1779	789–3823
Columbicola columbae								
2–5 September	2199	661	2124	1237–3329	2422	864	2595	1192–3843
15 September	753	517	640	220–1610	1443	804	1288	586–3306

extensive handling of hosts and prolonged exposure to bright light. Handling may have facilitated the transmission of hyperparasites such as *Trenomyces circinans*, a detrimental fungal ectoparasite of *C. columbae* (Eichler *et al.* 1972). Exposure to bright light probably killed large numbers of lice by desiccation (Clayton, unpublished data).

Despite the overall decrease in louse loads due to sampling, there was a significant effect of restored preening on both species of lice. The percentage reduction in the loads of experimental birds exceeded that of controls for both *C. columbae* (Wilcoxon $T=42$, $p=0.01$, one-tailed) and *C. bidentatus* ($T=45$, $p=0.004$). Furthermore, preening had a significantly greater impact on *C. bidentatus* than *C. columbae.* The marginal impact of restored preening (percentage reduction on experimental bird minus percentage reduction on paired control) was significantly greater for *C. bidentatus* than *C. columbae* ($T=32$, $p=0.025$, one-tailed). *C. columbae* showed greater resistance to host preening than *C. bidentatus*, as predicted.

The greater resistance of *C. columbae* to preening is striking, given that this species spends most of its time on the wings (Fig. 14.7), which rock doves preen more than all other regions combined (Fig. 14.11). This result strongly suggests that one adaptive function of *C. columbae*'s avoidance response is escape from host preening. This is not to say that escape is the only adaptive function of the avoidance response; it may also improve the tenacity of lice during host flight and serve other functions.

Conclusions

Adaptive function

The goal of these experiments was to explore the adaptive functions of avian grooming and ectoparasite avoidance. The first experiment verifies

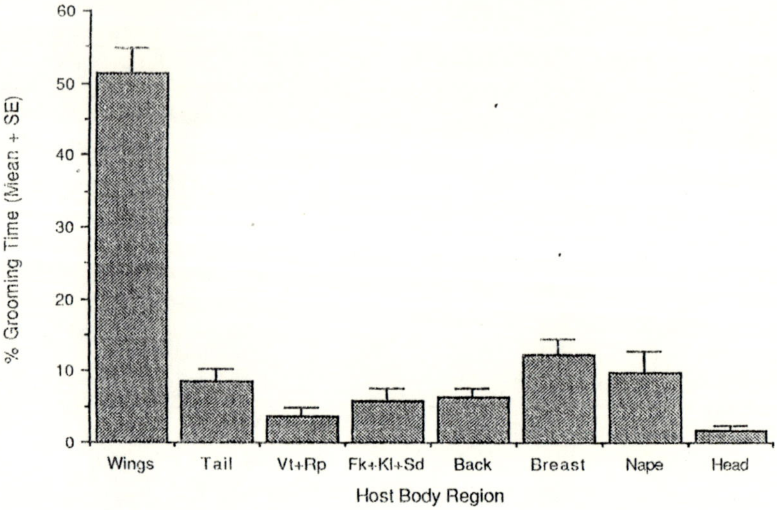

Fig. 14.11. Distribution of grooming by body region. Data are 24 rock doves—12 with large louse loads, and 12 with no lice. Time spent grooming did not differ significantly between the two groups for any region. All regions were groomed by preening, except the head which was groomed by scratching. See Clayton (1990*b*) for details.

the role of preening in louse control. The second experiment suggests that louse control is adaptive by demonstrating that lice cause extensive plumage damage. The third experiment provides comparative evidence that the avoidance response of *C. columbae* is adaptive because it facilitates escape from host preening. Taken together, these results suggest that host preening and louse avoidance are reciprocally adapted traits subject to host–parasite coevolution.

Although experiment I demonstrates the effectiveness of preening for louse control, it does not address the mechanics of control. Presumably preening controls lice by removing adults and nymphs from the plumage, but it may also control lice through the direct removal of nits attached to the feathers. Nits are often attached to regions that the host cannot groom, such as the gulum (Nelson and Murray 1971; Marshall 1981). *C. columbae* nits, which are plainly visible in large numbers on the underwing coverts of rock doves, were not visibly affected by preening in experiment III. However, the direct impact of grooming on the distribution and abundance of nits, as well as the distribution of adults and nymphs, requires detailed study. Also needed is a comparative analysis of the regional efficiency of grooming. The fact that rock doves preen their wings more than other regions may reflect the fact that the remiges are harder to reach than other feathers (Simmons 1964, 1985). The greater impact of preening on *C. bidentatus*, relative to *C. columbae*, could conceivably reflect variation in the efficiency of host defence, rather than variation in the ability of lice to escape host defence.

Experiment II documented the relationship between louse load and feather weight, the latter being a useful index of plumage insulation. Louse-imposed feather damage may reduce the thermoregulatory ability of the host, which would presumably have a negative impact on host fitness (see below). The metabolic cost of lice could be measured by comparing the thermoregulatory ability of birds with different louse loads. If possible, the metabolic cost of grooming behaviour also should be measured to determine more completely the impact of lice on host energetics.

Experiment III provided comparative data suggesting that the avoidance response of *C. columbae* is adaptive for escape from host preening. Although comparative data can serve as circumstantial evidence, they cannot demonstrate adaptive function. The function of the avoidance response could be demonstrated by measuring its direct contribution to escape ability, preferably in conjunction with experimental manipulation of the avoidance response.

Reciprocal selection and coevolutionary response

Functional and comparative analyses can identify the members of an adaptive complex, but they cannot provide direct evidence of selection or adaptation (Arnold 1983). The adaptive function of a trait may be due to chance, or indirect selection (Williams 1966; Gould and Vrba 1982; Endler 1986). For example, the louse-control function of preening may be an incidental effect of preening for general feather maintenance (Simmons 1964, 1985). Adaptation can be inferred most convincingly by documenting selection on adaptive, heritable traits. Likewise, coadaptation can be inferred by documenting reciprocal selection on the adaptive, heritable traits of interacting species (Schemske 1983; Gould 1988). Manly (1985), Endler (1986), and Price and Boag (1987) review recently derived techniques for the measurement of selection.

Clayton (1989) demonstrated a selective effect of lice on wild rock doves subjected to experimentally increased louse loads. The experimental increases mimicked natural inceases on rock doves with inhibited preening resulting from minor bill deformities. Although experimental increases had no effect on the reproductive success of established pairs of rock doves in the field or aviary, they reduced the winter survival of wild individuals, presumably owing to the thermoregulatory cost of feather damage. Feather damage was also apparently responsible for reduced courtship display by lousy males during captive mate choice trials, trials in which lousy males obtained significantly fewer mates than louse-free males (Clayton, 1990*b*). These results show that experimentally increased louse loads impose direct selection on rock doves via reduced survival, and indirect selection via their influence on host sexual selection.

Lice probably exert selection against natural bill deformities, given that the louse loads of deformed birds are much higher than those of birds with normal bills. Lice may also exert selection on the preening ability of birds with normal bills. This hypothesis could be tested by measuring the covariation of preening and louse load among hosts with normal bills. Before attempting this, however, it will be necessary to develop repeatable methods for measuring minute variations in preening ability. If preening ability covaries with bill morphology, as shown for avian foraging ability (Boag and Grant 1981), it may be feasible to test the selection hypothesis by measuring the covariation of bill morphology and louse load.

Although the selective effect of lice on rock doves has been demonstrated, the hypothesized reciprocal effect of host grooming on lice has not been tested. Grooming may eliminate lice purely at random, in which case it can have no direct effect on the evolution of louse avoidance. By contrast, if grooming eliminates lice selectively, it can directly influence the evolution of avoidance, assuming avoidance is a heritable trait (see below). The intensity of grooming-imposed selection is likely to vary among species of lice with different avoidance mechanisms. For example, it is possible that preening exerts selection on wing lice, but not on body lice which may rely on the density of the abdominal plumage to escape preening purely by chance. Comparing the intensity of preening-imposed selection on *C. columbae* versus *C. bidentatus* would provide a compelling test of this hypothesis.

A final issue that must be addressed in studies of coevolution is genetic response. Selection acts at the phenotypic level and may or may not cause evolution, which is change at the genotypic level (Haldane 1954; Lande and Arnold 1983). Selection can lead to an evolutionary response only if it influences heritable traits. For example, if grooming behaviour has no heritable component it will not coevolve with louse avoidance, regardless of the intensity of louse-imposed selection on the host. The stereotyped variation in grooming among closely related species of birds (e.g. McKinney 1965; Schreiber 1977) suggests that grooming has a genetic component. However, interspecific comparisons of this kind convey little information regarding the narrow-sense heritability required for evolution (Hailman 1986). Heritable variation can be measured using quantitative genetic techniques (Boag and van Noordwijk 1987; Falconer 1989).

In conclusion, the opportunity for coevolution can be thoroughly assessed by documenting reciprocal selection between interacting species, and the potential for coevolutionary response to selection. First, however, one must identify realistic targets of selection by demonstrating the adaptive functions of apparent coadaptations. The evidence presented in this chapter suggests specific adaptive functions for avian grooming and ectoparasite avoidance, and confirms these traits as excellent candidates for coevolutionary analysis. Further assessment of the opportunity for coevolution of grooming and avoidance requires: (1) direct measurement of grooming-imposed selection on ectoparasites; and (2) quantitative genetic analysis of host grooming and ectoparasite avoidance. Rock doves and their chewing lice constitute a realistic system for performing these steps in the near future.

Acknowledgements

Funds were provided by NSF Grant BSR-8612575, NIH Grant GM07197, the Frank M. Chapman Memorial Fund, Sigma Xi, the Van Tyne Fund, Kaytee Products Inc, and L. Getz. I am grateful to N. Burley, C. Cooper, D. Droge, P. Mankin, and D. Wake for providing facilities or accommodations at the University of Illinois, and to J. Humphrey, K. Vulinec, and Jean and Jim Nicholson for animal caretaking and assistance in the laboratory. This work benefited directly from discussions with many people over the years, including D. Alstad, J. Choe, F. Cuthbert, J. Fitzpatrick, K. Hamann, J. Humphrey, M. Johnston, K. Karoly, J. Kethley, L. Lemke, B. Nelson, T. Peterson, P. Phillips, R. Price, D. Stotz, R. Timm, J. Willis, and particularly D. Schemske. Statistical advice was furnished by R. Chappell, M. Morgan, S. Shuster, and M. Wade, and P. Phillips assisted with the preparation of figures. Comments on the manuscript were provided by S. Arnold, J. Fitzpatrick, R. Lande, M. Lloyd, D. Schemske, and an anonymous reviewer. The encouragement and assistance of K. Clayton was essential to all phases of this research.

References

Arnold, S.J. (1983). Morphology, performance and fitness. *American Zoologist*, **23**, 347–61.

Ash, J.S. (1960). A study of the Mallophaga of birds with particular reference to their ecology. *Ibis*, **102**, 93–110.

Barnett, L.B. (1970). Seasonal changes in temperature acclimatization of the house sparrow, *Passer domesticus*. *Comparative Biochemistry and Physiology*, **33**, 559–78.

Baum, H. (1968). Biology and ecology of the feather lice of blackbirds. (In German.) *Angewandte Parasitologie*, **9**, 129–75.

Blagoveschchenskii, D.I. (1959). Mallophaga. Part 1. Introduction. (In Russian.) *Nasekomye pukhoedy 1, Fauna SSSR*, n.s. no. 72, pp. 1–202.

Boag, P.T. and Grant, P.R. (1981). Intense natural selection in a population of Darwin's finches (Geospizinae) in the Galapagos. *Science*, **214**, 82–5.

Boag, P.T. and van Noordwijk, A.J. (1987). Quantitative genetics. In *Avian genetics* (ed. F. Cooke and P.A. Buckley), pp. 45–78. Academic Press, London.

Boyd, E.M. (1951*a*). The external parasites of birds: a review. *Wilson Bulletin*, **63**, 363–9.

Boyd, E.M. (1951*b*). A survey of parasitism of the starling *Sturnus vulgaris* L. in North America. *Journal of Parasitology*, **37**, 56–84.

Brooke, M. de L. (1985). The effect of allopreening on tick burdens of molting eudyptid penguins. Auk, **102**, 893–5.

Brown, C.R. and Brown, M.B. (1986). Ectoparasitism as a cost of coloniality in cliff swallows *Hirundo pyrrhonota*. *Ecology*, **67**, 1206–18.

Brown, N.S. (1971). A survey of the arthropod ectoparasites of pigeons (*Columba livia*) in Boston. *Journal of Parasitology*, **57**, 1379–80.

Brown, N.S. (1972). The effect of host beak condition on the size of *Menacanthus stramineus* populations of domestic chickens. *Poultry Science*, **51**, 162–4.

Calder, W.A. and King, J.R. (1974). Thermal and caloric relations of birds. *Avian Biology*, 4, 259–413.

Choe, J.C. and Kim, K.C. (1988). Microhabitat preference and coexistence of ectoparasitic arthropods on Alaskan seabirds. *Canadian Journal of Zoology*, **66**, 987–97.

Clay, T. (1949). Some problems in the evolution of a group of ectoparasites. *Evolution*, **3**, 279–99.

Clay, T. (1951). The Mallophaga as an aid to the classification of birds with special reference to the structure of feathers. *Proceedings of the Xth International Ornithological Congress*, (ed. S. Hörstadius) pp. 207–15. Almqvist and Wiksell, Uppsala.

Clay, T. (1957). The Mallophaga of birds. In *First symposium on host specificity among parasites of vertebrates* (ed. J.G. Baer), pp. 120–58. Institut de Zoologie, Université de Neuchâtel, Neuchâtel.

Clayton, D.H. (1989). Coevolutionary ecology of the rock dove *Columba livia* and its chewing lice (Mallophaga: Ischnocera). Unpublished D.Phil. thesis, University of Chicago.

Clayton, D.H. (1990*a*). Host specificity of *Strigiphilus* owl lice (Ischnocera: Philopteridae), with the description of new species and host associations. *Journal of Medical Entomology*, **27**, 257–65.

Clayton, D.H. (1990*b*). Mate choice in experimentally parasitized rock doves: lousy males lose. *American Zoologist*, **30**, 251–62.

Cook, E.F. (1954). A modification of Hopkins' technique for collecting ectoparasites from mammal skins. Entomological News, **65**, 35–7.

Derylo, A. (1974*a*). Studies on the economic harmfulness of Mallophaga. I. The influence of invasion of lice on the state of health in hens and turkeys. (In Polish.) *Medycyna Weterynaryjna*, **30**, 353–7.

Derylo, A. (1974*b*). Studies on the economic harmfulness of Mallophaga. II. Influence of lice infestation on egg laying and hatching of hens. (In Polish.) *Medycyna Weterynaryjna*, **30**, 406–10.

DeVaney, J.A. (1976). Effects of the chicken body louse, *Menacanthus stramineus*, on caged layers. *Poultry Science*, **55**, 430–5.

Dubinin, V.B. (1947). Investigation of the adaptation of ectoparasites. II. Ecological adaptations of feather-mites and Mallophaga. (In Russian.) *Parazitologicheskii Sbornik, Moscow*, **9**, 191–222.

Duffy, D.C. (1983). The ecology of tick parasitism on densely nesting Peruvian seabirds. *Ecology* **64**, 110–19.

Eichler, W., Zlotorczyka, J., Ludwig, H.W., and Stenram, H. (1972). The pigeon louse *Columbicola columbae columbae.* (In German.) *Angewandte Parasitologie*, **13**, 1–18.

Emerson, K.C. (1960). A new species of *Physconelloides* (Mallophaga) from Texas. *Journal of the Kansas Entomological Society*, **33**, 122–4.

Emerson, K.C. (1972). *Checklist of the Mallophaga of North America.* Desert Test Center, Dugway Proving Ground, Dugway, Utah.

Endler, J.A. (1986). *Natural selection in the wild.* Princeton University Press, Princeton, New Jersey.

Esch, G.W., Gibbons, J.W., and Bourque, J.E. (1975). An analysis of the relationship between stress and parasitism. *American Midland Naturalist*, **93**, 339–53.

Falconer, D.S. (1989). *Introduction to quantitative genetics* (3rd edn). Wiley, New York.

Futuyma, D.J. (1986). *Evolutionary biology*. Sinauer, Sunderland, Massachusetts.

Futuyma, D.J. and Slatkin, M. (ed.) (1983). *Coevolution.* Sinauer, Sunderland, Massachusetts.

Goodwin, D. (1983). *Pigeons and doves of the world.* Cornell University Press, Ithaca, New York.

Gould, F. (1988). Genetics of pairwise and multispecies plant-herbivore coevolution. In *Chemical mediation of coevolution* (ed. K.C. Spencer), pp. 13–56. Academic Press, San Diego.

Gould, S.J. and Lewontin, R.C. (1979). The spandrels of San Marco and the Panglossian paradigm: a critique of the adaptationist programme. *Proceedings of the Royal Society of London*, **B 205**, 581–98.

Gould, S.J. and Vrba, E.S. (1982). Exaptation—a missing term in the science of form. *Paleobiology*, **8**, 14–15.

Hailman, J.P. (1986). The heritability concept applied to wild birds. In *Current ornithology* Vol. 4 (ed. R.F. Johnston), pp. 71–95. Plenum Press, New York.

Haldane, J.B.S. (1954). The measurement of natural selection. *Proceedings of the IXth International Congress of Genetics*, **1**, 480–7.

Ilienko, A.L. (1959). Peculiarities of the ecology of *Passer domesticus* L. affecting their infestation with Gamasid mites. (In Russian.) *Journal of Zoology*, **38**, 1060–8.

Janzen, D.H. (1980). When is it coevolution? *Evolution*, **34**, 611–12.

Kartman, L. (1949). Preliminary observations on the relation of nutrition to pediculosis of rats and chickens. *Journal of Parasitology*, **35**, 367–74.

Keirans, J.E. (1975). A review of the phoretic relationship between Mallophaga (Phthiraptera: Insecta) and Hippoboscidae (Diptera: Insecta). *Journal of Medical Entomology*, **12**, 71–6.

Keler, S. (1939). Baustoffe zu einer Monographie der Mallophagen. II. Teil: Uberfam. Nirmoidea (I). Nova Acta (Leopoldina. Abhandlungen) der Kaiserlich Leopoldinisch–Carolinisch Deutschen Akademie der Naturforscher, *N.F.* **8**, 1–254.

Kim, K.C. (1972). Louse populations of the Northern Fur Seal *Callorhinus ursinus*. *American Journal of Veterinary Research*, **33**, 2027–36.

Lande, R. and Arnold, S.J. (1983). The measurement of selection on correlated characters. *Evolution*, **37**, 1210–26.

Ledger, J.A. (1969). Ectoparasite load in a laughing dove with a deformed mandible. *Ostrich*, **41**, 191–4.

Lemke, L.A. (1982). Distribution and population density of the northern fowl mite *Ornithonyssus sylviarum* (Canestrini and Fanzago 1877), on mature laying hens and evaluation of a visual sampling technique. Unpublished M.S. thesis, Pennsylvania State University.

Levi, W.M. (1957). *The pigeon*. R.L. Bryan, Columbia, South Carolina.

Lincoln, R.J., Boxshall, G.A. and Clark, P.F. (1982). *A dictionary of ecology, evolution and systematics*. Cambridge University Press, Cambridge.

Lodmell, D.L., Bell, J.F., Clifford, C.M., Moore, G.J., and Raymond, G. (1970). Effect of limb disability on lousiness in mice. V. Hierarchy disturbance on mutual grooming and reproductive capacities. *Experimental Parasitology*, **27**, 184–92.

Manly, B.F.J. (1985). *The statistics of natural selection on animal populations*. Chapman and Hall, London.

Marshall, A.G. (1981). *The ecology of ectoparasitic insects*. Academic Press, London.

Martin, M. (1934). Life history and habits of the pigeon louse *Columbicola columbae*. *Canadian Entomologist*, **66**, 6–16.

Matthysse, J.G., Jones, C.J. and Purnasiri, A. (1974). Development of northern fowl mite populations on chickens, effects on the host, and immunology. *Search Agriculture*, **4**, 3–39 (Entomology 13). Cornell University Agricultural Experiment Station, Ithaca, New York.

McGroarty, D.L. and Dobson, R.C. (1974). Ectoparasite populations on house sparrows in northwestern Indiana. *American Midland Naturalist*, **91**, 479–86.

McKinney, F. (1965). The comfort movements of Anatidae. *Behaviour*, **25**, 120–220.

Moss, W.W. and Camin, J.H. (1970). Nest parasitism, productivity and clutch size in purple martins. *Science*, **168**, 1000–3.

Murray, M.D. (1990). Influence of host behaviour on some ectoparasites of birds and mammals. In *Parasitism and host behaviour* (ed. C.J. Barnard and J.M. Behnke), pp. 286–311. Taylor and Francis, London.

Nelson, B.C. and Murray, M.D. (1971). The distribution of Mallophaga on the domestic pigeon *Columba livia*. *International Journal of Parasitology*, **1**, 21–9.

Nelson, W.A. (1962). Development in sheep of resistance to the ked *Melophagus ovinus* II. Effects of adrenocorticotrophic hormone and cortisone. *Experimental Parasitology*, **12**, 45–51.

Nelson, W.A. (1984). Effects of nutrition of animals on their ectoparasites. *Journal of Medical Entomology*, **21**, 621–35.

Nelson, W.A., Keirans, J.E., Bell, J.F., and Clifford, C.M. (1975). Host–ectoparasite relationships. *Journal of Medical Entomology*, **12**, 143–66.

Nelson, W.A., Bell, J.F., Clifford C.M., and Keirans, J.E. (1977). Interaction of ectoparasites and their hosts. *Journal of Medical Entomology*, **12**, 389–428.

Neter, J., Wasserman, W., and Kutner, M.H. (1983). *Applied linear regression models*. Richard D. Irwin Inc, Homewood, Illinois.

Nitecki, M.H. (ed.) (1983). *Coevolution*. University of Chicago Press, Chicago.

Noble, E.R. and Noble, G.A. (1982). *Parasitology*. Lea and Febiger, Philadelphia.

Pomeroy, D.E. (1962). Birds with abnormal bills. *British Birds*, **55**, 49–72.

Price, P.W. (1980). *Evolutionary biology of parasites*. Princeton University Press, Princeton, New Jersey.

Price, R.D. and Clayton, D.H. (1983). A review of the genus *Psittaconirmus* (Mallophaga: Philopteridae) from south pacific parrots. *International Journal of Entomology* (formerly *Pacific Insects*), **25**, 56–70.

Price, T.D. and Boag, P.T. (1987). Selection in natural populations of birds. In *Avian genetics* (ed. F. Cooke and P.A. Buckley), pp. 257–88. Academic Press, London.

Rand, A.L. (1959). Pity poor pigeon: host to a community. *Chicago Natural History Museum Bulletin*, **30**, 6–7.

Rollinson, D. and Anderson, R.M. (1985). *Ecology and genetics of host–parasite interactions*. Academic Press, London.

Rothschild, M. and Clay, T. (1952). *Fleas, flukes and cuckoos*. Collins, London.

SAS Institute (1985). *SAS/STAT™ guide for personal computers*, Version 6 edition. SAS Institute, Cary, North Carolina.

Schemske, D.W. (1983). Limits to specialization and coevolution in plant–animal mutualisms. In *Coevolution* (ed. M.H. Nitecki), pp. 67–110. University of Chicago Press, Chicago.

Schorger, A.W. (1952). Introduction of the domestic pigeon. *Auk*, **69**, 462–3.

Schreiber, R.W. (1977). *Maintenance behavior and communication in the brown pelican*, Ornithological Monographs, no. 22. American Ornithologists' Union, Washington, DC.

Shields, W.M. and Crook, J.R. (1987). Barn swallow coloniality: a net cost for group breeding in the Adirondacks? *Ecology* **68**, 1373–86.

Siegel, S. (1956). *Nonparametric statistics*. McGraw-Hill, New York.

Simmons, K.E.L. (1964). Feather maintenance. In *A new dictionary of birds* (ed. A. C. Thompson), pp. 278–86. McGraw-Hill, New York.

Simmons, K.E.L. (1985). Comfort behaviour. In *A dictionary of birds* (ed. B. Campbell and E. Lack), pp. 101–5. T.A.D. Poyser, Carlton.

Simmons, K.E.L. (1986). *The sunning behaviour of birds*. Bristol Ornithological Club, Bristol.

Spencer, K.C. (ed.) (1988). *Chemical mediation of coevolution*. Academic Press, San Diego.

Stenram, H. (1956). The ecology of *Columbicola columbae* L. (Mallophaga). *Opuscula entomologica*, **21**, 170–90.

Tendeiro, J. (1965). Estudos sobre Malofagos. Revisao monografica do genero *Columbicola* Ewing (Ischnocera, Philopteridae). (In Portuguese.) *Memórias da Junta de Investigações Ultramar*, 2nd series, **32**, 1–460.

Thompson, J.N. (1982). *Interaction and coevolution*. Wiley, New York.

Thompson, J.N. (1989). Concepts of coevolution. *Trends in Ecology and Evolution*, **4**, 179–83.

Waage, J. (1979). The evolution of insect/vertebrate associations. *Biological Journal of the Linnean Society*, **12**, 187–224.

Williams, G.S. (1966). *Adaptation and natural selection, a critique of some current evolutionary thought*. Princeton University Press, Princeton, New Jersey.

Zar, J.H. (1984). *Biostatistical analysis*. Prentice-Hall, Englewood Cliffs, New Jersey.

Zeuner, F.E. (1963). *History of domesticated animals*. Hutchinson & Co, London.

15 *Behavioural changes in breeding American kestrels infected with* Trichinella pseudospiralis

MICHELE D. SAUMIER, MANFRED E. RAU, and DAVID M. BIRD

Introduction

Although raptors are exposed to a great number of parasitic infections (Keymer 1972; Cooper and Greenwood 1981), little is known about whether such parasites affect the competitive fitness of these birds. Natural infections with the nematode *Trichinella pseudospiralis* are known to cause severe debilitation and death in Cooper's hawks (Wheeldon *et al.* 1983). Laboratory studies suggest that essentially subclinical infections with this parasite can elicit profound behavioural changes in American kestrels (*Falco sparverius*) (Saumier *et al.* 1988). The major effects of infection in non-breeding female kestrels involve a reduction in mobility. Severe and long-lasting effects appear as early as 15 days post-inoculation and coincide with the release of larvae into the tissues of the host. *Trichinella pseudospiralis* infections also severely impair the reproductive success of captive American kestrels (Saumier *et al.* 1986). This is manifested in a high proportion of broken eggs, increased embryo mortality, and nest abandonment. The present study explores the behavioural dimension of this failure and examines the behaviour of infected nestlings as well as nestling–parent interaction. Conceivably, behavioural patterns may be as sensitive to the effects of parasitic infections as are some secondary sexual characteristics. Finally, the study raises the question whether contact with infected individuals may alter the behaviour of other members of the family group.

Methods

American kestrels were bred and maintained in captivity according to the procedures of the Macdonald Raptor Research Centre of McGill University as described by Bird (1985). Birds were housed in cages measuring 2.4 m × 1.2 m × 2.4 m ($L \times W \times H$). The cages consisted of a plywood frame covered, top and bottom, with wire mesh, and were equipped with shelters, rope and wood perches, as well as a plywood nestbox. One-way glass windows gave a view of the interior of the pen and the nest box. A door provided access to the pen from a darkened, central observation corridor. Males were allowed to establish themselves in the cage for several days before being randomly paired with females. In the pens, kestrels were exposed to natural daylight and temperatures. Birds were fed freshly-thawed, 1-day-old chicks enriched with SA-37 vitamin-mineral supplement (Rogar Co, Montreal).

At the time of pairing (mid-April), 13 randomly selected pairs of birds were inoculated by gavage with 100 larvae of *Trichinella pseudospiralis* per gram of body weight. A control group of 26 pairs was sham inoculated with 0.85 per cent saline. Half of the clutches of control parents subsequently received 100 larvae per gram body weight at the age of 2–5 days.

At the end of the experiment, infected adult birds were euthanized with CO_2, and individual total muscle burdens of larvae were obtained by artificial digestion of the skinned and eviscerated carcasses. Tissues were digested in 0.5 per cent pepsin for 2 h, filtered through 3 layers of cheese-cloth, and washed in 0.85 per cent saline (modified after Tanner 1968).

Activity budget

Information was gathered on the manner in which the behaviour of the kestrels was partitioned among various activities in time, as well as on the relationship of these activities to the various stages of infection. An activity budget was obtained by recording the specific activities of individual kestrels during 15-min sessions at 15-s intervals (total of 60 observations per session). For convenience, the mean frequencies of each behaviour (out of 60) were converted to percentages. The method used was modified after Wiens *et al.* (1970) and Altmann (1974). All kestrels within the same flight cage (one family group) were observed simultaneously. Individual birds were identified with coloured plastic bands on both legs, as well as a single numbered aluminium band. At the age of 2–5 days, young nestlings were marked individually with ink on the down of different body regions. The behavioural data were processed through PROC RANK (SAS 1982) to obtain the ranks of the response variables, and parametric procedures were applied to these ranks (Conover and Iman 1981). Since sample sizes were unbalanced, the General Linear Model (GLM) procedure (SAS 1982) was used instead of ANOVA. The GLM model examined the main effect of class variables (treatment, sex, time of the day) as well as crossed effects (interactions of sex and treatment, treatment and time of the day, sex and treatment and time of the day, sex and time of the day). Duncan's Multiple Range test (DMR) was used to determine significant differences between the means of the ranked behavioural data. When interactions between the factors included the treatment, and were statistically significant, these are presented in the results. Thus, interactions between sex and treatment are designated as (S*T), between treatment and time of day (T*TD) or both (S*T*TD). Levels of activity are presented as means (*X*) of unranked data. Standard errors (SE) provide an indication of the variability about these means. The above two statistics were not used to determine significant differences between means, but rather provided a convenient format for quantifying the levels at which various behaviours occurred. Sample sizes (*N*) represent the number of 15-min bird sessions of observation.

The behavioural activities of adult kestrels were recorded throughout the periods of courtship, incubation, and parental care. Their offspring were observed throughout their nestling and juvenile phases. The courtship period was considered to be the interval between pairing and the time at which three eggs had accumulated in the nest. For kestrels, the task of incubation does not begin until the clutch is complete (Sherman 1913), or nearly complete (Porter and Wiemeyer 1972). Captive kestrels lay an average clutch of four eggs (Bent 1938). Incubation follows, and ends when the first egg hatches. Thereafter, the parents feed their nestlings, and the period of parental care commences (Hurrell 1974; Ratcliffe 1980). The parental care period ceases when all young have fledged. Young birds were observed during the period of parental care. Since nestlings are inactive and are usually covered by the brooding female during the first 10 days after hatching, they were not observed before this time. The interval, from day 10 post-hatching until fledging, was termed the nestling period.

Test schedule

Sampling was scheduled to document the activities of infected birds during every daylight hour. Cages to be observed were selected at random, but care was taken to distribute the observation sessions throughout the various periods of the day. Each day was separated into five observation periods: 6.00–9.00 hours (early morning), 9.00–12.00 hours (late morning), 12.00–15.00 hours (early afternoon), 15.00–18.00 hours (late afternoon), and 18.00 hours until dusk (evening). In early spring and late fall, the early morning and evening periods of observation were curtailed because of darkness. The above time intervals encompass the daily rhythms in behavioural patterns described by Smith *et al.* (1972) and Joly (1977). Sampling was not undertaken during periods of heavy rain, due to the reduced activity of the birds under such conditions. The behavioural repertoire of kestrels was divided into locomotory, perching, social, grooming, food handling, and breeding behaviours as well as miscellaneous behaviours of uncertain function.

Behaviour descriptions

Locomotory behaviours

- *Flying*. In captivity, kestrels fly from perch to perch, to the nest box, or in tight horizontal loops.
- *Walking*. Birds walk on the ground or the platforms. This process involves alternate steps with an occasional hop.
- *Exercising*. Birds walk or perch while beating their wings, without becoming airborne (Sherrod 1983). This behaviour pattern, as well as mock attacks on small, inanimate objects by young kestrels. (Balgooyen 1976), was considered to be a form of exercising.
- *Perching behaviours*. Perching behaviours were divided into five categories on the basis of the bird's posture and alertness.
- *High perching*. Birds stand with both feet on an elevated structure, such as a wood or rope perch or on top of the next box, and appear to be keenly aware of their surroundings.
- *Floor perching*. Birds stand on the floor of the flight cage and appear to be alert.
- *Nest Perching*. Birds behave as above but are standing in the nest box (except when brooding or nest inspecting).
- *Resting*. Birds either lie on their bellies or support their weight on one leg with one foot tucked into the belly plumage. They appear to remain aware of their surroundings (Joly 1977).

Social behaviours

- *Aggressive displays*. Characteristically, aggressive displays involve a chase, with or without calls, and occasionally a jab with the talons as described by Cade (1955) and Mueller (1971).
- *Calling*. Adult kestrels possess a repertoire of three distinct calls (Willoughby and Cade 1964). Vocalizations characteristically introduce or accompany other behavioural activities. When vocalization accompanied primary behavioural patterns, only the latter were recorded.

Grooming behaviours

- *Preening*. Kestrels use their bill to nibble, stroke, and quiver their feather coats, modes described in detail by Joly (1977) and Lefebvre and Joly (1982).
- *Scratching*. Kestrels use their feet to groom various body regions, particularly the head.
- *Comfort movements*. Comfort movements include stretching and bill-wiping, as described by Joly (1977).
- *Feather fluffing*. This activity consists of shaking the feather coat as described by Joly (1977).

Food handling behaviours

- *Feeding*. In the present study, birds were considered to be feeding only when they were actually tearing or ingesting prey.
- *Food transfer*. Food transfer between adults is usually followed by the presentation of food to the young by the female during the parental care period.

Breeding behaviours

- *Courtship feeding*. The male presents food to the female. The behaviour consists of a complex of calls, postures, and coordinations, and serves to strengthen the pair bond (Balgooyen 1976).
- *Solicitation*. The female presents her back to the male, flicks her tail, and assumes a horizontal posture, thus offering a landing platform for the male during copulation (Roest 1957).
- *Copulation*. Mating can occur regardless of whether the female was previously soliciting. This activity is well described by Willoughby and Cade (1964) and Balgooyen (1976).
- *Nestbox inspection*. This behaviour is most frequently performed by the male, and consists of foot scraping and loud chittering (Olendorff 1968).
- *Incubation*. The process of sitting on eggs is generally performed by the female. The male assists the female, permitting her to leave the eggs to feed, preen, and stretch (Balgooyen 1976).
- *Brooding*. This behaviour refers to the act whereby the parents shield their young with their wings and body. Kestrels assume the brooding position by leaning forward and carefully stepping around

the nestlings in a shuffling movement, thus placing their feet below the chicks (Ratcliffe 1980).

Miscellaneous behaviours

- *Tail flicking.* This behaviour consists of an up-and-down movement of the tail and is considered to be a prelude to copulation (Bishop 1925; Roest 1957).
- *Head bobbing.* This up-and-down movement of the head occurs when birds are intensely interested in whatever they are watching (Sherrod 1983).
- *Gaping.* This behaviour can be compared to a short, forced yawn, and is usually repeated every few seconds.
- *Panting.* This behaviour resembles hard breathing with a partially open beak (Sherrod 1983).

The present study concentrates on those behavioural processes that are affected by the *T. pseudospiralis* infections. Behavioural differences between infected and control birds that are attributable to the infections are presented along with any interactions with the sex of the birds and/or the time of day. Where treatment effects are dependent on the time of day, only the means for the time period in which significant differences occurred are presented in the text.

Results

The birds initiated courtship behaviour as soon as they were paired on 10–13 April. Among control birds courtship extended, on average, over 32.27 ± 1.45 days ($N=26$), but lasted significantly longer among infected pairs (40 ± 2.53 days, $N=8$); ($p=0.0200$). Incubation was first observed on 3 May and its duration was not significantly different ($p=0.200$) for controls (30.32 ± 0.51 days, $N=22$) and infected birds (31 ± 1.7 days, $N=3$). The first eggs hatched on 2 June, marking the beginning of the period of parental care and the nestling period. This segment of the bird's life cycle continued an average of 27.90 ± 0.33 days ($N=19$) for control parents, and 29.00 ± 0.71 days for infected pairs ($N=2$). These differences were not statistically significant ($p=0.400$). The last young kestrel fledged on 30 July.

Courtship period

Trichinella pseudospiralis infections influenced the flying activity of captive American kestrels. Infected kestrels were observed to fly less

Table 15.1. Courtship activity budget of *Trichinella pseudospiralis* infected kestrels. Data presented as percentage of time spent in each behaviour (mean ± standard error)

Flying	C		3.16 ± 0.44^{a}	$N=162$ (52)	Aggressive	C	F	EM 0.00 ± 0.00^{b}	$N=17$ (16)
(t, $p=0.0306$)	I		1.43 ± 0.28^{b}	$N=104$ (26)	displays			LM 0.00 ± 0.00^{b}	$N=28$ (26)
					(S*T*TD,			EA 0.33 ± 0.33^{b}	$N=20$ (20)
					$p=0.0040$)			LA 0.00 ± 0.00^{b}	$N=16$ (16)
High perching	C	F	73.90 ± 2.80^{a}	$N=81$ (26)			M	Behaviour not observed	
(S*T, $p=0.0028$)		M	50.46 ± 3.44^{b}	$N=81$ (26)		I	F	EM 0.00 ± 0.00^{b}	$N=13$ (13)
	I	F	42.25 ± 5.46^{b}	$N=52$ (13)				LM 0.36 ± 0.19^{a}	$N=14$ (13)
		M	39.27 ± 4.81^{b}	$N=52$ (13)				EA 0.00 ± 0.00^{b}	$N=12$ (12)
								LA 0.00 ± 0.00^{b}	$N=13$ (12)
							M	Behaviour not observed	
Floor perching	C	F	0.37 ± 0.28^{b}	$N=81$ (26)					
(S*T, $p=0.0001$)		M	2.98 ± 1.34^{b}	$N=81$ (26)	Scratching	C		EM 0.78 ± 0.19^{b}	$N=34$ (32)
	I	F	35.71 ± 6.29^{a}	$N=52$ (13)	(T*TD,			LM 1.10 ± 0.25^{b}	$N=56$ (52)
		M	0.64 ± 0.35^{b}	$N=52$ (13)	$p=0.0276$)			EA 0.58 ± 0.18^{b}	$N=40$ (40)
								LA 0.83 ± 0.29^{b}	$N=32$ (32)
						I		EM 0.38 ± 0.17^{b}	$N=26$ (26)
								LM 1.43 ± 0.24^{a}	$N=28$ (26)
								EA 0.28 ± 0.13^{b}	$N=24$ (24)
								LA 0.58 ± 0.21^{b}	$N=26$ (24)

Behaviour	Treatment	Sex	Mean	N	Behaviour	Treatment	Mean	N
Nest perching	C		1.20 ± 0.44^{a}	$N=162$ (52)	Copulation	C	0.56 ± 0.07^{a}	$N=162$ (52)
(T, $p=0.0219$)	I		0.05 ± 0.05^{b}	$N=104$ (26)	(T, $p=0.0003$)	I	0.22 ± 0.07^{b}	$N=104$ (26)
Sleeping		F	Behaviour not observed					
(S*T*TD,	C	M	EM 0.00 ± 0.00^{b}	$N=17$ (16)	Gaping	C	EM 0.39 ± 0.35^{ab}	$N=34$ (32)
$p=0.0201$)			LM 0.00 ± 0.00^{b}	$N=28$ (26)	(T*TD, $p=0.0035$)		LM 0.00 ± 0.00^{b}	$N=56$ (52)
			EA 0.00 ± 0.00^{b}	$N=20$ (20)			EA 0.00 ± 0.00^{b}	$N=40$ (40)
			LA 0.10 ± 0.10^{ab}	$N=16$ (16)			LA 0.00 ± 0.00^{b}	$N=32$ (32)
						I	EM 0.00 ± 0.00^{b}	$N=26$ (26)
	I	F	EM 1.03 ± 1.03^{ab}	$N=13$ (13)			LM 0.89 ± 0.66^{a}	$N=28$ (26)
			LM 0.00 ± 0.00^{b}	$N=14$ (13)			EA 0.00 ± 0.00^{b}	$N=24$ (26)
			EA 0.00 ± 0.00^{b}	$N=12$ (12)			LA 0.13 ± 0.09^{ab}	$N=26$ (24)
			LA 1.28 ± 1.04^{a}	$N=13$ (12)				
		M	EM 0.00 ± 0.00^{b}	$N=13$ (13)				
			LM 0.83 ± 0.83^{ab}	$N=14$ (13)				
			EA 0.00 ± 0.00^{b}	$N=12$ (12)				
			LA 0.00 ± 0.00^{b}	$N=13$ (12)				

p, probability obtained with GLM.
T, treatment; S, sex; TD, time of day; *, interactions.
C, control; I, infected. F, female; M, male.
EM, early morning; LM, late morning; EA, early afternoon; LA, late afternoon.
N, number of 15-min bird sessions of observation; ( ), number of individual birds observed.
Means with the same superscript are statistically indistinguishable using Duncan's multiple range test.

frequently than the control birds (T, $p=0.0306$) (Table 15.1). These effects were particularly noticeable among females. Infected females spent much of their time perching on the wire mesh floor. In contrast, control females were rarely seen to floor perch (S*T, $p=0.0001$). Control females occupied high perches significantly more frequently than their mates, and significantly more frequently than the infected females and infected males (S*T, $p=0.0028$). Furthermore, control birds were observed to stand more frequently in the nest box than were infected birds (T, $p=0.0219$). No significant differences were found between infected and uninfected pairs with respect to their walking and exercising activities.

Gaping was more common among infected individuals than among controls (T*TD, $p=0.0035$). This highly significant difference was apparently only during the late morning when gaping was characteristically most frequent (Table 15.1). In addition, infected birds scratched more frequently during late morning than did the controls (T*TD $p=0.0276$). Infected females slept more frequently during late afternoon than did female controls (S*T*TD, $p=0.0201$). The frequencies of resting, calling, preening, comfort movements, feather fluffing, feeding, and head bobbing were not found to be significantly different for control and infected pairs. Panting was not observed during the courtship period.

Aggressive displays were rarely witnessed during the courtship period, but statistical analysis revealed that infected females were more aggressive towards their mates, specifically during late morning, than were uninfected females (S*T*TD, $p=0.0040$). Infected birds flicked their tails more frequently during late morning ($3.83+0.95$, $N=56$), than did the control pairs ($4.10+1.04$, $N=28$) (T*TD, $p=0.0110$). Due to the infrequency of this behaviour, it is not presented in Table 15.1. Copulation took place more frequently among controls than among infected pairs (T, $p=0.0003$). The frequencies of solicitation, nestbox inspection, and courtship feeding were not found to differ significantly for infected and control pairs.

Incubation period

Seven infected females failed to produce complete clutches of six eggs and were thus excluded from observation during the incubation period. This reduced the sample size to six infected pairs. Since all of the uninfected females produced six eggs or more, the control sample size remained at 26 pairs.

The *T. pseudospiralis* infections had profound effects on the activity of female kestrels during the incubation period. In the late morning and late afternoon infected females spent most of their time perching on the floor of the flight pens. In contrast, during the same two periods, uninfected females were rarely observed to floor perch. These differences between infected and control birds were highly significant (S*T*TD, $p=0.0101$; Table 15.2). The frequencies of flying, walking, exercising, high perching, nest perching, resting, and sleeping did not differ significantly for controls and infected birds.

Panting was common during incubation, and was more frequently observed among infected birds than among controls during early afternoon, the warmest times of the day (T*TD, $p=0.0304$). Furthermore, infected females were observed to feed more frequently than their uninfected counterparts (S*T, $p=0.0140$). The frequencies of calling, preening, scratching, feather fluffing, head bobbing, comfort movements, tail flicking, and gaping were not significantly different for infected and control pairs.

Two breeding behaviours were affected by the *T. pseudospiralis* infection. Nestbox inspection was performed significantly more frequently by infected males than by the corresponding controls during late morning (S*T*TD, $p=0.0001$; Table 15.2). Furthermore, control females were observed to incubate more frequently than were infected females (S*T, $p=0.0021$). Concomitantly, infected males spent more than 30 per cent of their time incubating eggs during the late morning, whereas control males spent only 2–7 per cent of the time in this activity. Aggressive displays, solicitation, and courtship feeding in controls and infected birds were statistically indistinguishable.

Parental care period

Pairs which failed to produce young, or lost their broods due to cannibalism, were eliminated from this group. This reduced the controls to 19 pairs, and the infected group to three pairs. Since birds infected with *T. pseudospiralis* hatch fewer young (Saumier *et al.* 1986), some of the most severely affected pairs may thus have been removed. No differences were found in the frequencies of flying and walking between infected and control birds. Exercising was not observed during this period in either infected or uninfected parents.

Regardless of the time of day or their sex, control birds occupied elevated perches more frequently than did infected individuals (T, $p=0.0410$; Table 15.3). Furthermore, infected birds were not observed to scratch during the parental care period ($N=24$), whereas controls did (T, $p=0.0482$). During the early afternoon, panting was more frequent among infected birds than among the controls (T*TD, $p=0.0090$). Parenthetically, panting was not observed in the early morning and early evening when the temperatures were relatively cool. Tail flicking was

Table 15.2. Incubation activity budget of *Trichinella pseudospiralis* infected kestrels. Data presented as percentage of time spent in each behaviour (mean ± standard error)

Behaviour					Behaviour				
Floor perching	C	F	EM 0.36 ± 0.36[b]	$N=28$ (23)	Nest inspection	C	F	EM 0.12 ± 0.12[b]	$N=28$ (23)
(S*T*TD, $p=0.0101$)			LM 0.00 ± 0.00[b]	$N=24$ (23)	(S*T*TD,			LM 0.00 ± 0.00[b]	$N=24$ (23)
			EA 0.06 ± 0.06[b]	$N=28$ (24)	$p=0.0001$)			EA 0.06 ± 0.06[b]	$N=28$ (24)
			LA 0.43 ± 0.30[b]	$N=27$ (23)				LA 0.00 ± 0.00[b]	$N=27$ (23)
			EE 0.00 ± 0.00[b]	$N=12$ (12)			M	EE 0.00 ± 0.00[b]	$N=12$ (12)
		M	EM 0.83 ± 0.57[b]	$N=28$ (23)				EM 0.00 ± 0.00[b]	$N=28$ (23)
			LM 0.49 ± 0.36[b]	$N=24$ (23)				LM 0.00 ± 0.00[b]	$N=24$ (23)
			EA 0.30 ± 0.21[b]	$N=28$ (24)				EA 0.12 ± 0.12[b]	$N=28$ (24)
			LA 1.60 ± 1.48[b]	$N=26$ (23)				LA 0.00 ± 0.00[b]	$N=26$ (23)
			EE 0.24 ± 0.24[b]	$N=14$ (13)				EE 0.00 ± 0.00[b]	$N=14$ (13)
	I	F	EM 17.22 ± 17.23[b]	$N=3$ (3)		I	F	Behaviour not observed	
			LM 60.00 ± 30.56[a]	$N=3$ (3)			M	EM 0.00 ± 0.00[b]	$N=3$ (3)
			EA 0.00 ± 0.00[b]	$N=6$ (4)				LM 3.33 ± 2.55[a]	$N=3$ (3)
			LA 64.72 ± 20.50[a]	$N=6$ (4)				EA 0.00 ± 0.00[b]	$N=6$ (4)
			EE 5.55 ± 5.56[b]	$N=3$ (3)				LA 0.00 ± 0.00[b]	$N=6$ (4)
		M	EM 0.00 ± 0.00[b]	$N=3$ (3)				EE 0.00 ± 0.00[b]	$N=3$ (3)
			LM 5.00 ± 5.00[b]	$N=3$ (3)					
			EA 1.67 ± 1.67[b]	$N=6$ (4)					
			LA 0.00 ± 0.00[b]	$N=6$ (4)					
			EE 0.00 ± 0.00[b]	$N=3$ (3)					

Behaviour			Mean	N	Behaviour		Mean	N
Feeding	C	F	1.81 ± 0.67^{c}	$N=119$ (26)	Panting	C	EM 0.00 ± 0.00^{b}	$N=56$ (46)
($S^{*}T$, $p=0.0140$)		M	5.07 ± 1.44^{a}	$N=120$ (26)	($T^{*}TD^{*}$, $p=0.0304$)		LM 0.83 ± 0.46^{b}	$N=48$ (48)
	I	F	3.02 ± 1.50^{b}	$N=21$ (6)			EA 0.03 ± 0.03^{b}	$N=56$ (48)
		M	3.75 ± 3.40^{ab}	$N=21$ (6)			LA 0.85 ± 0.61^{b}	$N=53$ (46)
							EE 0.00 ± 0.00^{b}	$N=26$ (4)
Incubation	C	F	63.79 ± 3.47^{a}	$N=119$ (26)		I	EM 0.00 ± 0.00^{b}	$N=6$ (6)
($S^{*}T$, $p=0.0021$)		M	3.88 ± 1.26^{c}	$N=120$ (26)			LM 0.00 ± 0.00^{b}	$N=6$ (6)
	I	F	39.85 ± 10.42^{b}	$N=21$ (6)			EA 5.56 ± 3.63^{a}	$N=12$ (8)
		M	9.53 ± 6.57^{c}	$N=21$ (6)			LA 15.56 ± 10.51^{ab}	$N=12$ (8)
							EE 0.00 ± 0.00^{b}	$N=6$ (6)

p, probability obtained with GLM.
T, treatment; S, sex; TD, time of day; *, interactions.
C, control; I, infected. F, female; M, male.
EM, early morning; LM, late morning; EA, early afternoon; LA, late afternoon.
N, number of 15-min bird sessions of observation; ( ), number of individual birds observed.
Means with the same superscript are statistically indistinguishable using Duncan's multiple range test.

Table 15.3. Parental care activity budget of *Trichinella pseudospiralis*-infected kestrels. Data presented as percentage of time spent in each behaviour (mean ± standard error)

Behaviour	Treatment	Time of day	Mean ± SE	N
High perching (T, $p=0.0410$)	C		59.15 ± 3.31^{a}	$N=136$ (28)
	I		49.94 ± 7.58^{b}	$N=24$ (6)
Scratching (T, $p=0.0482$)	C		0.37 ± 0.08^{a}	$N=136$ (28)
	I		0.00 ± 0.00^{b}	$N=24$ (6)
Panting ($T^{*}TD$, $p=0.0090$)	C	EM	0.00 ± 0.00^{d}	$N=26$ (20)
		LM	5.14 ± 3.41^{cd}	$N=24$ (24)
		EA	5.39 ± 4.09^{cd}	$N=26$ (22)
		LA	1.50 ± 1.04^{cd}	$N=30$ (26)
		EE	0.00 ± 0.00^{d}	$N=30$ (22)
	I	EM	0.00 ± 0.00^{d}	$N=2$ (2)
		LM	14.73 ± 11.71^{bc}	$N=6$ (4)
		EA	27.23 ± 10.97^{a}	$N=6$ (6)
		LA	20.29 ± 12.69^{ab}	$N=6$ (6)
		EE	0.00 ± 0.00^{d}	$N=4$ (4)

p, probability obtained with GLM.
T, treatment; S, sex; TD, time of day; *, interactions.
C, control; I, infected. F, female; M, male.
EM, early morning; LM, late morning; EA, early afternoon; LA, late afternoon.
N, number of 15-min bird sessions of observation; ( ), number of individual birds observed.
Means with the same superscript are statistically indistinguishable using Duncan's multiple range test.

rarely observed during this period, but analysis revealed that the infected males flicked their tail marginally more frequently during late morning ($0.55+0.55$, $N=3$) than did the corresponding controls (0, $N=12$) ($S^{*}T^{*}TD$, $p=0.0457$). Due to the infrequency of this behaviour, it is not presented in Table 15.3. Breeding behaviours were greatly reduced in frequency during the period of parental care. Incubation was succeeded by brooding, which did not differ between infected and control individuals. Other behaviours which did not appear to be affected by the infection were floor perching, nest perching, resting, sleeping, calling, preening, comfort movements, feather fluffing, feeding, food transfer, solicitation, copulation, gaping, and head bobbing. Aggressive displays were not observed during the parental care period.

During the parental care period, uninfected birds rearing infected young behaved differently from control birds rearing normal young. During late morning, gaping was observed in normal parents with infected young whereas normal parents with normal young were not observed to gape ($T^{*}TD$, $p=0.006$; Table 15.4). Similarly, male kestrels with infected young called significantly more frequently during late afternoon than the corresponding controls ($S^{*}T^{*}TD$, $p=0.0259$). No differences in the frequency of flying, walking, high perching, floor

Table 15.4. Parental care activity budget of uninfected kestrels raising *Trichinella pseudospiralis*-infected young. Data presented as percentage of time spent in each behaviour (mean ± standard error)

Behaviour	Group	Sex	Time	Mean ± SE	N
Nest perching	P y c	F		12.68 ± 4.62^{a}	$N=46$ (10)
(S*T, $p=0.0163$)		M		0.33 ± 0.16^{b}	$N=46$ (10)
	P y i	F		0.23 ± 0.17^{b}	$N=22$ (5)
		M		0.30 ± 0.18^{b}	$N=22$ (5)
Copulation	P y c		EM	0.19 ± 0.13^{b}	$N=18$ (14)
(T*TD, $p=0.0165$)			LM	0.00 ± 0.00^{b}	$N=16$ (16)
	C	F	EA	0.00 ± 0.00^{b}	$N=18$ (16)
			LA	0.00 ± 0.00^{b}	$N=20$ (18)
			EE	0.00 ± 0.00^{b}	$N=20$ (18)
	P y i	M	EM	0.00 ± 0.00^{b}	$N=8$ (6)
			LM	0.00 ± 0.00^{b}	$N=8$ (8)
			EA	0.42 ± 0.27^{a}	$N=8$ (6)
			LA	0.00 ± 0.00^{b}	$N=10$ (8)
			EE	0.33 ± 0.22^{a}	$N=10$ (8)
Gaping	P y c		Behaviour not observed		
(T*TD, $p=0.0006$)	P y i		EM	0.00 ± 0.00^{b}	$N=8$ (6)
			LM	0.42 ± 0.27^{a}	$N=8$ (8)
			EA	0.00 ± 0.00^{b}	$N=8$ (6)
			LA	0.00 ± 0.00^{b}	$N=10$ (8)
			EE	0.00 ± 0.00^{b}	$N=10$ (8)
Calling	P y c	F	EM	0.00 ± 0.00^{c}	$N=9$ (7)
(S*T*TD,			LM	0.21 ± 0.21^{c}	$N=8$ (8)
$p=0.0259$)			EA	0.37 ± 0.37^{c}	$N=9$ (8)
			LA	0.67 ± 0.67^{c}	$N=10$ (9)
			EE	0.00 ± 0.00^{C}	$N=10$ (9)
		M	EM	1.30 ± 0.77^{abc}	$N=9$ (7)
			LM	1.46 ± 1.46^{bc}	$N=8$ (8)
			EA	0.00 ± 0.00^{c}	$N=9$ (9)
			LA	0.00 ± 0.00^{c}	$N=10$ (9)
			EE	0.33 ± 0.22^{bc}	$N=10$ (9)
	P y i	F	Behaviour not observed		
		M	EM	0.00 ± 0.00^{c}	$N=4$ (3)
			LM	0.750 ± 6.96^{ab}	$N=4$ (4)
			EA	0.00 ± 0.00^{c}	$N=4$ (3)
			LA	10.00 ± 7.66^{a}	$N=5$ (4)
			EE	3.00 ± 3.00^{bc}	$N=5$ (4)

p, probability obtained with GLM.
T, treatment; S, sex; TD, time of day; *, interactions.
Pyc, parents with uninfected young; Pyi, parents with infected young; I, infected. F, female; M, male.
EM, early morning; LM, late morning; EA, early afternoon; LA, late afternoon.
N, number of 15-min bird sessions of observation; ( ), number of individual birds observed.
Means with the same superscript are statistically indistinguishable using Duncan's multiple range test.

perching, resting, sleeping, preening, scratching, comfort movement, feather fluffing, feeding, food transfer, panting, tail flicking, and head bobbing were found between normal parents raising uninfected young and normal parents raising infected young. Exercising and aggressive displays were not observed in either of these parental groups.

During the parental care period, copulation was observed significantly more frequently among control parents caring for infected young, specifically during the early afternoon and evening (T*TD, $p=0.0165$; Table 15.4). Control parents with normal nestlings were observed to copulate only during early morning and significantly less frequently than the preceding group (T*TD, $p=0.0165$). In addition, uninfected females rearing normal young perched more frequently in the nest than did uninfected females rearing infected young (S*T, $p=0.0163$). Nest perching in uninfected adult males caring for normal nestlings did not differ from that of their counterparts which cared for infected nestlings. Brooding and solicitation by uninfected parents did not appear to be affected by the presence of infected nestlings. Nestbox inspection was not observed during this period.

Nestling period

Uninfected male nestlings reared by infected parents exercised more frequently than their counterparts reared by control parents (S*T, $p=0.0231$; Table 15.5). Such differences were not significant among uninfected female nestlings cared for by infected parents. The frequency of exercising among infected male nestlings did not differ significantly from that of their uninfected counterparts when both were reared by normal parents. The frequency of exercise among infected female nestlings reared by normal parents did not differ from their uninfected counterparts. Uninfected young raised by infected parents were observed to perch only marginally less frequently than were uninfected nestlings reared by normal parents (T, $p=0.0487$). Furthermore, the frequency of perching among infected nestlings of both sexes did not differ significantly from that among their uninfected counterparts when both were reared by normal parents. Calling was affected by the infection. The uninfected young of infected parents were observed to call significantly more frequently during late afternoon than were uninfected young of normal parents and infected young raised by normal parents (T*TD, $p=0.0003$). There was no difference in aggressive displays for any of the nestling groups.

The frequency of panting among female nestlings was affected by whether they were reared by infected or normal parents. During the early afternoon uninfected female nestlings raised by infected parents panted more frequently than uninfected female nestlings raised by normal parents (S*T*TD, $p = 0.0002$). Infected female nestlings raised by normal parents panted with a frequency of 60.00 ± 30.56 per cent ($N = 3$). This differs significantly from their uninfected counterparts which were not observed to pant ($N = 13$), (S*T*TD, $p = 0.0002$). Male nestlings did not exhibit such differences. The frequencies of preening, scratching, comfort movement, feather fluffing, tail flicking, head bobbing, feeding, and gaping were not affected by the parasitic infection.

The frequency of sleeping among nestlings was also affected by whether they were reared by infected or control parents. Uninfected young reared by infected parents slept significantly more frequently than their counterparts reared by normal parents, and also more frequently than the infected nestlings (T*TD, $p = 0.0003$). This difference was apparent only during the late morning observation period. There was no difference in resting for any of the nestling groups.

The mean total larvae recovered was $310\,560.92 \pm 76\,024.69$ for the 13 infected adult females and $116\,099.08 \pm 23\,745.17$ ($N = 13$) for the adult males. These represent, in our experience, relatively moderate infections. None of the infected birds died nor manifested clinical disease.

Discussion

Experimental *T. pseudospiralis* infections altered the behavioural processes of the adult kestrel host. Not all of these effects are seen in both sexes, during all periods of the host's life cycle, or during all times of the day. Infected birds exhibited effects that were consistent with the pathological tissue changes described from avian hosts infected with this parasite (Miroshnichenko 1978), as well as with the well-documented symptoms accompanying *T. spiralis* infections in the human host (Gould 1970). Thus, during the courtship period, infected birds of both sexes were observed to fly less frequently than uninfected controls. This effect may be attributable to muscle damage. In addition, infected kestrels also gaped more frequently than the controls, specifically during late morning. Conceivably, gaping is a manifestation of lung damage inflicted by migrating larvae. Larval migration is a transitory phenomenon and characterizes the early stages of the infection. In the present study, this phase fell within the courtship period of the birds. Since the observed differences in the frequency of gaping between infected and control birds were highly significant and since this behaviour did not show significant differences during subsequent periods of observation, they are considered to be transitory manifestations of infection. Similarly, infected birds of both sexes scratched more frequently (late

Table 15.5. Nestling activity budget of *Trichinella pseudospiralis*-infected young. Data presented as percentage of time spent in each behaviour (mean ± standard error)

Behaviour	Young	Parasite	Time	Mean ± SE	N
Perching	NY	(NP)		52.93 ± 4.30^{a}	$N=78$ (20)
(T, $p=0.0487$)	IY	(NP)		51.91 ± 5.14^{ab}	$N=56$ (18)
	NY	(IP)		36.25 ± 8.46^{b}	24 (7)
Calling	NY	(NP)	EM	0.00 ± 0.00^{b}	$N=13$ (13)
(T*TD,			LM	1.78 ± 0.93^{ab}	$N=15$ (15)
$p=0.0003$)			EA	0.00 ± 0.00^{b}	$N=18$ (16)
			LA	0.00 ± 0.00^{b}	$N=16$ (16)
			EE	0.21 ± 0.21^{b}	$N=16$ (16)
	IY	(NP)	EM	0.13 ± 0.13^{b}	$N=13$(13)
			LM	0.00 ± 0.00^{b}	$N=11$(11)
			EA	0.00 ± 0.00^{b}	$N=7$(7)
			LA	0.00 ± 0.00^{b}	$N=16$(16)
			EE	1.67 ± 1.67^{b}	$N=9$(9)
	NY	(IP)	EM	0.00 ± 0.00^{b}	$N=2$(2)
			LM	$0.00^{b} \pm 0.00^{b}$	$N=5$(5)
			EA	$0.00^{b} \pm 0.00^{b}$	$N=5$(5)
			LA	2.78 ± 1.36^{a}	$N=9$(8)
			EE	$0.00^{b} \pm 0.00^{b}$	$N=3$(3)

Behaviour	Young	Parasite	Sex	Time	Mean ± SE	N
Exercising	NY	(NP)	F		0.65 ± 0.27^{b}	$N=59$ (14)
(S*T, $p=0.0231$)			M		0.09 ± 0.09^{b}	$N=19$ (6)
	IY	(NP)	F		0.43 ± 0.17^{ab}	$N=27$ (8)
			M		0.23 ± 0.14^{b}	$N=29$ (10)
	NY	(IP)	F		0.00 ± 0.00^{v}	$N=15$ (4)
			M		0.93 ± 0.40^{a}	$N=9$ (3)
Panting	NY	(NP)	F	EM	0.00 ± 0.00^{e}	$N=10$ (10)
(S*T*TD,				LM	7.12 ± 3.90^{cde}	$N=11$ (11)
$p=0.0002$)				EA	0.00 ± 0.00^{e}	$N=13$ (11)
				LA	13.21 ± 8.10^{cde}	$N=13$ (13)
				EE	0.00 ± 0.00^{e}	$N=12$ (13)
			M	EM	0.00 ± 0.00^{e}	$N=3$ (3)
				LM	0.00 ± 0.00^{e}	$N=4$ (4)
				EA	20.00 ± 20.00^{cde}	$N=5$ (5)
				LA	31.11 ± 24.76^{abcd}	$N=3$ (3)
				EE	0.00 ± 0.00^{e}	$N=4$ (3)

Behaviour	Young	Parents	Sex	Time	Mean ± SE	N
Sleeping (T*TD, $p=0.0003$)	NY	(NP)		EM	0.00 ± 0.00^{b}	$N=13$ (13)
				LM	0.00 ± 0.00^{b}	$N=15$ (15)
				EA	23.43 ± 8.99^{b}	$N=18$ (16)
				LA	1.35 ± 1.35^{b}	$N=16$ (16)
				EE	0.00 ± 0.00^{b}	$N=16$ (16)
	IY	(NP)		EM	2.56 ± 2.57^{b}	$N=13$ (13)
				LM	0.00 ± 0.00^{b}	$N=11$ (11)
				EA	12.38 ± 12.39^{b}	$N=7$ (7)
				LA	6.04 ± 2.85^{b}	$N=16$ (16)
				EE	8.33 ± 4.31^{b}	$N=9$ (9)
	NY	(IP)		EM	0.00 ± 0.00^{b}	$N=2$ (2)
				LM	30.00 ± 8.62^{a}	$N=5$ (5)
				EA	3.33 ± 3.33^{b}	$N=5$ (5)
				LA	0.00 ± 0.00^{b}	$N=9$ (8)
				EE	0.00 ± 0.00^{b}	$N=3$ (3)
	IY	(IP)	F	EM	0.00 ± 0.00^{e}	$N=7$ (7)
				LM	6.39 ± 6.39^{de}	$N=6$ (6)
				EA	60.00 ± 30.56^{abc}	$N=3$ (3)
				LA	0.00 ± 0.00^{e}	$N=7$ (7)
				EE	0.00 ± 0.00^{e}	$N=4$ (4)
			M	EM	0.00 ± 0.00^{e}	$N=6$ (6)
				LM	6.67 ± 6.67^{e}	$N=5$ (5)
				EA	11.67 ± 11.67^{cde}	$N=4$ (4)
				LA	0.74 ± 0.74^{a}	$N=9$ (9)
	NY	(IP)	F	EM	0.00 ± 0.00^{e}	$N=1$ (1)
				LM	0.00 ± 0.00^{e}	$N=3$ (3)
				EA	100.0 ± 0.00^{a}	$N=3$ (3)
				LA	60.67 ± 22.20^{ab}	$N=5$ (4)
				EE	0.00 ± 0.00^{e}	$N=3$ (3)
			M	EM	0.00 ± 0.00^{e}	$N=1$ (1)
				LM	20.83 ± 20.83^{bcde}	$N=2$ (2)
				EA	0.00 ± 0.00^{e}	$N=2$ (2)
				LA	0.00 ± 0.00^{e}	$N=4$ (4)

p, probability obtained with GLM.
T, treatment; S, sex; TD, time of day; *, interactions.
NY (NP), normal young raised by normal parents; IY (NP), infected young raised by normal parents; NY (IP), normal young raised by infected parents;
F, female; M, male.
EM, early morning; LM, late morning; EA, early afternoon; LA, late afternoon.
N, number of 15-min bird sessions of observation; ( ), number of individual birds observed.
Means with the same superscript are statistically indistinguishable using Duncan's multiple range test.

morning) than did the controls during the courtship period. These differences were highly significant during the early stages of the infection, and at best only marginally significant during the subsequent periods. Conceivably, scratching is a response to the itching sensation described by Gould (1945) in cases of human trichinellosis.

The data suggest that *T. pseudospiralis* infections may create some thermal stress. On hot days, kestrels pant to maintain thermal homeostasis (Welty 1979). During the incubation and parental care periods (June and July), particularly in the early afternoon, infected birds of both sexes were observed to pant more frequently than the controls. The pathological basis for this effect is obscure, but may be related to the pyrogenic response to the invading parasites (Gould 1945). The consequences of infection seem to be much more severe in female than in male birds. Thus, during the courtship period, infected females were observed to sleep in the daytime, especially during late afternoon, and to feed more during the incubation period than did uninfected controls. Both activities may be associated with the demands of tissue repair.

During the courtship period, infected individuals flicked their tail more frequently (late morning) than their uninfected counterparts. Since tail flicking may be considered to be a prelude to copulation and since copulation among infected pairs was significantly less frequent than among controls, the elevated frequency of tail flicking among infected males may indicate a lowered threshold for this activity. During the courtship period, infected females exhibited an increase in the frequency of aggressive displays (late morning). This suggests that they may be less receptive to mating than uninfected females. A lack of receptivity has been documented from female mice infected with *T. spiralis*. Edwards and Barnard (1987) report that infected female mice avoid, or act aggressively towards males that attempt copulation.

Since infected birds which failed to produce complete clutches or to produce young at all (presumably the most severely affected individuals) had been removed from the experimental group, the observed changes in the behaviour of the remaining infected pairs are conceivably an underestimate of the effects of *T. pseudospiralis* infection.

Parasite-associated changes in kestrel behaviour extend to other activities as well. Infected birds of both sexes were seen to perch less frequently in their elevated nest boxes during the courtship period than did the corresponding controls. Infected females spent a greater amount of time on the floor of the flight cage both during the courtship and incubation period (late morning and late afternoon). Similarly, high perching was significantly more frequent among control females during the courtship period, than it was among their infected counterparts. This was also true for female and male controls during the period of parental care. Indeed, infected females tend to lay their eggs on the floor rather than in the nest box (Saumier *et al.* 1986). Such eggs usually break or are abandoned by the females. Furthermore, infected females were observed to incubate significantly less frequently than their uninfected counterparts. It is conceivable that the elevated prevalence of nest inspection during the incubation period exhibited by their mates during late morning constitutes a response compensatory to the lack of care shown by the infected females. Indeed, at certain times of the day, the males of infected pairs spent more time incubating then the males of control pairs. Infected pairs produce significantly fewer offspring than their uninfected counterparts (Saumier *et al.* 1986).

The altered behaviour of infected parents may affect the behaviour of their offspring. Uninfected nestlings reared by infected parents manifested signs of restlessness. Thus, both male and female nestlings called more frequently during late afternoon, and young, uninfected males reared by infected parents exercised more than their counterparts reared by normal birds. Calling by nestling kestrels is a sign of hunger or stress (Jones 1981). Conceivably, nestlings reared by infected parents suffered some degree of nutritional stress. However, since female nestlings are generally more aggressive and tend to be the first to get fed (Sherman 1913), this stress may be more severe among young males. Indeed, nestling females were observed to feed more frequently, whether infected or uninfected. Hunger will increase the activity of small mammals (Nagy *et al.* 1983). Conceivably, it may have similar effects on this raptor species. Following a period of high activity, young kestrels usually become quiescent (Balgooyen 1976). Uninfected nestlings raised by infected parents were observed to sleep more during late morning and perch less than did their counterparts raised by control birds. This increase in the time spent sleeping may be the result of fatigue from begging for food. Also, possible differences in the ability to cope with heat stress, as manifested by abnormally frequent panting among the uninfected female nestlings raised by infected parents (early afternoon), may be due to hunger (Welty 1979). What may be nutritionally mediated effects of maternal infection on the behaviour of their offspring have been reported by Rau (1985) from mice harbouring *T. spiralis*.

Although the effects of the parental infection on the behaviour of the nestlings are pronounced, the infection of the young themselves had little impact. This may be due to the fact that the infection in the nestling falls within the enteric phase of the parasite's life cycle, and larvae are produced and disseminated into the tissues only when nestlings are about to fledge. According to Meerovitch *et al.* (1982), larvae of *T.*

pseudospiralis are first seen in the musculature of kestrels on day 20 post-infection. In *T. spiralis* infections of mice behavioural changes accompany the release of migrating larvae (Zohar and Rau 1986). No behavioural changes were attributable to the enteric phase of the infection (Zohar and Rau 1984). It is conceivable that, as with *T. spiralis*, this phase of *T. pseudospiralis* infections is relatively benign.

The presence of infected nestlings also alters the behaviour of the normal parents. Females rearing infected nestlings spent less time in the nest with these young, and both such parents gaped more frequently during late morning than did parents rearing uninfected nestlings. In addition, the male parents rearing infected young called more during late afternoon than their counterparts. Such parents copulated during early afternoon and evening at a time in their reproductive cycle when their uninfected counterparts had largely abandoned this activity. The mechanisms by which these behavioural changes are induced in the uninfected bird are still obscure.

Trichinella pseudospiralis infections interfere with a wide variety of behavioural processes in the kestrel host. The behavioural changes induced by *T. pseudospiralis* in kestrels resemble those induced in mice by *T. spiralis* (Rau 1983; Rau and Putter 1984). Both host species experience significant reductions in mobility as a result of infection with their respective parasites. However, mice infected with *T. pseudospiralis* exhibit quite different behavioural modifications. Instead of a reduction in activity, *T. pseudospiralis*-infected mice, particularly heavily infected individuals, demonstrated a mild increase in activity (Rau 1984). Both *T. spiralis* in mice and *T. pseudospiralis* in kestrels elicit severe inflammatory responses in the musculature (Larsh 1963; Miroshnichenko 1978, respectively): *T. pseudospiralis* in mice does not (Bessonov *et al.* 1978). Indeed, *T. pseudospiralis* has an anti-inflammatory action in the latter host (Al Karmi and Faubert 1981). Conceivably, many of the behavioural changes manifested during the chronic phase of *T. pseudospiralis* infections in kestrels, as well as *T. spiralis* and *T. pseudospiralis* infections in mice, are attributable, at least in part, to the character and severity of the host responses elicited.

Acknowledgements

The assistance of S. Khanizadeh is gratefully acknowledged. This study was supported by the Natural Sciences and Engineering Research Council of Canada, the Fonds FCAR pour l'aide a la recherche, and the Canadian National Sportsmen's fund.

References

Al Karmi, T.O and Faubert, G.M. (1981). Comparative analysis of mobility and ultrastructure of intramuscular larvae of *Trichinella spiralis* and *Trichinella pseudospiralis*. *Journal of Parasitology*, **67**, 685–91.

Altmann, J. (1974). Observational study of behaviour: sampling methods. *Behaviour*, **49**, 227–67.

Balgooyen, T.G. (1976). Behavior and ecology of the American kestrel (*Falco sparverius* L.) in the Sierra Nevada of California. *University of California Publications in Zoology*, **103**, 1–83.

Bent, A.C. (1938). *Life histories of North American birds of prey*. US National Museum Bulletin, New York.

Bessonov, A.S., Penkova, R.A., and Gumenshchikova, V.P. (1978). *Trichinella pseudospiralis* Garkavi 1972, Morphological and Biological characteristics and host specificity. In *Trichinellosis. Proceedings of the Fourth International Conference on Trichinellosis*, Aug. 26–28, 1976 (ed. C.W. Kim and Z. Pawlowski), pp. 79–93. University Press of New England, Hanover, New Hampshire.

Bird, D.M. (1985). Evaluation of the American kestrel (*Falco sparverius*) as a laboratory research animal. In *The contribution of laboratory animal science to the welfare and animals: past, present and future*, 8th Symposium of the International Council for Laboratory Animal Science (ICLAS), Canadian association for laboratory animal science (CALAS). Vancouver, 31 July–5 August, 1983 (ed. J. Archibald, J. Ditchfield, and H.C. Rowsell), pp. 3–9. Gustav Fisher Verlag, New York.

Bishop, S.C. (1925). Notes on the mating habits of the sparrow hawk. *Auk*, **42**, 268–9.

Cade, T.J. (1955). Experiments on winter territoriality of the American kestrel, *Falco sparverius*. *Wilson Bulletin*, **67**, 5–17.

Conover, W.J. and Iman, R.L. (1981). Ranks transformation as a bridge between parametric and non-parametric statistics. *American Statistician*, **35**(3), 124–7.

Cooper, J.E. and Greenwood, A.G. (ed.) (1981). *Recent advances in the study of raptor diseases. Proceedings of the International Symposium on Diseases of Birds of Prey*. Chiron Publications Ltd, London.

Edwards, J.C. and Barnard, C.J. (1987). The effects of *Trichinella* on intersexual interactions between mice. Animal Behaviour, **35**, 533–40.

Gould, S.E. (1945). *Trichinosis*. Charles C. Thomas, Springfield, Illinois.

Gould, S.E. (1970). *Trichinosis in man and animals*. Charles C. Thomas, Springfield, Illinois.

Hurrell, L.H. (1974). On breeding the sparrow-hawk in captivity. In *A hawk for the bush* (ed. J.G. Mavrogordato), pp. 29–44. Spearman, London.

Joly, R. (1977). Cycles d'activité et morphologie des comportements individuels de la Crecerelle americaine (*Falco sparverius*). Unpublished MS thesis, Université de Montréal.

Jones, C.G. (1981). Abnormal and maladaptive behaviour in captive raptors. In *Recent advances in the study of Raptor diseases. Proceedings of the International Symposium on Diseases of Birds of Prey*, 1–3 July, 1980 (ed. J.E. Cooper and A. G. Greenwood), pp. 53–9. Chiron Publications Ltd, London.

Keymer, J.F. (1972). Diseases of birds of prey. *Veterinary Records*, **90**, 579–94.

Larsh, J.E. (1963). Experimental trichiniasis. *Advances in Parasitology*, **1**, 241–51.

Lefebvre, L.R. and Joly, R. (1982). Organisation rules and timing in kestrel grooming. Animal Behaviour, **30**, 1020–8.

Meerovitch, E., Chadee, K., and Bird, D.M. (1982). Experimental infection of American kestrels, *Falco sparverius* with *Trichinella pseudospiralis* Garkavi, 1972, and *T. spiralis*. *Canadian Journal of Zoology*, **60**, 3150–2.

Miroshnichenko, L.S. (1978). Pathology of skeletal muscles of birds infected with *Trichinella pseudospiralis*. *Wiadomości Parazytologiczne*, **24**, 91–6.

Mueller, H.C. (1971). Displays and vocalizations of the sparrow hawk. *Wilson Bulletin*, **83**, 249–54.

Nagy, Z.M., Porada, K.J., and Anderson, J.A. (1983). Undernutrition by rearing in larger litters delays the development of reflexive, locomotor and memory processes in mice. *Journal of Comparative Physiology and Psychology*, **91**, 682–6.

Olendorff, R.R. (1968). A contribution to the breeding behavior of the American kestrel in captivity. *Raptor Research*, **4**, 77–92.

Porter, R.D. and Wiemeyer, S.N. (1972). Reproductive pattern in captive American kestrels (sparrow hawks). *Condor*, **74**, 46–53.

Ratcliffe, D. (1980). *The peregrine falcon*. Buteo Books, Vermillion. South Dakota.

Rau, M.E. (1983). The open-field behaviour of mice infected with *Trichinella spiralis*. *Parasitology*, **86**, 311–18.

Rau, M.E. (1984). The open-field behaviour of mice infected with *Trichinella pseudospiralis*. *Parasitology*, **88**, 415–19.

Rau, M.E. (1985). The effects of *Trichinella spiralis* infection of pregnant mice on the future behavior of their offspring. *Journal of Parasitology*, **71**, 774–8.

Rau, M.E. and Putter, L. (1984). Running responses of *Trichinella spiralis* infected CD-1 mice. *Parasitology*, **89**, 579–83.

Roest, A.I. (1957). Notes on the American sparrow hawk. *Auk*, **74**, 1–19.

SAS (1982). *SAS user's guide: statistics*. SAS Institute Inc, Cary, North Carolina.

Saumier, M.D., Rau, M.E., and Bird, D.M. (1986). The effect of *Trichinella pseudospiralis* infection on the reproductive success of captive American kestrels (*Falco sparverius*). *Canadian Journal of Zoology*, **64**, 2123–5.

Saumier, M.E., Rau, M.E., and Bird, D.M. (1988). The influence of *Trichinella pseudospiralis* infection on the behavior of non-breeding American kestrels. (*Falco sparverius*). *Canadian Journal of Zoology*, **66**, 1685–2.

Sherman, A.R. (1913). The nest life of the sparrow hawk. *Auk*, **30**, 406–18.

Sherrod, S.K. (1983). *Behavior of fledgeling peregrines*. Pioneer Impressions, Colorado.

Smith, D.G., Wilson, C.R., and Frost, H.H. (1972). The biology of a kestrel in Central Utah. *Southwestern Naturalist*, **17**, 73–83.

Tanner, C.E. (1968). Relationship between infecting dose, muscle parasitism and antibody response in experimental trichinosis in rabbits. *Journal of Parasitology*, **54**, 98–107.

Welty, J.C. (1979). *The life of birds*. Saunders College Publishing, Philedelphia.

Wheeldon, E.B., Dick, T.A., and Schulz, T.A. (1983). First report of *Trichinella spiralis* var. pseudospiralis in North America. *Journal of Parasitology*, **69**, 781–2.

Wiens, J.A., Martin, S.G, Holthaus, W.R., and Iwen, F.A. (1970). Metronome timing in behavioural ecology studies. *Ecology*, **51**, 350–2.

Willoughby, E.J. and Cade, I.J. (1964). Breeding behavior of the American kestrel (sparrow hawk). *Living Bird*, **3**, 75–96.

Zohar, A.S. and Rau, M.E. (1984). The effect of the intestinal phase of *Trichinella spiralis* on the open-field behavior of mice. *Journal of Parasitology*, **70**, 927–30.

Zohar, A.S. and Rau, M.E. (1986). The role of muscle larvae of *Trichinella spiralis* in the behavioural alterations of the mouse host. *Journal of Parasitology*, **72**, 464–6.

Part IV
Sexual selection

16 *Parasites and bright birds: new data and a new prediction*

MARLENE ZUK

Introduction

The function of male secondary sexual characters such as colourful plumage or elaborate courtship displays is controversial among evolutionary biologists, with some suggesting that females assess the genetic quality of potential mates by the development of these characters (Thornhill 1980; Heisler 1984; Nur and Hasson 1984; Trivers 1985; Andersson 1986) and others maintaining that conspicuous traits indicate nothing about male fitness outside of sexual attractiveness, and have evolved through genetic correlation with a female preference for the ornament (Fisher 1930; Lande 1981; Kirkpatrick 1982; Arnold 1983; cf. Pomiankowski 1987). In a specific version of this former type of 'good genes' hypothesis, Hamilton and Zuk (1982) proposed that male ornaments signal the bearer's health and freedom from parasites, and that resistant males within a species have brighter plumage and more vigorous displays than susceptible males. The relationship between host and parasite is likely to generate 'good genes' sexual selection because frequency-dependent interactions between host and parasite alleles readily produce stable limit cycles in the gene frequencies of both species, helping to maintain heritable genetic variation in fitness (Eshel and Hamilton 1984; Hamilton 1982). Species susceptible to parasites will therefore be most likely to evolve male ornaments as indicators of male resistance to common pathogens. Across species, then, Hamilton and Zuk (1982) predicted that the degree of ornamentation should be positively correlated with the proportion of individuals infected with parasites.

Using data on blood parasites in 109 North American passerine bird species, Hamilton and Zuk (1982) found a significant positive association between the brightness of male plumage in a species and its parasite level, defined as the proportion of infected individuals harbouring an infection. A similar correlation was found between parasite level and the complexity of male song, another at least partly sexually-selected character (Hamilton and Zuk 1982). Read (1987, 1988) found support for these findings in a survey of both European and North American passerines; he additionally controlled for the potentially confounding effects of body size, taxonomic group, nest type, male parental investment, and other ecological variables, discovering that the relationship between brightness and proportion of individuals in a species infected with blood parasites is, if anything, strengthened by the removal of these factors.

Ward (1988) also examined the correlation between brightness and parasite intensity, using freshwater fish rather than birds, and sexual dichromatism rather than absolute degree of male or female brightness. He concluded that: (1) fish families with more reported parasite genera tended to be more dichromatic; and (2) this relationship remained significant after the effects of several ecological variables were removed.

The hypothesis has also been tested intraspecifically. Hamilton and Zuk (1982) predicted that, within a species, females should prefer less-parasitized males, using the degree of development of secondary sexual characters to indicate the health of potential mates. Read (1988) reviewed the intraspecific work published before 1988; since that time, several additional studies have been undertaken (Zuk *et al.* 1990; Kirkpatrick *et al.*, Chapter 18, this volume).

The comparative analyses remain in some ways more problematical than the within-species tests, partly because the role of interspecific comparisons in evolutionary biology is itself controversial (Pagel and Harvey 1988). Here I present the results of a comparative study of haematozoa of tropical Central and South American birds and discuss the value of such studies for critically evaluating the Hamilton and Zuk (1982) hypothesis.

The tropical data set allowed the study of a much larger sample (526 species) than had previously been examined, and one containing many of the gaudy birds found in tropical habitats. Tropical species are also often suggested to differ from temperate species in many aspects of their ecology and behaviour (Keast and Morton 1980), and an examination of the generality of application of the Hamilton and Zuk prediction to tropical birds is useful in this context. In addition, the use of tropical birds allowed a comparison between migrant and resident species; because the Hamilton–Zuk hypothesis requires that parasites coevolve closely with their hosts, I predicted that resident species which live year round with the same set of parasites should show the relationship between brightness and parasite level more strongly than migrant species, which are more likely to be exposed to different sets of haematozoa in wintering and breeding habitats. I also employed a contingency analysis different from that of either Hamilton and Zuk (1982) or Read (1987, 1988) which I feel better takes into account the very different sample sizes for each species. Finally, as with any comparative study, the confounding effect of phylogeny must be dealt with, i.e. certain groups of species may show the association between brightness and infection level merely because of shared ancestry, and an overall positive correlation could be the result of non-independence of the species compared (Pagel and Harvey 1988; Felsenstein 1985; Clutton-Brock and Harvey 1984). I present two different approaches to deal with this difficulty.

Methods

The tropical bird data came from White *et al.* (1978), who examined over 35 000 individuals for the presence of five groups of blood parasites. Four different blood parasite genera (*Haemoproteus*, *Plasmodium*, *Trypanosoma*, and *Leucocytozoon*), as well as microfilarial worms, were noted in the survey; data from all five groups were pooled in the analysis reported here to obtain infection frequencies. Male brightness and female brightness scores, on a scale of 1 to 6 from very dull to very bright, were assigned before I had looked at the parasite data and, in the case of many tropical species, were made using pictures from field guides and collections of drawings or photographs (see Hamilton and Zuk 1982). Personal experience was also used in cases where I had seen the species in the field. Species for which adequate colour portrayals could not be found were excluded from the analysis. A 'sexual dichromatism score' was calculated as the difference between each species' male and female brightness scores.

A median test was employed to test for an association between the proportion of individuals within a species harbouring blood parasites and its degree of brightness. An overall median infection level for all species combined was calculated by dividing the number of infected individuals by the total number of birds examined within each species, and then taking the median of these frequencies for all 526 species. Each species was then scored separately as being above or below this grand median. The numbers of species in each brightness category were totalled and a median test performed (Zar 1974; SAS 1985) of the null hypothesis that equal numbers of species fell above and below the median infection level regardless of brightness.

The median test helps to correct for a difficulty with the survey data available from the literature, which includes species without regard for the number of individuals taken in each sample. As a result, many species have only one or two individuals examined, and therefore often must show an infection frequency of 100 or 0 per cent. These numbers are not accurately comparable in ordinary regression or correlation techniques to frequencies of 100 per cent resulting from, say, 25 of 25 birds being infected. Because the median test relies only on frequencies being above or below the median, however, the magnitudes are not exaggerated in cases of small sample size. An alternative strategy would be to drop those species which contain some arbitrary number of individuals or fewer, such as 10, but the resulting loss of power from decreased sample size is undesirable, and the problem still remains, albeit to a smaller extent.

The tropical bird species were scored as either migrant or resident, and the analyses were also performed separately within each group. Two additional tests examined the role of phylogeny in the association. First, the median test was performed separately for each bird family, and the resulting *p*-values combined using a consensus test (Liptak–Stouffer combined probability test; Folks 1984; Rice 1990) to determine if associations between coloration and parasite level had arisen independently in more than one family. This test is more suitable than Fisher's combined probability test for evaluating a set of independent tests of the same null hypothesis. A spurious correlation could result if one or two speciose families are either exceptionally colourful and also heavily infected or exceptionally dull and parasite-free. The families were also separated into two categories: those containing 70–100 per cent resident species, and those with 0–50 per cent residents. The consensus *p*-values were also calculated within each of these categories. A second approach attempted to determine the relative contributions of family classification and male plumage brightness in explaining variation in infection level. Using 'family' as a dummy class variable, family and male brightness were regressed together on infection level (the proportion of individuals in each species that harboured blood parasites), and the resulting Type III mean squares were compared.

Results

Male brightness was significantly associated with parasite load (Table 16.1; χ^2 approximation = 16.13, $p = 0.006$), as was female brightness, although not as strongly (Table 16.1; χ^2 approximation = 12.34, $p = 0.03$). Fewer dull species had infection frequencies above the median than were expected, whereas the number of bright species with infection frequencies above the median was higher than expected. Sexual dichromatism scores, on the other hand, were not significantly associated with median scores (Table 16.1; χ^2 approximation = 7.42, $p = 0.19$).

When the data were separated into migrants and residents, neither male nor female brightness scores showed significant positive relationships with parasite level in the 75 migrant species (Table 16.2; χ^2

Table 16.1 Plumage brightness and frequency of haematozoa infection in 526 neotropical bird species

Plumage brightness*	Number of species with frequency of infection above median for all species		
	Observed	Expected	N
Male			
1	0	1.5	3
2	79	94.8	190
3	87	83.3	167
4	41	38.4	77
5	35	30.4	61
6	21	14.5	29
Female			
1	3	5.5	11
2	125	132.8	266
3	85	83.3	167
4	23	20.5	41
5	19	16.9	34
6	8	4.0	8
Sexual dichromatism (male–female score)			
1	2	2.0	4
0	178	188.1	377
1	44	41.4	83
2	21	14.5	29
3	15	14.0	28
4	3	3.0	6

* Plumage brightness was scored on a scale of 1–6 in order of increasing brightness. Frequency of infection is defined as the number of individuals infected/individuals examined for each species.

approximation = 3.70 for male brightness and 3.08 for female brightness; $p = 0.45$ and 0.54 respectively). In the remaining 451 resident species, however, scores for both male and female brightness remained significantly associated with parasite load (Table 16.2; χ^2 approximation = 16.18 for male brightness and 11.41 for female brightness; $p = 0.006$ and 0.04, respectively). Sexual dichromatism scores remained unrelated to parasite level in both migrants and residents.

The median tests were performed within each of the 18 tropical bird families that had more than two species. The overall consensus p-value for the association between infection level and male plumage brightness score was 0.07 (Table 16.3). When the data were separated into a group of 13 families with all or mostly resident species and one of five families with all or mostly migrant species, consensus p-values were 0.039 and

Table 16.2 Plumage brightness and frequency of haematozoa infection in 526 neotropical migrant and resident bird species

Plumage brightness*	Number of species with frequency of infection above median for all species		
	Observed	Expected	*N*
Migrants			
male			
2	14	15.3	31
3	8	9.4	19
4	7	6.9	14
5	6	4.4	9
6	2	1.0	2
Female			
2	18	21.2	43
3	14	11.8	24
4	2	2.0	4
5	2	1.5	3
6	1	.5	1
Residents			
Male			
1	0	1.5	3
2	63	79.0	159
3	79	73.0	147
4	35	31.3	63
5	28	25.8	52
6	19	13.4	27
Female			
1	3	5.5	11
2	103	110.3	222
3	73	71.0	143
4	21	18.4	37
5	17	15.4	31
6	7	3.5	7

* Plumage brightness was scored on a scale of 1–6 in order of increasing brightness. Frequency of infection is defined as the number of individuals infected/individuals examined for each species (see text for details of analysis).

0.52 for the two categories, respectively. This result suggests that the overall trend for brighter species to have more individuals infected with parasites is significant within resident, but not migrant, families, as was true at the species level. The regression of family and male brightness on infection level yielded a Type III mean square of 0.72 for the male brightness variable, nearly four times the 0.23 value attributable to family.

Table 16.3 Assocation between male plumage brightness and frequency of haematozoa infection within 18 families of neotropical birds

	Number of species	Migrant species (%)	*p*
Coerebidae	7	0	0.68
Corvidae	10	50	0.27
Cotingidae	18	0	0.45
Dendrocolapidae	24	0	1.40
Formicariidae	34	0	0.52
Fringillidae	78	12	0.17
Furnariidae	32	0	0.72
Hirundinidae	7	100	0.15
Icteridae	35	14	0.15
Mimidae	7	28	0.34
Parulidae	39	54	0.62
Pipridae	14	0	0.03
Sylviidae	4	20	0.15
Thraupidae	71	4	0.06
Troglodytidae	18	22	0.38
Turdidae	22	23	0.50
Tyrannidae	87	15	0.91
Vireonidae	11	55	0.54

* Plumage brightness was scored on a scale of 1–6 in order of increasing brightness. Frequency of infection is defined as the number of number of individuals infected/individuals examined for each species. Families with 1 or 2 species were omitted. *p*-values were obtained from a median test of the null hypothesis that infection level varies independently of brightness (see text for details of analysis).

Discussion

The comparative method has received a great deal of attention recently, not all of which has been favourable (Cheverud *et al.* 1985; Felsenstein 1985; Pagel and Harvey 1988). The same might be said for Hamilton and Zuk's (1982) hypothesis about the role of parasites in sexual selection (Harvey and Partridge 1982; Read 1987, 1988). Given this background, combining the two may seem problematical. Several authors imply or suggest that an intraspecific, preferably experimental, test of the hypothesis is inherently more powerful than an interspecific one (Harvey and Partridge 1982; Borgia 1986; Read 1988; Jaenike 1988). I disagree. Despite the pitfalls associated with comparative studies, they nevertheless provide one of the few avenues for making generalizations about adaptation, avenues which are simply not available via experiments on single populations (Pagel and Harvey 1988). Both the comparative and experimental/observational methods can provide crucial evidence regarding the parasite–sexual selection hypothesis.

The Hamilton and Zuk hypothesis is one of the only models of sexual selection which makes an interspecific prediction at all (Bradbury and

Andersson 1987). While not a requirement for models of the evolution of female choice and male ornaments, and interspecific comparison provides an additional test to those from single-species study and, if the interspecific prediction is sufficiently novel and can be applied to several very different taxomonic groups, the hypothesis can be tested all the more rigorously. Support of the parasite–sexual selection interspecific prediction in such disparate groups as North American passerines (Hamilton and Zuk 1982; Read 1987, 1988), tropical birds (this chapter), and British and Irish freshwater fishes (Ward 1988) lends support to the idea that the association between brightness and parasites occurs for the reasons postulated by Hamilton and Zuk (1982), rather than from an unknown third correlate of the two variables (Read 1988). In addition, it should be noted that the various studies used different techniques for assessing 'brightness' with Ward's (1988) study using sexual dichromatism rather than absolute levels of coloration and Read (1987) using a version of Baker and Parker's (1979) technique of scoring body parts for colour.

The prediction that resident species should show a greater tendency for brighter species to harbour more parasites than migrant species was borne out by the analysis of tropical birds reported here. This finding further supports the idea that the correlation results from coevolution between host and parasite, and not from an alternative explanation such as brighter species being more gregarious and therefore more parasite-ridden than duller, more solitary species. Such an alternative explanation predicts no relationship between migratory status and the association between haematozoa and coloration. In addition, Read (1987) found no support for any 12 ecological and behavioural variables accounting for the correlation in European birds, and it is unlikely that these variables would reverse their importance or lack of it in tropical regions. Because of the paucity of data on the ecology and mating systems of many tropical bird species, an analysis such as Read's (1987) is not feasible for this study.

Male brightness was more strongly associated with parasite load than were either female brightness or sexual dichromatism as measured by the difference between male and female scores. If males are subject to more intense sexual selection than females, as is frequently supposed, then females would choose bright, parasite-free males more often than males would choose females, and the relationship between plumage cues and pathogen resistance would be stronger in males. Why the difference between males and females should not also reflect this difference in which sex chooses is not immediately clear; perhaps the degree to which females become drab or gaudy is limited by developmental genetic constraints, correlated to the brightness of the male of the species. Another explanation might be that increasing pressure from parasites may have also increased selection on both sexes being discriminating, which could reduce the differences in appearance between males and females. Alternatively, both sexual selection and natural selection may operate differentially on each sex, confounding a simple interpretation based on mate choice alone (Read, in press).

Both tests of the potentially confounding effects of phylogenetic similarity showed that shared ancestry alone is unlikely to account for the overall correlation between plumage brightness and infection level (Pagel and Harvey 1988). The combined *p*-value test (Folks 1984; Rice, in press), which provides a way to compare associations among several taxa, should prove helpful in alleviating this problem in comparative studies (Felsenstein 1985; Clutton-Brock and Harvey 1984). Multiple regression techniques provide a somewhat different approach to the same problem by allowing the evaluation of the relative contributions of several variables to explaining variation in the variable of interest.

Badly needed now are examinations of the combined effects of all parasites, both internal and external, on sexual selection within a species. Mallophagan parasite loads, for example, fluctuate widely with both season and host breeding status (Foster 1969), and the interactive effects of lice and blood parasites, or intestinal worms, are unknown. Intraspecific tests of the parasite–sexual selection hypotheses completed to date have included work on field crickets (Zuk 1987*a*,*b*), *Drosophila* (Jaenike 1988), guppies (Kennedy *et al.* 1987), lizards (Ressel and Schall 1989), mice (Edwards and Barnard 1987), and bowerbirds (Borgia 1986), as well as those in this volume, but these investigate only one parasite type in the host species.

Ecologists have generally given scant attention to the effects of parasites, particularly sublethal ones, on the ecology and behaviour of their hosts. Freeland (1976, 1981) suggested that parasites and pathogens influence the evolution of dominance and social behaviour, including mate choice, and Hausfater and Watson (1976) found that dominant male baboons actually had more worm eggs in their faeces than subordinate males did, a difference which may reflect greater food intake in the dominant individuals. The field remains largely unexplored. Nonetheless, it is likely that these sublethal pathogens have had a great effect on the evolution of many aspects of host ecology, morphology, and behaviour.

Acknowledgements

I am grateful to W.D. Hamilton for starting all this, and to Bill Rice and John T. Rotenberry for advice regarding data analysis. J. David Ligon classified the migrant and resident species. Randy Thornhill, Kristine

Johnson, Andrew F. Read, Sam Loker, John T. Rotenberry, and J. David Ligon made valuable comments on an earlier version of the manuscript.

References

Andersson, M. (1986). Evolution of condition-dependent sex ornaments and mating preferences: sexual selection based on viability differences. *Evolution*, **40**, 804–16.

Arnold, S.J. (1983). Sexual selection: the interface of theory and empiricism. In *Mate choice* (ed. P. Bateson), pp. 67–107. Cambridge University Press, Cambridge.

Baker, R.R. and Parker, G.A. (1979). The evolution of bird coloration. *Philosophical Transactions of the Royal Society of London*, **B287**, 63–130.

Borgia, G. (1986). Satin bowerbird parasites: a test of the bright male hypothesis. *Behavioral Ecology and Sociobiology*, **19**, 355–8.

Bradbury, J.W. and Andersson, M.B. (ed.) (1987). *Sexual selection: testing the alternatives*. Wiley, Chichester.

Cheverud, J.M., Dow, M.M., and Leutenegger, W. (1985). The quantitative assessment of phylogenetic constraints in comparative analyses: sexual dimorphism in body weight among primates. *Evolution*, **39**, 1335–51.

Clutton-Brock, T.H. and Harvey; P.H. (1984). Comparative approaches to investigating adaptation. In *Behavioural ecology: an evolutionary approach*, (2nd edn), (ed. J.R. Krebs and N.B. Davies), pp. 7–29. Sinauer, Sunderland, Massachusetts.

Edwards, J.C. and Barnard, C.J. (1987). The effects of *Trichinella* infection on intersexual interactions between mice. *Animal Behaviour*, **35**, 533–40.

Eshel, I. and Hamilton, W.D. (1984). Parent–offspring correlation in fitness under fluctating selection. *Proceedings of the Royal Society of London*, **B122**, 1–14.

Felsenstein, J. (1985). Phylogenies and the comparative method. *American Naturalist*, **125**, 1–15.

Fisher, R.A. (1930). *The genetical theory of natural selection*. Clarendon Press, Oxford.

Folks, J.L. (1984). Combination of independent tests. In *Handbook of statistics*, Vol. 4 (ed. P.R. Krishnaiah and P.K. Sen), pp. 113–22. Elsevier Science, New York.

Foster, M.S. (1969). Synchronized life cycles in the orange-crowned warbler and its Mallophagan parasites. *Ecology*, **50**, 315–23.

Freeland, W.J. (1976). Pathogens and the evolution of primate sociality. *Biotropica*, **8**, 11–24.

Freeland, W.J. (1981). Parasitism and behavioural dominance among male mice. *Science*, **213**, 461–2.

Hamilton, W.D. (1982). Pathogens as causes of genetic diversity in their host populations. In *Population biology of infectious disease agents*, Dahlem Konferenz 1982, (ed. R.M. Anderson and M. May), pp. 269–96. Verlag-Chemie, Weinheim.

Hamilton, W.D. and Zuk, M. (1982). Heritable true fitness and bright birds: a role for parasites? *Science*, **218**, 384–7.

Hausfater, G. and Watson, D.F. (1976). Social and reproductive correlates of parasite ova emissions by baboons. *Nature*, **262**, 688–9.

Harvey, P.H. and Partridge, L. (1982). Bird coloration and parasites—a task for the future? *Nature, London*, **300**, 480–1.

Heisler, I.L. (1984). A quantitative genetic model for the origin of mating preferences. *Evolution*, **38**, 1283–95.

Jaenike, J. (1988). Parasitism and male mating success in *Drosophila testacea*. *American Naturalist*, **131**, 774–80.

Keast, A. and Morton, E.S. (ed). (1980). *Migrant birds in the Neotropics: ecology, behavior, distribution and conservation*. Smithsonian Institution Press, Washington, DC.

Kennedy, C.E.J., Endler, J.A., Poynton, S.L., and McMinn, H. (1987). Parasite load predicts mate choice in guppies. *Behavioural Ecology and Sociobiology*, **21**, 291–5.

Kirkpatrick, M. (1982). Sexual selection and the evolution of female choice. *Evolution*, **36**, 1–12.

Lande, R. (1981). Models of speciation by sexual selection. *Proceedings of the National Academy of Sciences, USA*, **78**, 3721–5.

Nur, N. and Hasson, O. (1984). Phenotypic plasticity and the handicap principle. *Journal of Theoretical Biology*, **110**, 275–97.

Pagel, M.D. and Harvey, P.H. (1988). Recent developments in the analysis of comparative data. *Quarterly Review of Biology*, **63**, 13–40.

Pomiankowski, A. (1987). The costs of choice in sexual selection. *Journal of Theoretical Biology*, **128**, 195–218.

Read, A.F. (1987). Comparative evidence supports the Hamilton and Zuk hypothesis on parasites and sexual selection. *Nature*, **327**, 68 70.

Read, A.F. (1988). Sexual selection and the role of parasites. *Trends in Ecology and Evolution*, **3**, 97–102.

Ressel, S. and Schall, J.J. (1989). Parasites and showy males, malarial infection and color variation in fence lizards. *Oecologia*, **78**, 158–64.

Rice, W.R. (1990). A consensus combined *P*-value test and the family-wide significance of component tests. *Biometrics*. **46**, 303–8.

SAS Institute (1985). *SAS/STAT guide for personal computers*, Version 6 Edition. SAS Institute Cary, North Carolina.

Thornhill, R. (1980). Competitive, charming males and choose females: Was Darwin correct? *Florida Entomologist*, **63**, 5–30.

Trivers, R. (1985). *Social evolution*. Benjamin Cummings, Menlo Park, California.

Ward, P.I. (1988). Sexual dichromatism and parasitism in British and Irish freshwater fish. *Animal Behaviour*, **36**, 1210–15.

White, E.M., Greiner, E.C., Bennett, G.F., and Herman, C.M. (1978). Distribution of the hematozoa of Neotropical birds. *Revista de Biologia Tropicale*, **26**, 43–102.

Zar, J.H. (1974). *Biostatistical analysis*. Prentice-Hall, Englewood Cliffs, New Jersey.

Zuk, M. (1987*a*). The effects of gregarine parasites, time of day, and body size on spermatophore production and sexual selection in field crickets. *Behavioral Ecology and Sociobiology*, **21**, 65–72.

Zuk, M. (1987*b*). Variability in attractiveness of male field crickets (Orthoptera: Gryllidae) to females. *Animal Behaviour*, **35**, 1240–9.

Zuk, M., Thornhill, R., Johnson, K., and Ligon, J.D. (1990). Parasites and mate choice in red jungle fowl. *American Zoologist*, **30**, 235–44.

17 *Parasites, sexual ornaments, and mate choice in the barn swallow*

ANDERS PAPE MØLLER

Introduction

The significance of extraordinary secondary sexual characters like the conspicuous tails of some bird species has puzzled naturalists for centuries (see Darwin 1871 for an early review). A large number of hypotheses have been advanced to account for such ornaments (see review in Bradbury and Andersson 1987). One of the most recent hypotheses is the suggestion by Hamilton and Zuk (1982) that females choose mates by means of the size of sexual ornaments. Such secondary sexual traits are suggested to indicate genetic resistance by males to parasites and/or diseases. If sexual ornaments indicate resistance to parasites, female choice of long-tailed males will result in genetically determined resistance being in linkage disequilibrium with genes for sexual ornaments. Female choice may continue to be advantageous because genetic variation is maintained by coadaptational cycles of hosts and their parasites (Hamilton and Zuk 1982).

The evidence for this hypothesis is scarce although interspecific studies seem to support the hypothesis (Hamilton and Zuk 1982; Read 1987, 1990; Zuk, Chapter 16, this volume); plumage brightness of male passerine birds was positively correlated with the prevalence of blood parasites in both a North American and a European sample, and the positive correlation remained after controlling for phylogeny and a number of other confounding variables. Intraspecific studies (review in Read 1988) also seem to support its predictions, although most evidence is based on correlations rather than experimental demonstrations. However, very few intraspecific studies have tested the critical predictions from the Hamilton–Zuk hypothesis and the experimental evidence is very limited indeed.

A first step towards testing the Hamilton–Zuk hypothesis is to acquire observational data useful for evaluation of the predictions. Highly social or gregarious birds may be appropriate choices because a large number of transmission possibilities will render such species likely candidates for long-term coevolution between a number of parasites and their hosts.

Colonially breeding bird species should thus be appropriate study objects.

In this chapter I provide data on the abundance of ectoparasites of the semi-colonial passerine, the nominate subspecies of the barn swallow *Hirundo rustica rustica*. These data are used to test the predictions that: (1) parasite numbers are repeatable traits on particular individuals; (2) parasite loads of mated and unmated birds differ; (3) individual swallows with many parasites take a longer time to acquire a mate; (4) swallows mate assortatively with respect to parasite loads; (5) females prefer males with few parasites as extra-pair copulation partners; and (6) parasite loads affect the size of male tail ornaments.

Study animals

Barn swallows are 20-g insectivorous passerines that feed on the wing and usually nest in small colonies. Males have longer (on average 16 per cent) and more variable outermost tail feathers than females, whereas sexual size dimorphism in other characters is slight (less than 2 per cent) (Møller 1988*a*). The number of pairs per breeding site in my study area ranges from 1 to 48. Population size varies markedly from one year to another. Males attempt to attract a female by singing and displaying their long tails, and long-tailed individuals are more successful in acquiring a mate than short-tailed ones (Møller 1988*a*). Tail length increases with age, but female barn swallows still prefer long-tailed males when controlling for age effects experimentally or statistically (Møller 1988*a*, 1990). A number of males remain unmated, on average 12.5 per cent (out of 25–200 males each year) during 1977–88, and they pursue various alternative reproductive strategies like extra-pair copulations and infanticide (Møller 1988*b*). However, females also prefer long-tailed males as extra-pair copulation partners. Such copulations are not forced on female barn swallows since they are able to either accept or reject attempts (Møller 1988*a*). Both members of mated pairs build the nest; females incubate and brood alone, while both sexes feed the offspring. Most pairs raise two clutches per year. Nests are reused extensively both within and between years.

The most abundant swallow ectoparasite in the study area is the tropical fowl mite *Ornithonyssus bursa* which occurs in variable, but usually large numbers (Gjelstrup and Møller 1986). *O. bursa* is a haematophagous mite living both on birds and in their nests. Mites are found on the body beneath feathers, particularly on the head, where they feed specifically on swallow blood. I have recorded mites in barn swallow nests at all times of the year. The entire life cycle of the mite lasts only 5–7 days and each female produces 1–7, most frequently 2–5 eggs per clutch (Sikes and Chamberlain 1954). Outside of the reproductive season of swallows, mites can be found both on the hosts and in their nests.

The number of mites found in nests was not dependent on the presence of old nests. I removed all old nests from a number of colonies shortly after the breeding season and simultaneously cleaned the nest scars to remove any remaining nest material and parasites. Mite infestation levels in these experimental colonies were compared with controls. A total of seven such experiments did not reveal any significant differences in infestation level (A. P. Møller, in preparation). Mite infestation levels on adults upon arrival were, however, positively correlated with later infestation levels of nests during the same reproductive season ($r=0.60$, $t=8.37$, $df=70$, $p<0.001$; A. P. Møller, in preparation). Mite populations in nests thus apparently primarily originate from propagules inoculated by the adult barn swallows. However, I removed all old nests from the parasite study sites in order to avoid any potential bias because of mite infections from old nests.

A number of species of Mallophaga also occur on barn swallows. The two most frequently recorded species are *Hirundoecus malleus* and especially *Myrsidea rustica* (A. P. Møller, unpublished). They are ectoparasites living year round on their hosts and feeding on feathers. Mite infestation rates of male swallows were not significantly correlated with Mallophaga infestation rates ($r=0.18$, $N=84$, not significant), neither did analyses of data from single years reveal any significant relations. Infestations by one species of ectoparasite apparently did not promote infestations by another (Marshall 1981).

Methods

I studied barn swallows in the Kraghede area (57° 12′N, 10° 00′E), Denmark, during the years 1971–88. A large fraction of the population was colour-banded during the years 1983–88. Barn swallows were caught at night on the breeding sites shortly after arrival and body measurements were taken according to standard procedures after each capture. Daily observations of the behaviour of most individuals were made throughout the breeding season. Data on the mating status of males and their involvement in extra-pair copulations were acquired in the course of these regular observations. Reproductive data were also recorded throughout the breeding season. More detailed information can be found in Møller (1987).

Parasites were sampled both on adult barn swallows during 1984–87 and in their nests during 1982 and 1984–87. *O. bursa* was sampled on birds by gently pushing forwards the feathers on their heads and simultaneously counting the number of mites. Few mites were obtained when I placed 20 barn swallows caught at night in a jar with chloroform vapour (Fowler and Cohen 1983), the head of the birds being outside the jar during sampling. Mite abundance estimated from chloroform extractions and from counts of mites on the heads of barn swallows were positively correlated ($r=0.84$, $t=6.57$, $df=18$, $p<0.001$). Most mites were present on the head (86 per cent of the total number of mites extracted and counted), apparently because swallows cannot easily remove parasites from there. I made repeated counts of the number of mites on the same individual barn swallows during successive captures within the same season and in different seasons. In addition, I recorded the number of *O. bursa* in nests by placing my hand on the rim of the nest in the morning during the day after fledging. After 10 s I estimated the number of mites on my hand to either 0, 10, 100, 1000, or 10 000. These estimated numbers were of a similar magnitude to those counted after extraction from nests placed in Berlese funnels for 72 hours. The correlation between estimated and counted numbers was strongly positive ($r=0.98$, $t=71.10$, $df=41$, $p<0.001$), although numbers estimated were lower than those extracted (log 10 (estimated number) = 0.90(0.02) log 10 (extracted number)—0.005(0.03), $F=2596.76$, $df=1,42$, $p<0.001$).

Mallophages were also extracted from a few barn swallows placed in jars with chloroform vapour (Fowler and Cohen 1983). Because mallophages were killed by this sampling technique and because it was very time-consuming, I decided instead to count the number of holes made by mallophages in the tail and the wing feathers of barn swallows as an approximation of their abundance. Mallophage abundance as estimated from extractions and from counts of holes in feathers were significantly positively correlated ($r=0.62$, $t=3.35$, $df=18$, $p<0.01$), indicating that holes in feathers were reliable estimates of mallophage abundance. Since it takes mallophages quite a long time to make holes in feathers, these holes are estimates of their long-term abundance. I counted the number of these holes after each capture in the same season and in different seasons.

I used statistical procedures as described in Sokal and Rohlf (1981), Siegel (1956), and Becker (1984). A pair of counts for each bird formed the data base for the repeatability calculations. Sample sizes for repeatabilities are thus the number of such pairs. All statistical tests are two-tailed.

Table 17.1 Repeatability (R) of parasite loads (SE in brackets) on male barn swallows within and between years for non-standardized and standardized data (value adjusted to mean = 0, SD = 1). Calculations were based on pairs of counts from all the different birds

Parasite load	Within years			Between years		
	R	F	N	R	F	N
Non-standardized data						
Mites on bird	0.90(0.09)	19.22†	84	0.83(0.15)	10.40†	49
Mallophaga on bird	0.99(0.03)	187.75†	84	0.88(0.13)	16.17†	49
Mites in nest	0.39(0.18)	2.43†	49	0.79(0.16)	8.43†	49
Two clutches						
in same nest	−0.13(0.21)	0.78	17	—	—	—
in two nests	0.41(0.23)	2.41*	32	—	—	—
Standardized data						
Mites on bird	0.91(0.08)	22.35†	84	0.94(0.10)	31.32†	49
Mallophaga on bird	0.99(0.02)	341.38†	84	0.96(0.12)	18.66†	49
Mites in nest	0.42(0.18)	2.43*	49	0.80(0.15)	9.11†	49
Two clutches						
in same nest	−0.14(0.21)	0.76	17	—	—	—
in two nests	0.43(0.23)	2.43*	32	—	—	—

* $p<0.01$.
† $p<0.001$.

Results

Repeatability of parasite loads

Repeatabilities are useful measurements of the reliability of sampling methods (Becker 1984). If samples taken by the same method on the same individual barn swallow by the same person vary extensively between different days, the sampling technique has to be changed. Repeatabilities of mite and mallophage counts with at least 1 month's interval during the same season were very high, especially when taking into consideration sampling variance and changes in numbers of parasites due to emigration, immigration, reproduction, and death (Table 17.1). Standardization of each of the two samples of parasite counts to a mean of zero and a standard deviation of one did not increase repeatabilities (Table 17.1).

Repeatabilities between years are useful because they provide an upper limit to heritability, low repeatabilities indicating low heritabilities of resistance to parasites. Statistically significant heritability estimates can then only be obtained with very large sample sizes (Falconer 1983). High repeatabilities indicate potentially high heritabilities (Falconer 1983). High repeatabilities between years exceeding 0.80 were found for both

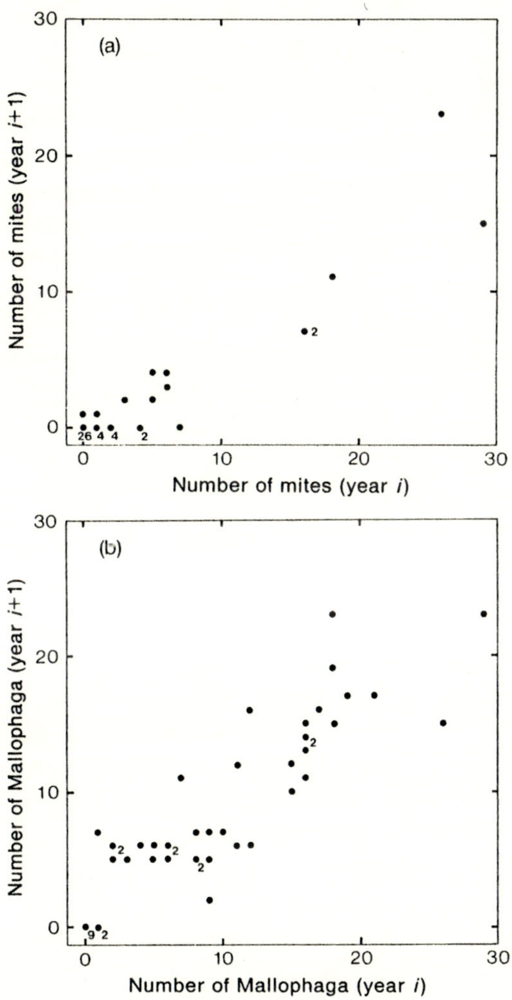

Fig. 17.1. (a) Repeatability in successive years of O. *bursa* mite loads on male barn swallows captured upon arrival at the breeding grounds. Numbers refer to number of individual swallows. (b) Repeatability in successive years of Mallophaga loads on male barn swallows captured upon arrival at the breeding grounds. Numbers refer to number of individual swallows.

mite and mallophage counts (Fig. 17.1, Table 17.1). Individual barn swallows that suffered from high parasite infestations in one year thus generally also had high infestation levels the following year. One might think that parasites might affect younger barn swallows in particular and that parasite numbers might decrease with age (see Scott and Edman, Chapter 10, this volume for discussion). This, however, was not the case as is evident from the standardized parasite counts, in which the annual mean value was adjusted to 0 with a standard deviation of 1. The recalculated repeatabilities did not increase markedly (Table 17.1) indicating that age effects were small.

Mite counts in different nests of the same individual within the same season had a significant repeatability, whereas counts in the same nests during different clutches had a low repeatability (Table 17.1). Numbers of mites increased when barn swallows reused the same nest during the same season, whereas newly occupied or built nests had mite loads similar to previous nests by the same individuals within the same season. Mite counts in different newly built first-clutch nests of the same barn swallows in successive years were highly repeatable confirming that individual barn swallows retained their parasite status between years. Standardization of the mite counts did not improve repeatabilities suggesting that age effects were negligible (Table 17.1).

Parasite loads of mated and unmated males

The Hamilton-Zuk hypothesis predicts that mated and unmated individuals should differ in parasite loads since females would choose mates from secondary sexual traits correlated with resistance against parasites and/or diseases. A number of barn swallow males remain unmated each year making a comparison of parasite loads of mated and unmated individuals possible (Møller 1988*b*).

Mites were totally absent from a larger fraction of the mated as compared to the unmated males (Table 17.2). Mallophage numbers also differed in the same way although to a smaller degree. Parasite loads were also significantly higher on unmated barn swallows as compared to mated conspecifics (Table 17.2).

Parasite loads and duration of the pre-mating period

Female barn swallows prefer long-tailed males over short-tailed ones (Møller 1988*a*). This preference is revealed by significantly shorter premating periods among long-tailed as compared to short-tailed males (Møller 1988*a*).

The length of the pre-mating period was significantly positively related to parasite loads (mites: $r=0.32$, $t=2.57$, $df=71$, $p<0.02$; Mallophaga: $r=0.24$, $t=2.08$, $df=71$, $p<0.05$). The duration of the premating period was 3.7 ± 3.3 days (mean $\pm$ SD); none of the analyses of data from single years yielded significant relationships, probably because

Table 17.2 Parasite loads and infestation rates of different groups of male barn swallows

Compared groups	Range	Mean	Median	z	Infestation rate (%)	G	N
O. bursa							
Mated	0–29	5.7	0	4.26‡	32	9.88†	74
Unmated	0–38	10.7	12		82		11
EPC	0–15	1.7	0	4.08‡	24	12.61‡	17
Others	0–36	7.1	7		71		68
Mallophaga							
Mated	0–27	9.4	7	4.01‡	82	3.86*	74
Unmated	6–23	13.5	12		100		11
EPC	0–16	6.2	5	3.82‡	59	9.10†	17
Others	0–26	11.3	11		91		68

t scores according to Mann–Whitney U-test. EPC, males engaged in successful extra-pair copulations.

* $p<0.05$.
† $p<0.01$.
‡ $p<0.001$.

of the small sample sizes. All correlation coefficients were, however, positive and of the same magnitude as for the pooled sample.

Assortative mating for parasite loads

Females may prefer mates that have sexual ornaments indicative of genetic resistance to parasites and/or diseases. In monogamous species, where males invest heavily in rearing offspring, it is likely that males also should prefer female mates with resistance to parasites. The overall outcome should be a tendency for assortative mating with respect to parasite loads.

There was a highly significant tendency for assortative mating with respect to mites ($r=0.76$, $t=9.85$, $df=71$, $p<0.0001$) and a less clear relationship with respect to Mallophaga ($r=0.27$, $t=2.36$, $df=71$, $p<0.02$; Fig. 17.2). All analyses for mites during single years were highly significant, whereas analyses for Mallophaga were only significant during 1 out of 4 years (even though all Spearman rank correlation coefficients were positive and of the same size). Tendencies for pair members to have mite and mallophage loads of the same magnitude conceivably may be due to transmission. Male barn swallows, however, are caught before the arrival of their mates, and females are caught shortly after their arrival before they have chosen a mate. Transmission

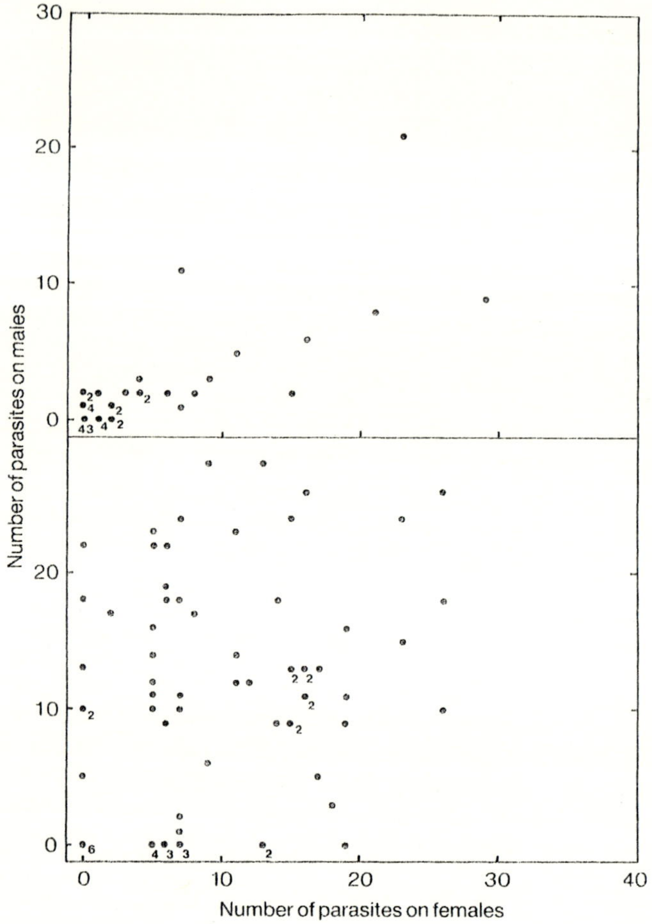

Fig. 17.2. Relations between *O. bursa* mite (upper) and Mallophaga (lower) loads of female barn swallows and their mates. Parasite loads were recorded upon arrival at the breeding grounds. Numbers refer to number of swallow pairs.

cannot account for the relationship between mallophage numbers on mates since holes in feathers are made during a very long period.

Extra-pair copulation partners and parasites

Barn swallows regularly engage in extra-pair copulations. All males attempt extra-pair copulations, especially when they do not have any mate guarding duties (Møller 1985). They approach females in their fertile period and especially choose those that are not well-guarded by male mates. Male barn swallows then hover over the female and try to mount. The female determines the outcome of these attempts, since she can either remain sitting and take part in the extra-pair copulation, or behave aggressively and fly away (Møller 1985). Females do not actively solicit copulations or extra-pair copulations. Old males and young females are engaged in extra-pair copulations more often then expected by chance (Møller 1985). Female barn swallows were found to choose extra-pair copulation partners with longer-than-average tails, especially if they were mated with short-tailed individuals (Møller 1988*a*). Females do not get any material benefits from extra-pair copulation partners.

Males that were successful in achieving extra-pair copulations more often lacked mites than did unsuccessful males, and this was also the case with respect to Mallophaga (Table 17.2). The number of mites and Mallophaga was much higher on unsuccessful as compared to successful males (Table 17.2).

Parasite loads and sizes of tail ornaments

Females should only be able to choose mates with resistance to parasites and/or diseases, if external characters are reliable predictors of parasite loads (Hamilton and Zuk 1982). The size of extravagant sexual ornaments should be especially indicative of such resistance.

I searched for correlations between parasite loads on male swallows when captured shortly after their arrival and the size of the sexually dimorphic outermost tail feathers. Mite loads were negatively correlated with the size of the outermost tail feathers (Fig. 17.3; $r = -0.26$, $t = 2.44$, $df = 82$, $p < 0.05$). Mallophage counts were even more strongly negatively correlated with the size of tails (Fig. 17.3; $r = -0.42$, $t = 4.19$, $df = 82$, $p < 0.001$). Analyses of parasite loads and tail sizes in single years gave similar results.

The relationships between parasite loads and the size of sexual ornaments have phenotypical and/or genetic reasons. The size of tail ornaments can change only once a year, namely during the annual moult in winter quarters. Parasite loads may have negative effects on the condition of swallows and thereby indirectly on tail size, since change in tail size from year i to year $i+1$ was significantly negatively related to mite loads on male barn swallows at arrival in year i ($r = -0.47$, $t = 4.16$, $df = 61$, $p < 0.001$) and to estimated number of mites in first nests in year i (Fig. 17.4; $r = -0.39$, $t = 3.31$, $df = 61$, $p < 0.01$).

Discussion

The Hamilton–Zuk hypothesis has generated a lot of debate about the role of parasites in sexual selection (e.g. Bradbury and Andersson 1987).

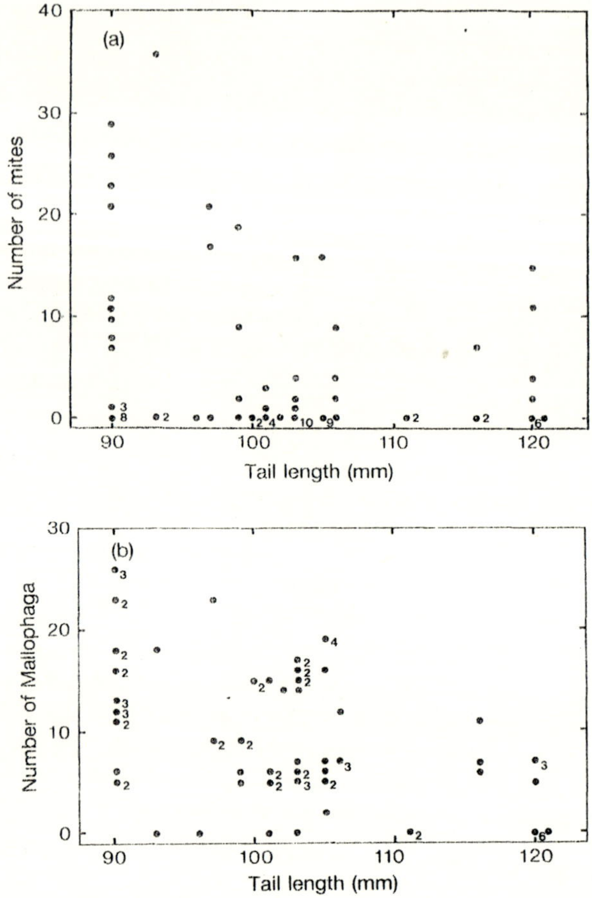

Fig. 17.3. (a) *O. bursa* mite loads recorded upon arrival at the breeding grounds in relation to tail length of male barn swallows. Numbers refer to number of individuals. (b) Mallophaga loads recorded upon arrival at the breeding grounds in relation to tail length of male barn swallows. Numbers refer to number of individual swallows.

Real data from studies of individually marked birds are scarce, but are indeed a prerequisite for a proper evaluation of the hypothesis. A preliminary idea of the ability of the hypothesis to account for parasite patterns can be obained from observations. Contrary to many other kinds of evolutionary questions, continuous coevolution between parasites and their hosts should result in distributions of parasites on hosts that can be used directly for a test of the hypothesis. Other traits (e.g. morphological ones) may have arisen during evolution, but they are not under selection today (Gould and Lewontin 1979). Observational studies

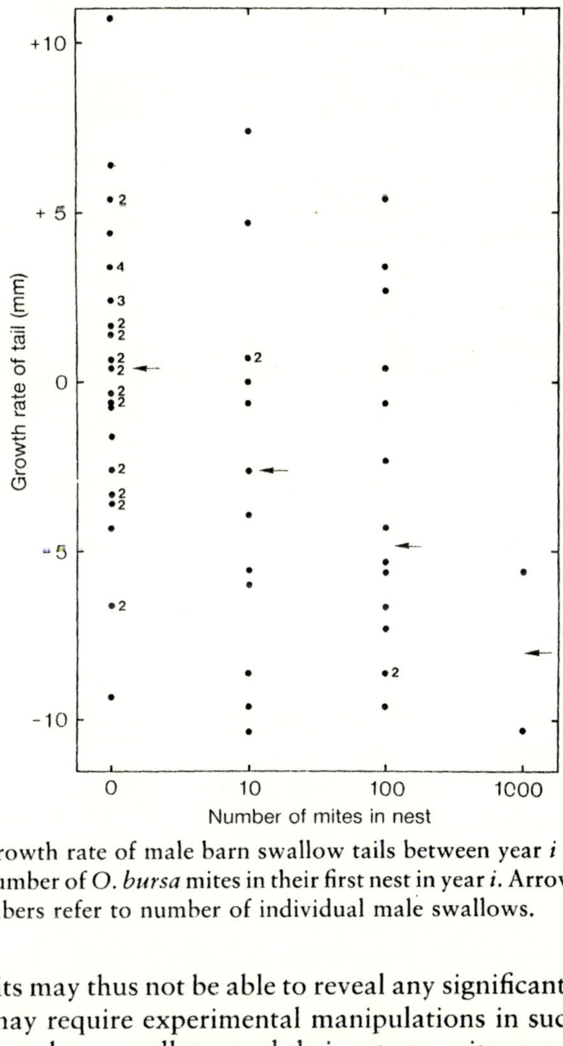

Fig. 17.4. Growth rate of male barn swallow tails between year *i* and year *i*+1 in relation to number of *O. bursa* mites in their first nest in year *i*. Arrows indicate mean values. Numbers refer to number of individual male swallows.

of such traits may thus not be able to reveal any significant relationships, and tests may require experimental manipulations in such cases.

My data on barn swallows and their ectoparasites are indicative of a role for parasites in the maintenance of long tails in males and thus in mate choice. The Hamilton–Zuk hypothesis (1982) predicts that: (1) there should be a relationship between intensity of parasite infection and some component of host fitness; (2) there should be heritable variation in parasite resistance; (3) the expression of sexual ornaments should covary with parasite loads; and (4) females should choose males with fewer parasites.

My data also show that *O. bursa* mites negatively affected the fitness of their hosts. Barn swallows with high parasite burdens in their nests had reduced breeding success, and fledglings from nests with high parasite loads were comparatively light (Møller 1987). Adult barn swallows actively avoided old nests with mites, and the tendency of pairs to build new nests for their second clutches was positively related to the mite loads in the nests of their first breeding episode (Møller 1987; see also Loye and Carroll, Chapter 12, this volume). The fitness costs of parasitism were experimentally demonstrated by manipulation of mite loads: nests were either fumigated, inoculated with mites, or kept as controls (A. P. Møller, in preparation). Many mites, experimentally introduced into barn swallow nests built for first clutches, significantly reduced nestling survival in both first and second clutches, reduced body condition of nestlings in both first and second clutches, and caused premature fledging in both first and second clutches. Mites also delayed second clutches, reduced their frequency, and reduced clutch size in second clutches (A. P. Møller, in preparation). Fitness costs of mite parasitism were therefore large. Mites are often vectors for blood parasites and diseases, but it remains to be established whether this was the case in *O. bursa* parasitizing barn swallows. In other bird species, *O. bursa* has been suspected to cause nest desertion and other kinds of fitness loss (Burgerjon 1964; Petersen 1979; Powlesland 1977, 1978). Mallophaga are not thought to have direct fitness costs to their hosts (Marshall 1981; see also Clayton, Chapter 14, this volume), but are regular vectors for other parasites and infectious diseases (Seegar *et al.* 1976). Whether this is the case for the barn swallow remains unknown.

There is no evidence for genetically determined variability of resistance to the parasites mentioned in the barn swallow. High repeatabilities of parasite loads between years may, however, suggest that resistance to parasitism might be heritable (e.g. Wakelin and Blackwell 1988). The lack of age effect on parasite loads in barn swallows is unusual (see Scott and Edman, Chapter 10, this volume). This may indicate a long-lasting specific association between the parasites and the barn swallow resulting in a long evolutionary history of coadaptation. Additional data on heritability of resistance to parasites has been accumulated by experimentally manipulating *O. bursa* parasite loads and simultaneously performing cross-fostering of offspring. These data indicate a strong genetic, but a weak environmental, effect on the abundance of mites on barn swallow offspring reared together (A. P. Møller, in preparation).

Tail ornaments of male swallows were found to vary with parasite burdens in the predicted direction since males with many parasites had shorter tails as compared to parasite-free individuals. Male swallows did not have tails of the same size each year, and the change in tail size was significantly related to parasite loads of the birds and their nests. Large parasite burdens were related to a decrease in tail length after the moult, suggesting a direct relationship between the ability to cope with parasites and the phenotype. This relationship should be amenable to an experimental test, since males with high parasite burdens should be affected and subsequently decrease the size of their tails during moult. This, in turn, should have fitness consequences for the male barn swallow because female choice is directly related to the size of male sexual ornaments (Møller 1988*a*). I have set up such an experiment and am waiting for the swallows to return from their winter quarters as this is being written.

There is evidence that swallow females were choosing mates on the basis of their parasite loads, since unmated males more often had parasites and more parasites than mated ones. Female barn swallows apparently also rejected as extra-pair copulation partners males with parasites, especially those with many parasites. All males attempted extra-pair copulations, irrespective of their mating status and parasite loads. A simple alternative could be that females were choosing males from the size of their tails. Since tail length and parasite loads are negatively correlated, males preferred by females will, on average, have few parasites. Because of the negative correlation between parasite burdens and tail ornament size and changes in size, respectively, it is likely that both traits could be affected by genes in linkage disequilibrium. By preferring males with long tails and as a result obtaining partners with few or no parasites, genes for long tails and for parasite resistance will get coupled. Even if tail length and parasite resistance are not coupled, and even if tail length is not heritable, the expression of the tail character may still indicate resistance to parasites as predicted by the Hamilton–Zuk hypothesis. An alternative explanation for the correlations between tail length, parasite loads, and mate choice could be that long-tailed males picked up fewer parasites because of behavioural differences. However, mite loads of nests were positively correlated with mite loads of barn swallow nest owners at arrival (A. P. Møller, in preparation), suggesting that nests are infected from the adult birds. Since long-tailed male barn swallows are more successful in acquiring extra-pair copulations than short-tailed males (Møller 1988*a*; unpublished), they also encounter more females. The probability of becoming infected by mites should thus be higher for long-tailed than for short-tailed males. Despite these frequent encounters of females by long-tailed males, such male barn swallows had fewer parasites than other males. Female preferences for males with long tails as a measure of parasite resistance should be tested experimentally.

Any progress in the test of the Hamilton–Zuk hypothesis will rely on experimental manipulation of parasite loads. Good and consistent measures of the size of parasite burdens and their relationships with sexual ornaments and behaviour of individually marked animals are a must as a platform from which experiments can be designed.

Acknowledgements

This study was supported by a grant from the Swedish Natural Science Research Council. I am grateful to W. D. Hamilton, J. Loye, S. Ulfstrand, and two anonymous referees for suggestions on a previous draft of this chapter.

References

Becker, W.A. (1984). *Manual of quantitative genetics.* Academic Enterprises, Pullman, Washington.

Bradbury, J.W. and Andersson, M.B. (ed.) (1987). *Sexual selection: testing the alternatives.* J. Wiley, Chichester.

Burgerjon, J.J. (1964). Some census notes on a colony of South African cliff swallows *Petrochelidon spilodera* (Sundevall). *Ostrich*, **35**, 77–85.

Darwin, C. (1871). *Sexual selection and the descent of man.* J. Murray, London.

Falconer, D.S. (1983). *Introduction to quantitative genetics* (2nd edn). Longman, London.

Fowler, J.A. and Cohen, S. (1983). A method for the quantitative collection of ectoparasites from birds. *Ringing and Migration*, **4**, 185–9.

Gjelstrup, P. and Møller, A.P. (1986). A tropical fowl mite, *Ornithonyssus bursa* (Berlese, 1988) (Macronyssidae, Gamasida) in Danish swallow (*Hirundo rustica*) nests; with a review of mites and ticks from Danish birds. *Entomologiske Meddelelser*, **53**, 119–25.

Gould, S.J. and Lewontin, R.C. (1979). The spandrels of San Marco and the Panglossian paradigm: a critique of the adaptationist programme. *Proceedings of the Royal Society of London*, **B205**, 581–98.

Hamilton, W.D. and Zuk, M. (1982). Heritable true fitness and bright birds: a role for parasites. *Science*, **218**, 384–7.

Marshall, A.G. (1981). *The ecology of ectoparasitic insects.* Academic Press, London.

Møller, A.P. (1985). Mixed reproductive strategy and mate-guarding in a semi-colonial passerine, the swallow *Hirundo rustica. Behavioral Ecology and Sociobiology*, **17**, 401–8.

Møller, A.P. (1987). Advantages and disadvantages of coloniality in the swallow, *Hirundo rustica. Animal Behaviour*, **35**, 819–32.

Møller, A.P. (1988*a*). Female choice selects for male sexual tail ornaments in the monogamous swallow. *Nature*, **332**, 640–2.

Møller, A.P. (1988*b*). Infanticidal and anti-infanticidal strategies in the swallow *Hirundo rustica. Behavioral Ecology and Sociobiology*, **22**, 365–71.

Møller, A.P. (1990). Male tail length and female mate choice in the monogamous swallow *Hirundo rustica. Animal Behaviour*, **39**, 458–65.

Petersen, G.W. (1979). Infestations of *Ornithonyssus bursa*, a haematophagous mite, on starling over the non-breeding season. *New Zealand Journal of Zoology*, **6**, 319–20.

Powlesland, R.G. (1977). Effects of the haematophagous mite *Ornithonyssus bursa* on nestling starlings in New Zealand. *New Zealand Journal of Zoology*, **4**, 85–94.

Powlesland, R.G. (1978). Behaviour of the haematophagous mite *Ornithonyssus bursa* in starling nest boxes in New Zealand. *New Zealand Journal of Zoology*, **5**, 395–9.

Read, A.F. (1987). Comparative evidence supports the Hamilton and Zuk hypothesis on parasites and sexual selection. *Nature*, **328**, 68–70.

Read, A.F. (1988). Sexual selection and the role of parasites. *Trends in Ecology and Evolution*, **3**, 97–102.

Seegar, W.S., Schiller, E.L., Sladen, W.J.L., and Trpis, M. (1976). A Mallophaga, *Trinoton anserinum*, as a cyclodevelopmental vector for a heartworm parasite of waterfowl. *Science*, **194**, 739–41.

Siegel, S. (1956). *Nonparametric statistics for the behavioral sciences*. McGraw-Hill Kogakusha, Tokyo.

Sikes, R.K. and Chamberlain, R.W. (1954). Laboratory observations on three species of birds mites. *Journal of Parasitology*, **40**, 691–7.

Sokal, R.R. and Rohlf, F.J. (1981). *Biometry* (2nd edn). W.H. Freeman, San Francisco.

Wakelin, D. and Blackwell, J.M. (ed.) (1988). *Genetics of resistance to bacterial and parasitic infection*. Taylor and Francis, London.

18 *Phenotypic correlates of blood parasitism in the common grackle*

CARL E. KIRKPATRICK, SCOTT K. ROBINSON, and URIEL D. KITRON

Introduction

In a novel hypothesis, Hamilton and Zuk (1982) suggested that the susceptibility of North American passerine birds to infections by haemotropic parasites is associated with the expression of secondary sexual characteristics. They further suggested that the expression of these characteristics may influence mate choice. This hypothesis has stimulated an enormous amount of research on the relationships of parasite infections to plumage brightness both between species (Read 1987; Zuk, Chapter 16, this volume) and within species (e.g. in this volume: Møller, Chapter 17; Burley *et al.* Chapter 19; Johnson and Boyce, Chapter 20; Spurrier *et al.*, Chapter 21).

If the Hamilton and Zuk hypothesis is correct, then at least five conditions must be met (Read 1988): (1) the host must be infected with parasites that reduce fitness; (2) the parasite must alter the appearance of a sexually dimorphic trait of importance in female choice; (3) susceptibility to parasitism must be heritable; (4) females must prefer the morphology of the uninfected males; and (5) there are at least two species of parasites that cycle, and resistance to one is not correlated with resistance to another.

The original purpose of this study was to examine the second of these predictions in a free-living population of common grackles (*Quiscalus quiscula*) (Passeriformes: Icterinae) that were found to be naturally infected with several of the haemotropic parasite genera examined by Hamilton and Zuk (1982) and Read (1987). These results were intended to be the first step in a more comprehensive test of Hamilton and Zuk's model in grackles, but, as we describe below, the initial results uncovered an enormous amount of variation that complicated our interpretations of the data. The purpose of this chapter is therefore twofold. First, we present the results of our exploratory study into the correlations between grackle phenotypes and morphology. Second, and perhaps more usefully, we describe sources of natural variation in parasites and morphology within populations that may complicate interpretations of field data.

Methods

Birds

A grackle population in an orchard in Urbana, Illinois (40°6′N, 88°12′1W), was studied from 12 March to 26 June, 1987, and from 21 June to 30 June 1988. Grackles began arriving at the study site in late February, and they remained throughout the summer. Fledglings began appearing in the traps in late May (1987) or early June (1988). Adult male and female grackles are readily distinguishable according to plumage. Sex-related external characteristics of fledglings have not been identified.

In the morning, grackles were captured in corn-baited walk-in or Potter traps within the orchard. Captured birds were placed in boxes and transported to a nearby laboratory. There, birds were banded with uniquely numbered, US Fish and Wildlife Service metal leg bands, weighed, measured, and (in 1988) examined for nits (louse eggs).

Blood samples (0.2 ml) were taken aseptically by jugular venipuncture using a disposable tuberculin syringe fitted with a 27-gauge needle. Haemostasis was assured by digital pressure over the venipuncture site for approximately 30 s. Blood smears were made and quickly dried with the aid of an aerosol duster. The smears were fixed in absolute methanol, stained in Giemsa stain solution, and permanently coverslipped. Blood was also placed into heparinized, 50-μl capillary tubes and centrifuged in a microhaematocrit centrifuge for 5 min.

In addition, plumage coloration of adult grackles of both sexes captured during 1988 was determined for the most iridescent regions of the body (i.e. head, neck, and breast) by examination under diffused fluorescent lighting. All of the birds were examined under the same bank of fluorescent bulbs at a fixed distance. The relative proportions of six primary colours at each body region were evaluated according to an arbitrary 0 to 5 scoring system, in which 0 indicated the absence of a colour and 5 indicated the predominance of a colour. All colours were scored by the same person during the 1988 field season to assure consistency. A composite score for each colour on each bird was determined by adding the scores from each body region. We did not assign a brightness score; scores for green and bronze were previously shown to be directly correlated with brightness scores (S. Robinson, unpublished data).

After banding, examination, and blood sampling, birds were returned to the study site and released. Some of the birds captured and sampled during 1987 were recaptured during 1988, and some of the 1988 birds were recaptured later in the same season. Because of the temporal variation in parasite prevalences (described below), data from recaptures were treated the same as those from primary captures.

Examinations for parasites

Blood smears were microscopically examined for parasites by first scanning them under ×200 magnification for 5 min (one smear per bird). This was useful for detecting the larger parasites, such as *Leucocytozoon* spp. and microfilariae of filariid nematodes. Each smear was then examined for 5 min under ×400 magnification using dark-field illumination. This technique makes *Plasmodium* spp. and *Haemoproteus* spp. more conspicuous, because the haemozoin ('pigment') granules of these intraerythrocytic parasites are highly refractile under dark-field illumination. Then, each smear was examined for an additional 5 min under standard (bright-field) illumination at ×400 magnification to detect non-refractile parasites (e.g. *Trypanosoma* spp.). Finally, any protozoan parasite detected was examined under ×1000 (oil-immersion) magnification for specific identification.

Parasitaemia (i.e. the proportion of blood cells infected) of the protozoan parasites was determined by actually counting 2000 erythrocytes as recommended by Godfrey *et al.* (1987).

The plasma portion of the blood in centrifuged capillary tubes was microscopically examined under ×100 magnification to detect microfilariae and trypanosomes (Bennett 1962). Motile trypanosomes were found just above the 'buffy coat' (packed white cells), whereas microfilariae were distributed the length of the plasma portion of the tube. Microfilaraemias (i.e. the concentration of microfilariae in the blood) were determined by counting all of the microfilariae in the plasma portion of each capillary tube. A bird was considered to be infected with trypanosomes or microfilariae if positive by either of the tests employed (i.e. blood-smear examination or capillary-tube examination). Representative blood smears have been deposited in the International Reference Centre for Avian Hematozoa, Memorial University, St. John's, Newfoundland, Canada A1B 3X9 and have been assigned accession numbers 99439–99443.

Results

Prevalence and intensity of parasite infections

The majority of grackles were infected with *Haemoproteus quiscalus* and/or microfilariae of an unidentified filariid nematode (Table 18.1). These nematodes were either *Eufilaria hibleri* or *Chandlerella quiscali* or

Table 18.1 Monthly prevalences of haemotropic parasite infections in adult common grackles (*Quiscalus quiscula*) 1987–88

	Month	*n*	Percentage infected with* H	L	P	T	M
Male							
	March	54	52	11	0	17	59
	April	141	78	12	1	19	59
	May	81	75	16	1	30	68
	June	24	25	13	8	42	88
	Total	300	68	13	1	24	63
Female							
	March	33	52	6	0	12	42
	April	88	85	14	0	19	58
	May	103	81	12	1	38	83
	June	70	60	13	0	34	79
	Total	294	71	12	<1	28	70
	Grand total	594	71	13	1	26	66

* H, *Haemoproteus*; L, *Leucocytozoon*; P, *Plasmodium*; T, *Trypanosoma*; M, *microfilariae* of unidentified *filariid nematode*.

both (Granath 1981; Huizinga and Granath 1984). The microfilariae seen on stained blood smears appeared to be those of *C. quiscali*, based on measurement. Other parasites found included *Leucocytozoon majoris*, *Leucocytozoon fringillinarum*, *Trypanosoma avium*, and *Plasmodium* spp. (Table 18.1). During 1988, nits of the louse *Myrsidea fuscomarginata* (Mallophaga: Menoponidae) were observed on the feathers of the head and neck of some of the birds (Table 18.2). All the results described below are based on multiple comparisons. The Bonferroni inequality was used to adjust the *p*-values based on the number of comparisons. Results that are defined as significant take this adjustment into account.

There were no significant differences between sexes with respect to haemotropic parasite prevalences ($p > 0.05$, χ^2 test; Table 18.1); however, the prevalence of nits was significantly higher in males compared to females ($\chi^2 = 13.2$, $df = 1$, $p < 0.001$; Tables 18.1 and 18.2). Temporal variation in parasite prevalences was observed during both the 1987 and 1988 breeding seasons. In both males and females, *H. quiscalus* prevalence peaked during April and May, and both *T. avium* and microfilaria prevalences rose in May. *Leucocytozoon* spp. infections occurred rather constantly throughout the observation period; *Plasmodium* spp. prevalences were too low to evaluate their temporal distribution. While too few birds were examined for nits during March of 1988, prevalence of nit

Table 18.2 Monthly prevalences (1988) of *Myrsidea fuscomarginata* in adult common grackles (*Quiscalus quiscula*)

	Month	*n*	Number infested	Percentage infested
Male				
	March	4	3	75
	April	103	40	39
	May	49	16	33
	June	12	0	0
	Total	168	59	35
Female				
	March	6	0	0
	April	71	21	30
	May	50	5	10
	June	33	1	3
	Total	160	27	17
	Grand total	328	86	26

infestation appeared steady during April and May, with a decline in June.

Parasitaemias of *H. quiscalus* were estimated for the first 100 birds examined during the 1988 observation period. *Haemoproteus quiscalus* infections were detected in 73 of these birds. The median parasitaemia of these infections was <1 infected erythrocyte of 2000 erythrocytes examined (i.e. <0.05 per cent); the highest parasitaemia recorded was 19/2000 (1 per cent). Of the *Haemoproteus*-infected birds, 19 had parasitaemias ⩾ 1/2000; the median parasitaemia in this group was 4/2000 (0.2 per cent). Because the parasitaemias were mostly low and the range of values was so narrow, we elected not to evaluate further *H. quiscalus* parasitaemias.

Parasitaemias of microfilariae varied widely in both males and females (Table 18.3). The mean intensities (i.e. mean number of parasites/infected birds) rose about 10-fold between March and April, and they declined somewhat in June. Intensities were significantly higher in female versus male grackles ($Z = 2.19$, $p = 0.03$, Wilcoxon rank-sum test).

Among 59 nit-infested males, the mean intensity was 7.1 (range, 1–30) and among 27 females the mean intensity was 8.8 (range, 1–25) (Table 19.2). There was no significant difference between infested males and females with respect to nit burden ($Z = 0.21$, $p > 0.05$, Wilcoxon rank-sum test).

Table 18.3 Monthly intensities of microfilariae in adult common grackles (*Quiscalus quiscula*) 1987–88

	Month	*n*	Microfilaraemia* Mean ± SD	Median	Range
Male					
	March	22	18 ± 33	5	1–140
	April	80	131 ± 292	27	1–1807
	May	55	184 ± 303	35	1–1399
	June	21	163 ± 259	40	1–913
	Total	178	137 ± 277	26	1–1807
Female					
	March	8	18 ± 27	7	1–81
	April	51	183 ± 303	29	1–1148
	May	84	305 ± 642	94	1–5039
	June	54	230 ± 419	48	1–2367
	Total	197	241 ± 500	52	1–5039
	Grand total	375	192 ± 412	31	1–5039

* Number of microfilariae per capillary tube (approximately 40 μl of blood).

Haemotropic parasite prevalences in the 24 fledgling birds examined during 1987 and 1988 were: *H. quiscalus*, 100 per cent; *T. avium*, 17 per cent; and microfilariae, 13 per cent. No *Leucocytozoon* spp. or *Plasmodium* spp. infections were detected in fledglings.

Considering the capillary-tube method as the definitive test for trypanosome and microfilaria detection (C. E. Kirkpatrick, unpublished data), the sensitivity of blood-smear examination as a screening test for these parasites was calculated. Thus, the sensitivity (i.e. the proportion of infected birds detected by the test) of blood-smear examination was 45.1 per cent for microfilaria detection and 12.5 per cent for trypanosome detection. Blood smears never detected trypanosomes that were undetected by the capillary tube method.

Structural correlates of parasitism

Results of body mass, tail length, and wing length measurements are presented in Table 18.4. Males were significantly heavier than females, and tails and wings of males were significantly longer than those of females (in all cases, $p < 0.001$, Wilcoxon rank-sum test, Bonferroni inequality used). Therefore, males and females were evaluated separately for structural correlates of parasitism. Within sexes, significant positive correlations ($p < 0.001$ in all comparisons) were found between body

Table 18.4 Some mensural characteristics (1988) of adult common grackles (*Quiscalus quiscula*)

Factor	*n*	Mean ± SD	Median	Range
Male				
Mass (g)	298	119.6 ± 7.1	120	91–142
Tail length (mm)	287	125.5 ± 7.5	125	107–147
Wing length (mm)	287	138.9 ± 3.8	139	124–150
Female				
Mass (g)	289	92.2 ± 7.3	92	74–124
Tail length (mm)	274	109.3 ± 6.4	109	94–125
Wing length (mm)	274	125.9 ± 4.0	126	112–142

Table 18.5 Associations between mensural characteristics and parasite prevalences (1987 to 1988*). Kruskal–Wallis test statistics and significance level

Factor	H†	L	M	T	N
Female					
Body mass	0.09	5.67	0.48	0.49	10.4
	(0.77)	(0.02)‡	(0.49)	(0.49)	(0.001)§
Tail length	0.18	0.59	10.0	0.47	2.4
	(0.89)	(0.44)	(0.002)§	(0.49)	(0.12)
Wing length	1.30	3.61	1.07	0.02	1.7
	(0.25)	(0.06)	(0.30)	(0.88)	(9.19)
Male					
Body mass	3.16	0.09	0.19	0.01	2.07
	(0.08)	(0.93)	(0.66)	(0.94)	(0.15)
Tail length	1.99	0.62	9.8	0.27	2.51
	(0.16)	(0.43)	(0.02)§	(0.61)	(0.11)
Wing length	18.8	0.12	3.94	0.48	0.25
	(0.000)§	(0.73)	(0.047)	(0.49)	(0.61)

* Only 1988 data for nits.
† H, *Haemoproteus*; L, *Leucocytozoon*; M, microfilaria of unidentified filariid nematode; N, nits.
‡ Significant on the 0.05 level (after a Bonferroni adjustment).
§ Significant on the 0.01 level (after a Bonferroni adjustment).

mass and tail length (Spearman rank correlation, $r = 0.28$ for males, and $r = 0.22$ for females), between body mass and wing length ($r = 0.29$ for males; $r = 0.38$ for females), and between tail length and wing length ($r = 0.51$ for males; $r = 0.48$ for females). Kruskal–Wallis analysis was used to associate parasite prevalence with weight, tail length, and wing length (Table 18.5). Out of 30 comparisons, six were significant. Among females, birds infected with *Leucocytozoon* were heavier, and birds infected with microfilaria also had longer tails whereas birds infected

Table 18.6. Interaction (1988) of plumage colours of 198 adult male common grackles (*Quiscalus quiscula*)

	Bronze	Green	Blue	Purple
Green	0.31*†			
Blue	−0.63†	−0.21‡		
Purple	−0.51†	−0.67†	0.33†	
Rose	−0.32†	−0.64†	0.19‡	0.62†

* Spearman rank correlation coefficient.
† $p \leqslant 0.001$.
‡ $p \leqslant 0.01$.

with *H. quiscalis* had shorter wings. Among 1988 birds, nit-infested females tended to be heavier.

Spearman rank correlations supported these results. Microfilaraemia was positively correlated with tail length in both males and females (Spearman rank correlation; $r=0.22$, $p<0.001$, and $r=0.24$, $p<0.001$, respectively). In males, microfilaraemia was also positively correlated with wing length ($r=0.16$, $p<0.01$). Also, there was a positive correlation between nit counts and body mass of females ($r=0.25$, $p<0.01$).

Haemoproteus quiscalus parasitaemias, scored on the first 73 *Haemoproteus*-infected birds (as described above), did not correlate significantly with any of the three structural measurement, ($p>0.05$ in all cases).

Plumage colour correlates of parasitism

Six distinct colours were recognized in adult male and female grackles: brown, bronze, green, blue, purple, and rose. All of the colours, except brown, are iridescent. Fledgling grackles were completely non-iridescent; hence they were not colour scored. Brown coloration was detected on males so infrequently that it was not included in subsequent analyses. Within sexes, most of the colours were significantly correlated with one another, either negatively or positively (Tables 18.6 and 18.7).

Out of 40 possible (20 for each sex), associations between plumage colours and parasite prevalences, six significant associations (χ^2 contingency test) were identified. The more predominant that purple was on females, the more likely they were to be infected with *T. avium* ($p=0.003$). Prevalence of microfilariae was less among brown females ($p=0.001$). Greener males had lower *T. avium* prevalence)$p=0.03$), and rosier males had lower *H. quiscalus* prevalence ($p=0.02$) and lower microfilaria prevalence ($p=0.01$), yet higher louse prevalence ($p=0.03$).

In correlating nit burdens with colour scores, a significantly negative correlation with green on females was found ($r=-0.16$, $p<0.05$).

Table 18.7. Interaction (1988) of plumage colours of 173 adult female common grackles (*Quiscalus quiscula*)

Brown	Bronze	Green	Blue	Purple	
Green	−0.39*†				
Blue	−0.65†	0.63†			
Purple	−0.24‡	−0.30†	−0.12		
Rose	0.10†	−0.56†	−0.58†	0.43†	
	−0.11	−0.29†	−0.28†	0.22‡	0.47†

* Spearman rank correlation coefficient.
† $p \leqslant 0.001$.
‡ $p \leqslant 0.01$.

Among males, rose was positively correlated with nit burdens ($r=0.18$, $p<0.02$). Among birds infected with microfilaria, intensities in females were negatively correlated with brown coloration ($r=-0.25$, $p<0.01$) and positively correlated with bronze coloration ($r=0.24$, $p<0.01$). In females, the occurrence of these two colours was negatively correlated (Table 18.6).

On some birds, both male and female, a distinct collar of colour was observed on the nape. Among females, those with a collar were more likely to be infected with *T. avium* ($p=0.015$, χ^2 test) and with microfilariae ($p=0.02$, Fisher exact probability test).

Additional interactions and confounding factors

Of 399 adult birds captured during the 1988 study period, 53 (13.3 per cent) had been banded in previous years. These recaptured birds were considered to be > 1 year old; the age of birds in the previously unbanded group could not be determined, although some of them undoubtedly were also > 1 year old. Nevertheless, when prevalences of parasitic infections between previously banded ('older') and previously unbanded ('younger') birds were compared. *Leucocytozoon* spp. and microfilarial infections were more prevalent among older females ($T=5.6$ and 5.8, respectively, $p=0.02$ in both cases, Kruskal–Wallis test). Furthermore, microfilaraemias in older females were significantly higher than in younger females ($T=11.4$, $p<0.001$). In contrast, the burden of nits was higher among younger females ($r=4.2$, $p=0.04$).

The relationships between adult age and colour predominance among sexes were examined (Kruskal–Wallis test). Only among females were significant relationships found. Younger females were browner ($r=17.1$, $p<0.01$), whereas older females were more likely to have green coloration ($r=7.4$, $p=0.01$).

In comparing adult age with mensural characteristics, older birds had longer wings ($T = 5.8$, $p = 0.02$ for males; $T = 11.5$, $p < 0.001$ for females) as well as longer tails ($r = 16.0$ for females, $T = 13.6$ for males, $p < 0.001$ for both sexes) than younger birds.

Finally, correlations among colour predominance and mensural characteristics were examined. Among males, tail length was positively correlated with the colours bronze and green, and negatively correlated with the colour purple ($r = 0.17$, 0.14, and -0.17, respectively; $p < 0.05$ in all cases). Male body weight correlated positively with the colour bronze ($r = 0.18$, $p < 0.05$). Among females, tail length was positively correlated with the colours bronze, green, and blue, and negatively correlated with the colour brown ($r = 0.26$, 0.34, 0.22, and -0.54, respectively; $p < 0.01$ in all cases). Also among females, wing length was positively correlated with the colour green, and negatively correlated with the colour brown ($r = 0.23$ and -0.37, respectively; $p < 0.01$ in both cases). Furthermore, female body mass was positively correlated with the colour rose ($r = 0.18$, $p < 0.05$).

Discussion

The great majority of grackles examined in this study were found to be parasitized with one or more genera of haemotropic parasite (see also Atkinson and van Riper, Chapter 2, this volume). In particular, *H. quiscalus* and microfilariae were so commonly found, that these organisms may be considered virtually normal fauna of the common grackle. Compared with results from selected studies of others (Table 18.8), grackles at our study site appear to be more frequently infected with *H. quiscalus, T. avium*, and microfilariae. It is likely that the higher prevalences of these three parasites found in our grackle population are due, in part, to the more sensitive detection methods that we employed (i.e. dark-field microscopy for *H. quiscalus* and capillary tube examination for *T avium* and microfilariae). In contrast, *Plasmodium* spp. were less frequently detected in the grackles of our study. The reported prevalence of *Leucocytozoon* spp. varied markedly among studies.

Prevalence *per se* is a relatively poor indicator of effects of parasites on their hosts (May 1984; Anderson 1986), although changes in prevalence may provide valuable information (Kitron and Higashi 1985). Especially for metazoan parasites (e.g. microfilariae and nits), infection does not necessarily equal disease, the effect of parasite on host being density-dependent (Crofton 1971). The hypothesis that parasitism may be associated with secondary sexual characteristics is only tenable when: (1)

Table 18.8. Prevalences of haemotropic parasites common grackles (*Quiscalus quiscula*) surveyed at various locations

Locality	Number Examined	Number Infected	Percentage infected with* H	L	P	T	M	Reference
Illinois	633	586	71	13	1	26	66	Present study
Washington DC	90	56	24	47	16	2	6	Wetmore 1941
S. Carolina	301	131	21	7	26	1	0	Hunninen & Young 1950
New York	75	56	40	57	16	0	0	Manwell 1951
Ohio	26	23	50	62	8	0	0	Smith 1967
New Jersey	897	180	9	9	3	<1	1	Williams & Bennett 1978
Vermont	102	94	9	92	2	10	3	Barnard & Bair 1986
Various†	973	433	20	23	13	2	3	Greiner *et* al. 1975

* H, *Haemoproteus*; L, *Leucocytozoon*; P, *Plasmodium*; T, *Trypanosoma*; M, microfilariae of unidentified filariid nematode.

† Compilation of data from various North American localities; includes data from Washington DC, South Carolina, New York, and Ohio studies listed above.

the parasite, when present in a certain density, has a debilitating effect on its host; and (2) the correlated host characteristics can be shown to be associated with reproductive success. Moreover, possible cumulative or even synergistic effects of parasites on hosts and the possibility that mate choice depends on a specific combination of epigamic characters means that a correlation should be attempted between more complex measures of parasite burdens and host appearance.

Sources of variation in haemotropic parasite prevalences in birds, other than test methods employed, include locality, seasonal fluctuations, and annual fluctuations. These three sources of variation are all probably related to climatic conditions which, in turn, influence the species distribution and density of local vector populations (Bennett and Cameron 1974; Bennett *et al.* 1975; Atkinson *et al.* 1988; van Riper, Chapter 8, this volume). All of the parasite genera studied here are transmitted by haematophagous flies. It is possible that parasite prevalence variations may occur also because of grackle race or subspecies (e.g. bronzed versus purple race; Manwell 1951). Furthermore, previous studies did not consider the effects of age and sex distributions of host populations on parasite prevalences. Finally, it is possible that diurnal fluctuations in parasitaemias may occur; therefore, the time of day when birds are sampled is another potential source of variation (Gore *et al.* 1982; Duffy, Chapter 13, this volume).

Increasing prevalences of haemotropic parasites among temperate-zone, passerine birds during the spring have been shown previously (Bennett and Cameron 1974; Barnard and Bair 1986; Kirkpatrick and

Suthers 1988). In addition, we show that louse oviposition prevalence and concentrations of circulating microfilariae appear to fluctuate seasonally as well. Increased parasite prevalences in spring are thought to occur for two reasons: relapse of latent infections and acquisition of new infections (Beaudoin *et al.* 1971). In chronic infections (e.g. of *Plasmodium* spp.), parasite division and, hence, parasitaemias may become so low that parasites may elude detection, particularly by blood-smear examination (Manwell 1934). Spring increases in parasitaemias (i.e. relapses) appear to result from hormonal influences (Applegate and Beaudoin 1970).

At our study site, transmission of *H. quiscalus*, *T. avium*, and certain unidentified filariid nematodes can be inferred, because infections with these parasites were detected in fledglings. It is likely that *Leucocytozoon* spp. infections were not acquired locally, because they were not detected in fledglings and because the appropriate habitat for their simuliid vectors does not occur locally. Because *Plasmodium* spp. prevalences were so low in our grackle population, conclusions on the locality of transmission of these parasites cannot be made. However, more sensitive methods for detection of *Plasmodium* spp. infections may have revealed a higher prevalence and the existence of local transmission (Herman 1968).

We found several significant associations among parasite prevalences and intensities and certain plumage colour, mensural, age, and sex characteristics. Plumage coloration was associated with either higher or lower prevalence of certain parasites, depending on the colour, in both males and females. Some colours appeared to be associated with increased resistance to parasites (e.g. green in males), whereas other colours appeared to be associated with increased susceptibility (e.g. purple in females). Interestingly, among males, the colour rose was associated with lower *H. quiscalus* and microfilarial prevalences as well as higher nit prevalence and nit burden. Thus, the expression of a certain colour by a bird may be sending conflicting signals to a prospective mate: resistance to one genus of parasite and susceptibility to another.

When adult grackles were classified into two age groups, the associations of certain colours and mensural characteristics were found to be age-related. Generally, older birds tended to be larger and more colourful. Thus, some of the associations found between colour and parasitism and between size and parasitism may have been confounded by age. For example, a higher *H. quiscalus* prevalence was found among smaller males and among younger males. Smaller and browner females had lower microfilarial prevalences and burdens; younger females had similar characteristics, and they were also smaller and browner.

It is remarkable that significant correlations between age and other characteristics, including prevalences of some parasites, were found. The younger (i.e. unbanded) group was surely diluted to an unknown extent with older birds that had not been captured previously. Examination of a grackle population with a larger proportion of birds of known age may reveal both stronger and additional age-associated relationships.

Immature birds are considered to be more susceptible to haemotropic protozoan parasites than mature birds, probably due to the development of immunity following initial exposure (Herman 1968). This is borne out for *H. quiscalus* infection in our study, because all of the fledgeling birds examined were found to be infected, and older adults had a lower prevalence of this parasite. In contrast, microfilarial prevalence and burdens appeared to increase with age in our grackle population. This suggests that strong immunity to filarial nematodes does not develop and that repeated infections may be cumulative.

Although many of the associations between host phenotypic characters and parasitism were statistically significant, the correlations were often weak (i.e. r values were low). In addition, the large number of pairwise comparisons performed in this study would tend to reduce the acceptable level of significance (Matthews and Farewell 1985). Furthermore, host age may have been a confounding variable. Primarily, our study provides a cautionary tale. To test the Hamilton and Zuk hypothesis, not only need we examine intraspecific variations, but the choice of correlates is complex. Unless criteria are established which measure actual damage of the host on the one hand and reproductive outcome on the other, correlations are like to be weak and possibly only accidental. The large number of possible confounders which must be considered renders field studies even less likely to provide significant results without detailed, long-term population studies.

In conclusion, unequivocal support for Hamilton and Zuk's hypothesis did not emerge from our preliminary study of haemotropic parasites in common grackles. Previous studies within species of passerine birds did not find clear associations among haemotropic parasite prevalence or intensity and host body mass (Smith and Cox 1972; Bennett *et al.* 1988; Kirkpatrick and Suthers 1988). Yet additional studies of haemotropic parasitism in well-characterized, free-living populations of given avian species are warranted. We believe that the best test of Hamilton and Zuk's hypothesis is an examination of parasite-associated variation in epigamic characters, including behaviour, within an avian host species.

The results of our study emphasize that, in addition to epigamic characteristics and parasite prevalences, studies which seek to address the hypothesis using a similar host–parasite model system need to collect additional data. Such data include the distributions of age and sex classes in the population under study and parasite burdens as well as prealences. Age classification may need to discriminate beyond just 'hatching-year' and 'after-hatching-year', as indicated in this study. Temporal variations in parasite prevalences and intensities must be taken into account. Furthermore, it would be worthwhile to correlate epigamic characters, such as plumage brightness, with behaviour and reproductive success in the field.

Acknowledgements

We thank Jennifer Leak and Jennifer Bothwell for assistance in the field. Jennifer Bothwell scored the colours on all birds used in this study. We also thank Dr K.C. Emerson for identifying louse specimens and Dr G. F. Bennett for examining representative blood smears. Two anonymous reviewers provided many helpful comments. This work was supported by funds from the Research Board of the University of Illinois.

References

Anderson, R.M. (1986). The population dynamics and epidemiology of intestinal nematode infections. *Transactions of the Royal Society of Tropical Medicine and Hygiene*, **80**, 686–96.

Applegate, J.E. and Beaudoin, R.L. (1970). Mechanism of spring relapse in avian malaria: effect of gonadotropin and corticosterone. *Journal of Wildlife Diseases*, **6**, 443–7.

Atkinson, C.T., Forrester, D.J., and Greiner, E.C. (1988). Epizootiology of *Haemoproteus meleagridis* (Protozoa: Haemosporina) in Florida: seasonal transmission and vector abundance. *Journal of Medical Entomology*, **25**, 45–51.

Barnard, W.H. and Blair, R.D. (1986). Prevalence of avian hematozoa in central Vermont. *Journal of Wildlife Diseases*, **22**, 365–74.

Beaudoin, R.L., Applegate, J.E., Davis, D.E. and McLean, R.G. (1971). A model for the ecology of avian malaria. *Journal of Wildlife Diseases*, **7**, 5–13.

Bennett, G.F. (1962). The hematocrit centrifuge for laboratory diagnosis of hematozoa. *Canadian Journal of Zoology*, **40**, 124–5.

Bennett, G.F. and Cameron, M. (1974). Seasonal prevalence of avian hematozoa in passeriform birds of Atlantic Canada. *Canadian Journal of Zoology*, **52**, 1259–64.

Bennett, G.F., Cameron, M., and White, E. (1975). Hematozoa of the passeriforms of the Trantamar Marshes, New Brunswick. *Canadian Journal of Zoology*, **53**, 1432–42.

Bennett, G.F., Caines, J.R., and Bishop, M.A. (1988). Influence of blood parasites on the body mass of passeriform birds. *Journal of Wildlife Diseases*, **24**, 339–43.

Crofton, H.D. (1971). A quantitative approach to parasitism. *Parasitology*, **62**, 179–93.

Godfrey, R.D. Jr, Fedynich, A.M., and Pence, D.B. (1987). Quantification of hematozoa in blood smears. *Journal of Wildlife Diseases*, **23**, 558–65.

Gore, T.C., Noblet, G.P., and Noblet, R. (1982). Effects of pinealectomy and ocular enucleation on diurnal periodicity of *Leucocytozoon smithi* (Haemosporina) gametocytes in the peripheral blood of domestic turkeys. *Journal of Protozoology*, **29**, 415–20.

Granath, W.O. Jr (1981). *Eufilaria hibleri* sp. n. (Nematoda: Filaroidea) from the common grackle (*Quiscalus quiscula versicolor*). *Proceedings of the Helminthological Society of Washington*, **48**, 17–23.

Greiner, E.C., Bennett, G.F., White, E.M., and Coombs, R.F. (1975). Distribution of avian hematozoa of North America. *Canadian Journal of Zoology*, **53**, 1762–87.

Hamilton, W.D. and Zuk, M. (1982). Heritable true fitness and bright birds: a role for parasites? *Science*, **218**, 384–7.

Herman, C.M. (1968). Blood protozoa of free-living birds. *Symposia of the Zoological Society of London*, **24**, 177–95.

Huizinga, H.W. and Granath, W.O. Jr (1984). Seasonal prevalence of *Chandlerella quiscali* (Onchocercidae: Filaroidea) in brain of the common grackle) (*Quiscalus quiscula versicolor*). *Proceedings of the Helminthological Society of Washington*, **51**, 191–5.

Hunninen, A.V. and Young, M.D. (1950). Blood protozoa of birds at Columbia, South Carolina. *Journal of Parasitology*, **36**, 258–60.

Kirkpatrick, C.E. and Suthers, Suthers, H.B. (1988). Epizootiology of blood parasite infections in passerine birds from central New Jersey. *Canadian Journal of Zoology*, **66**, 2374–82.

Kitron, U.D. and Higashi, G.I. (1985). *Schistosoma haematobium* in Upper Egypt: analysis of dispersion patterns. *American Journal of Tropical Medicine and Hygiene*, **34**, 331–40.

Manwell, R.D. (1934). The duration of malarial infection in birds. *American Journal of Hygiene*, **19**, 532–38.

Manwell, R.D. (1951). Leucocytozoa and other blood parasites of the purple grackle, *Quiscalus quiscula quiscula*. *Journal of Parasitology*, **37**, 301–6.

May, R.M. (1984). Ecology and population biology. In *Tropical and geographical medicine* (ed. K.S. Warren and A.A.F. Mahmoud), pp. 152–66. McGraw Hill, New York.

Matthews, D.E. and Farewell, V.T. (1985). *Using and understanding medical statistics*. Karger, Basel.

Read, A.F. (1987). Comparative evidence supports the Hamilton and Zuk hypothesis on parasites and sexual selection. *Nature*, **328**, 68–70.

Read, A.F. (1988). Sexual selection and the role of parasites. *Trends in Ecology and Evolution*, **3**, 97–102.

Smith, D.G. (1967). A survey of the occurrence of blood parasites in the local bird population in the Hiram, Ohio, area. *Bulletin of the Wildlife Diseases Association*, **3**, 185–6.

Smith, V.W. and Cox, F.E.G. (1972). Blood parasites and the weights of palaearctic migrants in central Nigeria. *Ibis*, **114**, 105–6.

Wetmore, P.W. (1941). Blood parasites of birds of the District of Columbia and Patuxent Research Refuge vicinity. *Journal of Parasitology*, **27**, 379–93.

Williams, N.A. and Bennett, G.F. (1978). Hematozoa of some birds of New Jersey and Maryland. *Canadian Journal of Zoology*, **56**, 596–603.

19 *Bill colour and parasite levels of zebra finches*

NANCY BURLEY, SONIA C. TIDEMANN,
and KARL HALUPKA

Introduction

Freeland (1981) hypothesized that, for male mammals, dominance status is inversely correlated with parasite load in such a way that females mating with dominant males produce offspring with genes for pathogen resistance. Hamilton and Zuk (1982) developed this idea further and applied it to levels of coloration and song characteristics of birds. Basically a 'good genes' approach, this hypothesis posits that female preference for brightly coloured males is functional in that such coloration signals heritable parasite resistance and predicts that, within species, brighly coloured males will have lower parasite counts than dully coloured ones.

While the Hamilton and Zuk hypothesis is usually applied to female choice of male traits (e.g. Borgia 1986; Read 1987; Zuk 1987; Ward 1988; this volume: Zuk, Chapter 16; Møller, Chapter 17; Kirkpatrick *et al.*, Chapter 18; Johnson and Boyce, Chapter 20; Spurrier *et al.* Chapter 21), for many species of birds with biparental care it is likely that both sexes participate in mate choice (Burley 1977). In such species offspring quality should be of concern to males as well as females. It is parsimonious to predict for these species that levels of coloration show the same direction of correlation with pathogen resistance in both sexes.

Hamilton and Zuk also predicted an interspecific positive correlation between diversity of parasites and brightness of coloration. Their rationale for this prediction is that advertisement of good condition is most valuable in species that are chronically susceptible to a wide variety of parasites; thus, high parasite loads will promote the evolution of bright coloration through sexual selection. In making this prediction, however, Hamilton and Zuk ignored the effects that increased coloration level may have on organisms. In any good genes model, the indicator traits (i.e. coloration level) must be reliably correlated with condition. To be reliable, these traits must have some cost (e.g. Zahavi 1977; Kodric-Brown and Brown 1984). Hence, evolutionary amplification of indicator traits will either increase the net total cost of epigamic signals, or previously evolved indicator traits must lose their cost (and their signal value) as new or amplified indicators evolve. Since it seems unlikely that costs of traits are routinely lost, a positive correlation between coloration levels and parasite loads may result from the costs of producing and maintaining coloration levels. Thus, bright coloration, previously evolved in response to social factors, may impose a physiological cost that ultimately increases susceptibility to parasitic infection.

Some species of birds are highly dimorphic, with males being more colourful than females, despite similar levels of parental investment, and presumably despite bidirectional mate choice. In such situations we can predict, on the basis of the above reasoning, that males will have higher average parasite loads than females because of the costs associated with higher average coloration levels. This prediction does not contradict the notion that the expression of the colourful traits will be highest in males with the lowest relative loads; rather, individuals that manage to avoid parasitic infections and their associated costs will be better able to invest in bright coloration.

In sum, the Hamilton–Zuk hypothesis predicts positive correlations between parasite loads and level of avian coloration both within and among species. While the prevailing approach to testing this prediction has been correlative, the real purpose of the correlative approach is to permit a *preliminary* investigation of the potential merit of a hypothesis. It is in this spirit that this study was conducted. Ultimately, tests of correlation are inadequate to determine causation because alternative explanations for observed correlations usually can be envisioned (e.g. the alternative explanation for the positive interspecific correlation given above; see also Alker 1969; Bertalanffy 1968; Burley 1983).

We set out to examine the relationship between bill colour and parasite load in the zebra finch (*Taeniopygia guttata castanotis*). We felt that bill colour has the potential to be an important indicator trait because it is quite variable both within and between sexes (males tend to have redder and darker bills than females) and is used in mate choice by both sexes (Burley and Coopersmith 1987). The only plumage trait that shows comparable variability is the size of the male breast band, which does not appear to be used by females in mate choice (Coopersmith and Burley, manuscript). There is also evidence that bill colour varies with condition of birds (Burley, Price and Zann, manuscript). We explored as many parasites (haematozoa, endo-, and ectoparasites, as well as air sac mites) as feasible. This approach was taken because there is little documentation of the parasite fauna and relative pathogenicity of parasites of Australian estrildine finches. We therefore did not know which parasites were most likely to influence bill colour. We tested two

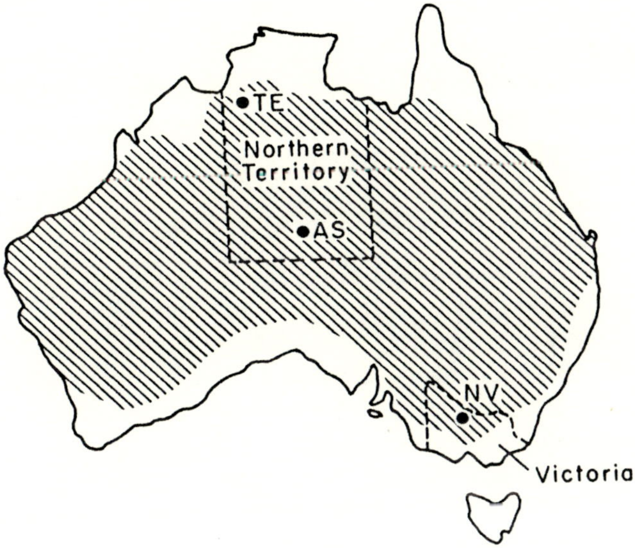

Fig. 19.1. Map of Australia illustrating location of study sites. AS, Alice Springs, Northern Territory; TE, Top End, Northern Territory; NV, northern Victoria.

predictions: (1) within each sex, there is an inverse correlation between bill colour and parasite counts; and (2) males have higher parasite counts than females. Zebra finches are a wide ranging species over the continent of Australia and we sampled them in three geographically and ecologically distinct regions (Fig. 19.1).

Ideally, bill colour would be measured and parasites recovered just prior to the time of pair formation. This is not possible for free-living zebra finches because they can form pair bonds at any time of year and they maintain pair bonds for long periods of time, possibly for life (Immelmann 1962). The importance of obtaining measurements prior to pair formation is diminished by the fact that birds remain paired for long periods of time: under these conditions, birds have most probably been selected to be attentive to indicator traits that signal long-term health characteristics of their bearers. Consistent with this expectation is the finding that the bill colour of individuals shows highly significant repeatability over time (Burley *et al.*, manuscript); that is, individual variability in bill colour persists into adulthood. Finally, the sorts of parasites envisioned to be most important to the Hamilton–Zuk hypothesis are those with chronic, rather than acute, effects. These effects are likely to be manifested during and quite possibly after the breeding season as well as at the time of pair formation.

Study sites

Data were collected at three locales (Fig. 19.1). Habitats, dates, and the types of data collected are described below. Bill colours were scored at all three locales.

Alice Springs, Northern Territory (N.T.)

Located in the semi-arid zone, Alice Springs receives an average of 250 mm of rain per year, most of which typically falls between October and March (Lazarides 1970). All data were collected in the open woodland habitat of the grounds of the Commonwealth Scientific and Industrial Research Organization just outside Alice Springs. Ectoparasite data were collected between December 1986 and January 1987, near the end of a breeding pulse that began in late August and peaked in early November (Burley *et al.* 1989). From December through April breeding occurred at a low level, but finches did not breed during the subsequent four months. Ectoparasite and faecal samples were collected at the same locale in July 1987.

Top End, N.T.

Blood samples were collected between August and October 1986 in low, open eucalypt woodland on several pastoral properties in the transition zone (for details, see Burley *et al.* 1989). In this zone, monsoonal rain (averaging 720 mm per year) typically falls between November and April (McAlpine 1976). Zebra finches breed during the dry season (May through October), and birds were breeding at the time data were gathered. Birds were collected to be examined for air sac mites at these and additional sites in the transition zone. Several other species of estrildines were also sampled for blood parasites and air sac mites at these and other study sites in the region at about the same time.

Northern Victoria

Study sites were located approximately 150 km north of Melbourne on the alluvial flats of the Goulbourn River, where the climate is temperate, with a mean annual rainfall of 450 mm. The area is flood-irrigated and subject to intensive agriculture, with only remnants of the original eucalypt woodland remaining. Faeces and ectoparasite samples were collected in March and April 1987, during the autumn breeding season.

Methods

Age and sex

Zebra finches display unambiguous sexual dimorphism from about the age of 40 days and are fully adult by 90 days of age. Age of immatures (40–89 days) can be approximated on the basis of bill colour, iris colour, secondary sexual plumage traits (males only), and wear and moult patterns of flight feathers (Zann 1985; Burley *et al.* 1989). Adult birds (90 days or more) cannot usually be aged by plumage traits reliably.

Bill colour measurements

Bill colours were scored by the Munsell® system using procedures described elsewhere (Burley and Coopersmith 1987). This system encodes hue, value (lightness–darkness), and chroma (dullness–brightness) into a complex numeral. To facilitate statistical analysis, this numeral was then converted to standard decimal notation using the formula

$$3(15-Z)+1.5(6-Y)+0.5(X-12)$$

where X = chroma, Y = value, and Z = hue as scored by the Munsell® system. This formula reflects differences in the sizes of perceptual intervals in the Munsell® system (Burley *et al.*, manuscript): differences in two adjacent hue standards are the largest, whereas differences in chroma standards are the smallest. Hue and value of bill colour are highly intercorrelated, while chroma varies relatively little among adult birds. In this transcribed notation, colours that are redder, brighter, and darker have higher scores than those that are less red, duller, and lighter. Bill colours that are red, bright, and dark are perceived as most striking by humans, and are preferred by females in heterosexual choice tests (Burley and Coopersmith 1987).

Ectoparasites

Birds were recovered from baited traps and transferred to individual children's socks. They were then taken to the laboratory, where each bird was scored for bill colour and then 'fumigated' for ectoparasites. To collect parasites, procedures similar to those reported elsewhere (e.g. Fowler and Cohen 1983; Wheeler and Threlfall 1986) were employed. A bird was suspended in a cylindrical chamber impregnated with chloroform vapour so that its entire body except its head was exposed to chloroform. This was accomplished by placing the bird's head through a hole in a piece of rubberized cloth, grasping the cloth and the bird's neck between the first and second fingers, and then securing the cloth to the chamber with a rubber band. To minimize the bird's movement, its toes were also held between the thumb and first finger of the same hand. The bird was held in this position for 110 s, at which time the toes were released and the bird shaken slightly to encourage movement. At 120 s the bird was removed from the chamber. A piece of filter paper at the bottom of the chamber was removed and examined under a dissecting scope for ectoparasites. Faeces and stray feathers were then removed from the sock, and it was thoroughly examined for ectoparasites.

The 120 s criterion was adopted because, in preliminary trials, there was little increase in number of ectoparasites recovered after 240 s over that recovered after 120 s. High ambient temperatures also made it necessary to remove birds from socks quickly. Hippoboscid flies were not always anaesthetized in 120 s, but their presence was easily detected. It is not expected that this method results in a measure of the total load borne by individual birds, but rather that the number recovered in a fixed interval is proportional to the total load (Fowler and Cohen, 1983).

Blood parasites

A drop of blood was collected from the brachial vein in a heparinized microcapillary tube. The blood was transferred to a clean glass slide, smeared immediately, air-dried, and fixed in absolute alcohol for 3 min. Blood smears were stained for 30 min using slow, buffered Giemsa. Thirty fields per slide were examined for blood parasites, in particular microfilariae and trypanosomes. All red cells were checked for malarial parasites.

Endoparasites

Faecal samples from same-sex individuals were pooled within sampling locale, with five individuals in each pool. Pooled samples (up to 0.5 g dry weight) were mixed with 10 ml of saturated NaCl and 5 drops of amyl alcohol, shaken, and filtered through cheesecloth. The filtrate (2 ml) and distilled water (2 ml) were centrifuged for 2 min at 2000 r.p.m. Three ml of supernatant were removed. The entire residue was then placed, one drop at a time, on a slide with one drop of Lugols iodine. The total number of eggs was counted at × 250 magnification.

Air sac mites

Birds were collected and dissected from the ventral side and the body cavity examined using a binocular microscope (× 10– × 64). Air sacs, lungs, and trachea were searched for the presence of mites.

Other procedures

Wing moult was scored by examining the primaries and secondaries of one wing for the relative age and wear of feathers, as well as for the presence of partially grown feathers. Weights were measured with a Pesola balance. Birds were also scored on an ordinal scale (0–8) for volume of crop contents. Weights were later corrected by running regressions of weight against crop content volume. Regression coefficients varied by location and date. At Alice Springs from December to January, $r^2=0.186$, $p<0.001$. In July $r^2=0.396$, $p<0.0001$. In northern Victoria $r^2=0.177$, $p<0.0001$). A bird's corrected weight is equal to its measured weight minus the product of its crop contents score times the regression coefficient for the locale and time period at which it was captured.

The frequency of ectoparasites on sex and age classes was evaluated using analysis of variance and *t*-tests. Spearman's *r* was used to measure correlation between parasite count and other attributes. Sex differences in endoparasite counts were evaluated using the Wilcoxon rank sum test. Generally, louse and mite counts were not intercorrelated and parametric partial correlation analysis did not reveal patterns in addition to those described below.

Results

Bill colour varied between sex and age classes in ways very similar to those reported for captive birds (Burley and Coopersmith 1987). Males tended to have redder and darker bills than females, and adults had higher scores than same-sex immatures (Fig. 19.2).

One species of mite, *Pellonyssus reedi* (Acarina: Macronyssidae), two undescribed species of Mallophaga (*Brueelia* sp. and *Myrsidea* sp.) and one species of hippoboscid fly (*Ornithophila metallica*) were recovered from 318 adult and 110 immature zebra finches sampled in Alice Springs and northern Victoria. Hippoboscids were very uncommon and are not included in the data analyses presented below. Ectoparasites were recovered from about 80 per cent of all birds sampled. The incidence of lice and mites varied considerably between locales and time periods (Fig. 19.3). Accordingly, separate analyses were performed for each sampling period. Immatures were not present during the second sampling period in Alice Springs (AS2).

The relative frequency of ectoparasites on sexes and age classes

Results of two-way analyses of variance (sex * age) for incidence of lice, mites, and total parasite count showed significant age effects in three of

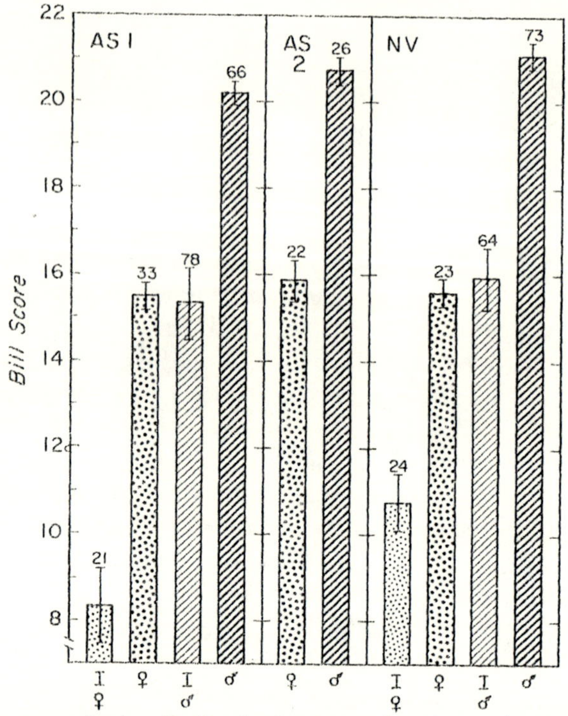

Fig. 19.2. Bill scores of zebra finches in three sampling intervals. Higher scores are redder/darker/brighter than lower scores. AS1, Alice Springs, December 1986–January 1987; AS2, Alice Springs, July 1987; NV, northern Victoria, March–April, 1987. I♀, immature females, age 40–89 days; ♀, adult females, >90 days; I♂, immature males; ♂, adult males.

six analyses (Table 19.1). Means for immatures were consistently higher than those for adults (Fig. 19.3), largely as a result of the higher louse scores obtained for immatures at both locales. In northern Victoria (NV), there were also significant age–sex interactions (Table 19.1), with immature males having the highest louse loads and mature males having the lowest (Fig. 19.3).

Due to the absence of juveniles at AS2 and the presence of interaction effects in NV, *t*-tests were performed for each age class at each site. For AS1, adult females tended to have higher mite counts than adult males ($t=1.86$, $df=78, 66$, $p=0.06$). Among immatures, males tended to have higher louse counts in NV ($t=-1.97$, $df=20,16$, $p=0.06$). No other comparisons approached significance.

To determine the relationship between louse and mite counts on individual birds, Spearman correlation analyses were run for each sex and age category at each sampling period. For adult females at NV, there was a positive correlation between counts ($r=0.282$, $p=0.04$). The only

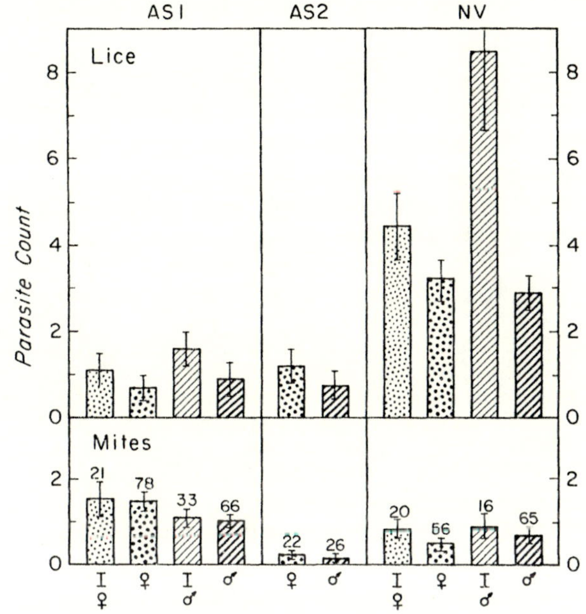

Fig. 19.3. Counts of ectoparasites of zebra finches in three sampling intervals. Sample sizes (number of birds sampled) are the same for louse counts as for mite counts. Legend as in Fig. 19.2.

Table 19.1 Results of two-way ANOVA for sex and age

	Lice		Mite		Σ par		
	F	*p*	*F*	*p*	*F*	*p*	*df*
AS1							
Model	3.06	0.029	1.51	0.212	1.29	0.278	3,194
sex	3.25	0.073	4.48	0.036	0.00	0.988	1
age	5.78	0.017	0.06	0.801	3.74	0.055	1
sex*age	0.16	0.690	0.00	0.974	0.13	0.714	1
NV							
Model	9.03	0.0001	0.72	0.543	9.38	0.0001*	3,153
sex	0.55	0.459	0.84	0.360	0.88	0.349	1
age	18.34	0.0001	1.16	0.283	19.94	0.0001*	1
sex*age	8.20	0.005	0.15	0.701	7.33	0.008*	1

AS1: Alice Springs, first sampling interval; NV: Northern Victoria. **Σ** par = sum of louse and mite ectoparasites. Sample sizes are in Fig. 19.3.

* Correlations with significant probability levels.

Table 19.2 Correlations between ectoparasite count and bill colour of adults

		Females			Males		
		Lice	Mites	Σ par	Lice	Mites	Σ par
AS1	r_s	0.044	0.349	0.290	0.277	0.050	0.180
	p	0.70	0.002*	0.01*	0.02*	0.69	0.15
AS2	r_s	0.117	−0.222	0.062	0.152	−0.322	0.070
	p	0.57	0.28	0.76	0.45	0.10	0.73
NV	r_s	0.466	0.259	0.497	0.083	0.020	0.057
	p	0.0003*	0.05*	0.0001*	0.51	0.87	0.65

r_s, Spearman test statistic, p, 2-tailed probability; AS1, Alice Springs, first sampling period; AS2, Alice Springs, second sampling period; NV, Northern Victoria; Σ par, sum of louse and mite counts.
* Correlations with significant probability levels. Sample sizes as in Fig. 19.3.

other correlations that approached significance were also positive (adult females during AS2: $r=0.299$, $p=0.14$; adult males during AS1: $r=0.168$, $p=0.18$).

Overall, immatures had higher louse counts than adults. No sex effects were significant.

The relationship between bill colour and parasite count

Separate correlation analyses were performed for each sampling period for each sex and age group to examine the relationship between bill colour and louse, mite, and total parasite count.

For adult females, eight of nine correlations were positive, five of them significantly so (Table 19.2), indicating that females with higher parasite counts had brighter, redder, darker bills. Significant positive correlations occurred for all three (louse, mite, and total) parasite categories, and during two sampling periods. Eight of nine correlations involving adult males were positive, one of them (bill colour versus louse count, AS1) significantly so (Table 19.2). Thus, for adults 16 of 18 correlations were positive ($\chi^2=9.389$, 1 *df* with Yates' correction, $p<0.005$), and six of these were statistically significant.

For immature females, four of six correlations were positive, and none was significant (Table 19.3). Three of six correlations were positive for immature males; one negative correlation (bill colour versus mite count, NV) was significant. Among immatures 7 of 12 correlations were positive, with one negative correlation reaching significance.

For all age and sex categories, correlations between louse count and bill colour were always (10 of 10 correlations) positive whereas mite count correlations were positive in five of the 10 analyses.

Table 19.3 Correlations between ectoparasite count and bill colour of immatures

		Females			Males		
		Lice	Mites	Σ par	Lice	Mites	Σ par
AS1	r_s	0.252	0.025	0.160	0.048	−0.272	−0.160
	p	0.27	0.91	0.49	0.80	0.14	0.39
NV	r_s	0.030	−0.144	−0.020	0.137	−0.595*	0.032
	p	0.90	0.54	0.93	0.61	0.02	0.91

r_s, Spearman test statistic; p, 2-tailed probability; AS1, Alice Springs, first sampling period; NV, Northern Victoria; Σ par, sum of louse and mite counts.
* Correlations with significant probability levels. Sample sizes as in Fig. 19.3.

Bill colour and age of immatures

Bill colour changes considerably between 40 and 90 days of age (Burley and Coopersmith 1987). Hence age is a potentially confounding variable in the relationship between bill colour and parasite count of immatures. To explore the relationship between ectoparasite count and age, one-way analyses of variance were performed with ages broken down into three subclasses among the immatures. Given sample size constraints, the sexes were lumped within sampling periods, because differences in parasite counts were smaller between sexes than between periods (Fig. 19.2). Analysis of variance was employed rather than a correlation test because the rate of change of secondary sexual characteristics, including bill colour (Burley and Coopersmith 1987; Burley *et al.* 1989; Burley, unpublished data), may not be uniform over development. Thus, if development rate affects load, we might expect to find peak ectoparasite loads in the middle of the immature period. There was, however, no relationship between ectoparasite count and age of immatures (p varied between 0.27 and 0.61).

Other variables

Two possible indicators of condition other than bill colour are activity of moult (zebra finches display continuous wing moult; Zann 1985) and weight. The relationship between ectoparasite count and activity of wing moult in adults showed no trends (0 of 18 correlations significant; 11 of 18 positive). Because wing moult is highly correlated with age of immatures, this variable is not reported for the younger age category.

To examine the relationship between weight and parasite counts, weights were first standardized for volume of crop contents (see Methods). Of 24 correlations performed (nine for each adult sex, six for each sex of immatures), only one was significant (weight versus mite count among adult males in NV: $r=0.316$, $p=0.025$) and several were nearly significant (lice versus weight for adult males in AS2: $r=-0.374$, $p=0.06$; sum of parasites versus weight for adult males in AS2: $r=-0.378$, $p=0.06$; mites versus weight for adult females in AS2: $r=0.390$, $p=0.07$). Fifteen of 24 correlations were positive.

Other parasites

Endoparasites

Nematode eggs were found. Because faecal samples were lumped, the only question that we can address here is whether adult males and females differed in counts. For NV, 10 sets of males averaged 5837 eggs/g of faeces, while nine sets of females averaged 2511 eggs/g (Wilcoxon Rank Sum Test, $U=59$, $p>0.10$). For AS1, six sets of males averaged 983 eggs/g, with five sets of females averaging 845 eggs/g (Wilcoxon Rank Sum Test, $U=15$, $p>0.10$). All pooled samples contained eggs. Thus, there is no apparent sex difference in numbers of nematode eggs passed.

Haematozoa

No blood parasites (microfilariae, trypanosomes, or other blood parasites) were found among 141 birds sampled. This total lack of blood parasites is consistent with the findings for 344 estrildines of eight different species sampled in the Top End at about the same time (star finch, *Neochmia ruficauda*, $n=77$; double-barred finch, *Taeniopygia bichenovii*, $n=41$; masked finch, *Poephila personata*, $n=29$; long-tailed finch, *Poephila acuticauda*, $n=79$; pictorella, *Heteromunia pectoralis*, $n=19$; chestnut-breasted mannikin, *Lonchura castaneothorax*, $n=19$; yellow-rumped mannikin, *Lonchura flaviprymna*, $n=10$; Gouldian finch, *Erythrura gouldiae*, $n=76$).

Air sac mites

No air sac mites were recovered from 31 dissected zebra finches. All of the above species except the yellow-rumped mannikin were also studied or air sac mites. Mites, *Sternostoma tracheacolum*, were found in pictorellas and Gouldian finches (Tidemann *et al.* manuscript), but in none of the others, including 306 individuals of the three species most closely related to zebra finches.

Discussion

Ectoparasites

Contrary to the central prediction of the Hamilton and Zuk hypothesis, ectoparasite counts tended to correlate positively with bill colour in adults. This trend was quite strong and, despite the low counts, which we feel probably underestimate the parasite load somewhat, the possibility that these correlations are biologically significant cannot be dismissed until both the direct effects of these parasites and their role as disease vectors (Krantz 1978; see also this volume Møller, Chapter 17; Johnson and Boyce, Chapter 20; Spurrier *et al.*, Chapter 21) are evaluated fully.

Five hypotheses to explain the positive correlation between bill colour and ectoparasite count are presented below. These hypotheses are not necessarily mutually exclusive.

1. Birds in better condition (i.e. those with redder bills) are more likely to breed. This hypothesis assumes that breeding birds have higher ectoparasite counts due to social contacts or to high parasite activity in active nests. This hypothesis is not supported by available data. In the laboratory, bill colour declines within and over successive breeding attempts (Burley *et al.*, manuscript). In Alice Springs in 1986, mean bill colour of birds captured both at the beginning (August 1986) and end (December 1986) of a breeding pulse showed absolute declines for both sexes (Burley *et al.*, manuscript). Thus, it is unlikely that, for birds sampled near the end of a breeding pulse (AS1 and NV), those having bred heavily would have higher bill scores.
2. Birds with redder bills have more social contacts from which they might obtain ectoparasites.
3. Parasites thrive on birds with redder bills. That is, if birds with redder bills are in better condition, such birds may provide a better environment for parasites (Hausfater and Watson 1976).
4. Birds with redder bills can afford to have more parasites and so spend less time controlling them.
5. 'Runaway' sexual selection has favoured the amplification of bill colour to such a degree that increased loads are a cost of bright coloration (see 'Introduction').

Hypotheses 3 and 4 are similar, but in 3 the active role in determining parasite distribution is ascribed to the parasite and in 4 it is attributed to the host.

At present, there is insufficient evidence to evaluate hypotheses 2–5, with only limited data bearing on 2 and none on the remaining hypotheses. Generally, intraspecific transmission of Mallophaga occurs during social contact (Turner 1971; Clayton, Chapter 14, this volume). Physical contact between adult zebra finches consists of copulation, preening, huddling, and joint incubation, all of which occur most commonly between mated individuals.

The nest may be an important site for transmission of mites (Camin 1963; this volume: Burtt *et al.*, Chapter 6; Møller, Chapter 17). Mites of the genus *Pellonyssus* are reported to lay their eggs in protected sites in the nests, with both protonymphs and adults feeding on the host (Clark and Yunker 1956). Nidicolous mites are known to parasitize multiple individuals of the host species within their lifetime (Camin 1963), thus potentially serving as important vectors of disease organisms. Zebra finches have more contact with nests than many birds. They roost in nests at night even outside the breeding season. Sometimes they adopt abandoned nests for this purpose and regularly build new nests on old nest platforms (Burley, unpublished observations).

A final consistent trend in the ectoparasite results is that immatures have higher louse and total parasite counts than adults. This could occur: (1) due to incomplete development of acquired immunologic resistance; (2) because they are less expert at removing parasites; (3) because they are less expert at other activities such as foraging, and so have less time to devote to removing parasites; (4) because they lack an efficient social (allopreening) partner; and (5) if juveniles are less sensitive to the presence of parasites during their extensive moult.

Other parasites

Evidence strongly suggests that air sac mites are not important parasites of zebra finches, at least during the Top End dry season. Evidence regarding haematozoans, however, is somewhat more equivocal (see this volume: Atkinson and van Riper, Chapter 2; Kirkpatrick *et al.*, Chapter 18). Kirkpatrick and Suthers (1988) reported that blood smear examination was inferior to blood culture techniques for the identification of trypanosomes. Similarly, more sensitive tests for malarial parasites (Herman *et al.* 1966) and microfilariae (Seegar 1979) have been suggested. In all cases, blood smears underestimate the prevalence of parasites. This consideration makes it impossible to rule out these haematozoa as possibly significant parasites. The fact that nearly 500 birds were tested without a single haematozoan being detected suggests, however, that they are relatively unimportant parasites of estrildines in the Top End dry season. By contrast, the substantial numbers and great variability in nematode egg counts suggest that further exploration of parasitism by nematodes might be profitable.

Evaluation of the Hamilton–Zuk hypothesis

No evidence was found to support the prediction that birds with bright bills had fewer parasites. We cannot conclusively falsify this hypothesis, however, for several reasons.

1. We did not census every conceivable parasite. Thus, while the hypothesis does not hold for the parasites we censused, it might hold for some other parasite.
2. We lack adequate information on the life cycle of all parasites and their pathogenic effects on hosts. This information is needed to determine if a given parasite is truly a good candidate pathogen with which to test the hypothesis.
3. The effects of parasites may be more pronounced at early life stages, or under environmental circumstances other than those in which they were sampled (see also this volume: Kirkpatrick *et al.*, Chapter 18; Rogers *et al.*, Chapter 7).

These considerations make it clear that the Hamilton–Zuk hypothesis is essentially non-falsifiable by the correlation procedures typically used to address it. A lack of (or positive) correlation may simply mean that the wrong parasite was selected for study, that the parasite's effects were felt at a different life stage, or some other variable intervened (e.g. Fowler *et al.* 1984). Supporters of the generality of the Hamilton–Zuk hypothesis may therefore wrongly feel it justifiable to ignore non-corroborative evidence. Another weakness with the correlation approach is the fact that a negative correlation between parasite abundance and brightness of coloration is of itself not an adequate test of the hypothesis. It is also necessary to show that there are mate preferences for bright coloration and that parasite resistance is heritable.

Our study is not singular in finding a positive correlation between parasite load and coloration level. Schall (1986) and Ressel and Schall (1989) have reported similar results for two species of lizards. The finding that multiple tests on diverse taxa show contradictory results is not surprising and underscores another pitfall of the correlational approach: without a good deal more research it is not possible to discount some sets of results and accept others. The best rationale for the correlational approach is that it is not time-consuming, yet to interpret results meaningfully, the 'time-consuming' research remains necessary.

Sex differences in parasite loads

A sex difference in the parasite loads of dichromatic species was not predicted by Hamilton and Zuk, nor are loads of the two sexes typically considered separately in assessments of the hypothesis (Hamilton and Zuk 1982; Read 1987). Sex differences in ectoparasite loads have, however, been reported (e.g. Ash 1960; Hall and Gross 1975; Nelson *et al.* 1975). Hamilton and Zuk hypothesized that the function of preferences for bright colours is the production of parasite-resistant offspring. It would be most ironic if an increase in parasite loads proves to be a cost of the evolutionary amplification of coloration level.

We must keep an open mind to the possibility that sexual selection is not a simple consequence of high parasite loads, but rather that it may amplify such loads. To discriminate among possibilities, tests of correlation will surely be inadequate.

Towards future research

For reasons outlined above, we recommend de-emphasizing the correlation approach to the testing of the Hamilton–Zuk hypothesis. An ideal alternative approach would be rigorous experimentation on a system in which the parasites' effect on the host and life history characteristics are well known, in which loads can be accurately measured and experimentally manipulated, and in which heritability of resistance and mate preferences for bright coloration can be determined. Although much is known about the mate preferences of zebra finches, not enough is presently known about the identity, prevalence, and pathogenicity of their parasites to currently permit such an approach. We expect, in the near future, to explore the incidence of nematode infection and to conduct preliminary experiments to ascertain the effect of drug therapy on parasite load, bill colour, and other aspects of the phenotype. We also plan to continue to look for sex differences in parasite loads.

Acknowledgements

We thank Bill Freeland, Jenella Loye, Rich Symanski, Ken Todd, Marlene Zuk, and two anonymous reviewers for helpful criticisms and the following people for parasite identifications: Don Colless (hippoboscids), Robert Domrow (mites), and Dale Clayton and Roger Price (lice). Marlene Calley examined blood smears. For hospitality in Northern Victoria N.B. thanks Richard Zann. This research was supported by a grant from the National Science Foundation (BSR 8516728 to N.B) and the Conservation Commission of the Northern Territory (S.C.T). Bird bands were provided by the Australian Bird and Bat Banding Scheme.

References

Alker, H.R. Jr (1969). A typology of ecological fallacies. In *Quantitative ecological analysis in the social sciences* (ed. M. Dogan and S. Rokkan), pp. 69–86. MIT Press, Cambridge, Massachusetts.

Ash, J.S. (1960). A study of the Mallophaga of birds with particular reference to their ecology. *Ibis*, **102**, 93–110.

Borgia, G. (1986). Satin bowerbird parasites: a test of the bright male hypothesis. *Behavioral Ecology and Sociobiology*, **19**, 355–8.

Bertalanffy, L.V. (1968). *General systems theory*. George Brazzler, New York.

Burley, N. (1977). Parental investment, mate choice, and mate quality. *Proceedings of the National Academy of Sciences USA*, **74**, 3476–9.

Burley, N. (1983). The meaning of assortative mating. *Ethology and Sociobiology*, **4**, 19–203.

Burley, N. and Coopersmith, C.B. (1987). Bill color preferences of zebra finches. *Ethology*, **76**, 13–151.

Burley, N., Zann, R.A., Tidemann, S.C., and Male, E.B. (1989). Sex ratios of zebra finches. *Emu*, **89**, 83–92.

Camin, J.H. (1963). Relation between host-finding behavior and life histories in ectoparasitic Acarina. In *Advances in acarology*, Vol. 1 (ed. J.A. Naegele). Comstock, Ithaca, New York, pp. 411–24.

Clark, G.M. and Yunker, C.E. (1956). A new genus and species of Dermanyssidae (Acarina: Mesostigmata) from the English sparrow, with observations of its life cycle. *Proceedings of the Helminthological Society of Washington*, **23**(2), 93–101.

Freeland, W.J. (1981). Parasitism and behavioral dominance among male mice. *Science*, **213**, 461–2.

Fowler, J.A. and Cohen, S. (1983). A method for the quantitative collection of ectoparasites from birds. *Ringing and Migration*, **4**, 185–9.

Fowler, J.A., Miller, C.J., and Cohen, S. (1984). Ectoparasite populations from breeding and wandering storm petrels. *Bird Study*, **31**, 126–130.

Hall, R.D. and Gross, W.B. (1975). Effect of social stress and inherited plasma corticosterone levels in chicken on populations of the northern fowl mite, *Ornithonyssus sylviarum*. *Journal of Parasitology*, **61**, 1096–1100.

Hamilton, W.D. and Zuk, M. (1982). Heritable true fitness and bright birds: a role for parasites? *Science*, **218**, 384–7.

Hausfater, G. and Watson, D.F. (1976). Social and reproductive correlates of parasite ova emissions by baboons. *Nature, London*, **262**, 688–9.

Herman, C.M., Kinsley, J.O. Jr, and Snyder, E.L. (1966). Subinoculation as a technique in the diagnosis of avian *Plasmodium*. *Avian Diseases*, **10**, 541–7.

Immelmann, K. (1962). Beiträge zu einer vergleichenden Biologie australischer Prachtfinken (Spermestidae). *Zoologische Jahrbücher (Syst.)*, **90**, 1–196.

Kirkpatrick, C. and Suthers, H. (1988). Epizootiology of blood parasite infections in passerine birds from central New Jersey. *Canadian Journal of Zoology*, **66**, 2374–82.

Kodric-Brown, A. and Brown, J.H. (1984). Truth in advertising: the kinds of traits favored by sexual selection. *American Naturalist*, **124**, 309–23.

Krantz, G.W. (1978). *A manual of acarology* (2nd edn). Oregon State University, Corvallis.

Lazarides, M. (1970). *The grasses of Central Australia*. Australian National University, Canberra.

McAlpine, J.R. (1976). Climate and water balance. *CSIRO Land Research Series*, **38**, 35–49.

Nelson, W.A., Keirans, J.E., Bell, J.F., and Clifford, C.M. (1975). Host–ectoparasite relationships. *Journal of Medical Entomology*, **12**, 143–66.

Read, A.F. (1987). Comparative evidence supports the Hamilton and Zuk hypothesis on parasites and sexual selection. *Nature*, **328**, 68–70.

Ressel, S. and Schall, J.J. (1989). Parasites and showy males: malarial infection and color variation in fence lizards. *Oecologia*, **78**, 158–64.

Schall, J.J. (1986). Prevalence and virulence of a haemogregarine parasite of the Aruban whiptail lizard, *Cnemidophorus arubensis*. *Journal of Herpetology*, **20**, 318–24.

Seegar, W.S. (1979). Comparison of four blood survey techniques for detecting microfilariae in avian blood. *Ibis*, **121**, 104–6.

Turner, E.C. Jr (1971). Fleas and lice. In *Infectious and parasitic diseases of wild birds* (ed. J.W. Davis, R.C. Anderson, L. Karstag, and D.O. Trainer), pp. 175–85. Iowa State University Press, Ames.

Ward, P.I. (1988). Sexual dichromatism and parasitism in British and Irish freshwater fishes. *Animal Behaviour*, **36**, 1210–15.

Wheeler, T.A.. and Threlfall, W. (1986). Observation of the ectoparasites of some Newfoundland passerines (Aves: Passeriformes). *Canadian Journal of Zoology*, **64**, 630–6.

Zahavi, A. (1977). The cost of honesty (further remarks on the handicap principle). *Journal of Theoretical Biology*, **53**, 205–14.

Zann, R. (1985). Slow continuous wing-moult of zebra finches *Poephila guttata* from southeast Australia. *Ibis*, **127**, 184–96.

Zuk, M. (1987). The effects of gregarine parasites, body size and time of day on spermatophore production and sexual selection in field crickets. *Behavioral Ecology and Sociobiology*, **21**, 65–72.

20 *Female choice of males with low parasite loads in sage grouse*

LINDA L. JOHNSON and MARK S. BOYCE

Introduction

A characteristic of lek species is large variance in mating success of males (Gibson and Bradbury 1986). On sage grouse leks, copulations are performed by only a few males each spring and many cocks never mate (Wiley 1978; Hartzler and Jenni 1988). Females appear to select the 'best' male, but it is difficult to envisage that preferred males are really genetically superior because selection should reduce heritability for characters highly correlated with fitness (Falconer 1981, Chapter 20). This is particularly troublesome in lekking species where no paternal benefits accrue from choosing a good mate. This is termed the lek paradox (Borgia 1979; Taylor and Williams 1982).

Various hypotheses have been proposed to explain the evolution and maintenance of lek behaviour, but currently few data exist to test the hypotheses (Avery 1984). The focus of our research is to provide empirical evidence to evaluate the Hamilton and Zuk (1982) hypothesis relating to coevolution of parasites and host as a mechanism for maintaining heritable variance (Boyce *et al.* 1988) in true fitness amongst males. The hypothesis proposes that secondary sexual characteristics may allow females to select mates that are relatively resistant to the parasite genotypes exploiting the host population at that time. By mating with such males, females obtain resistance genes for their offspring.

Female sage grouse have been shown to select males that strut frequently and vigorously and that attend the lek regularly (Gibson and Bradbury 1985). Strutting frequency and lek attendance may be cues which females use to identify genetic resistance to parasites among potential mates. It is expected that males with heavy parasite loads will not be able to display as vigorously or may attend leks less frequently. Resistance to parasites may therefore enhance a male's potential vigour and ability to attain master cock status.

Avian malaria, *Plasmodium pediocetii*, is a particularly intriguing candidate for the Hamilton and Zuk hypothesis because infected erythrocytes erupt in the morning in sage grouse releasing metabolic wastes and haematoxins into the bloodstream (Stabler and Kitzmiller 1976; Atkinson and van Riper, Chapter 2, this volume). This coincides with morning courtship display and suggests that males infected with avian malaria may feel poorly at precisely the time they would be strutting to attract females.

In this chapter, we show that females preferentially mate with males bearing low parasite loads as predicted by the Hamilton and Zuk hypothesis. Sage grouse offer a good opportunity to study lek behaviour since large populations occur in relatively undisturbed habitats near Laramie, Wyoming, and they are easy to capture, mark, and observe on leks.

Methods

Our study was conducted in south-eastern Wyoming on 11 leks located 25–100 km north of Laramie (Fig. 20.1). Data were collected between March and May during 1987 and 1988. Sage grouse were captured by night lighting and rocket netting as described by Giesen *et al.* (1982). After capture, each bird was placed in a burlap bag for approximately 30 min. This gave the bird time to calm down and made handling easier.

At the time of capture a series of morphological measurements were taken for each bird, including body mass, total length, tail length, keel length, bill depth, bill length, comb length, tarsus, and toe length. A condition index was defined to be (body mass)·(keel length)$^{-3}$.

A blood sample from the brachial vein was collected and two blood films were prepared and preserved using methyl alcohol. Slides were later prepared with Giemsa stain and examined under oil immersion ($\times$ 1000) for haematozoa including *Plasmodium, Haemoproteus, Leucocytozoon*, trypanosomes, and microfilariae.

Faecal samples were collected and examined for intestinal parasites. Faecal samples were centrifuged for two 5-min periods in water and the liquid was decanted. A coverslip was placed on the centrifuge tube and the sample was centrifuged for 5 min in saline solution (Pritchard and Kruse 1982). The coverslip was transferred to a microscope slide and examined for coccidia and other intestinal parasites under $\times$ 100 magnification.

Birds were inspected for ectoparasites, particularly in relatively unpreenable places such as the back of the head and air sacs. Samples of ectoparasites were collected for identification.

Each bird was marked with a unique colour combination of numbered patagial tags (plastic sheep tags) which allowed for identification during morning observation periods. Leks were observed daily throughout the

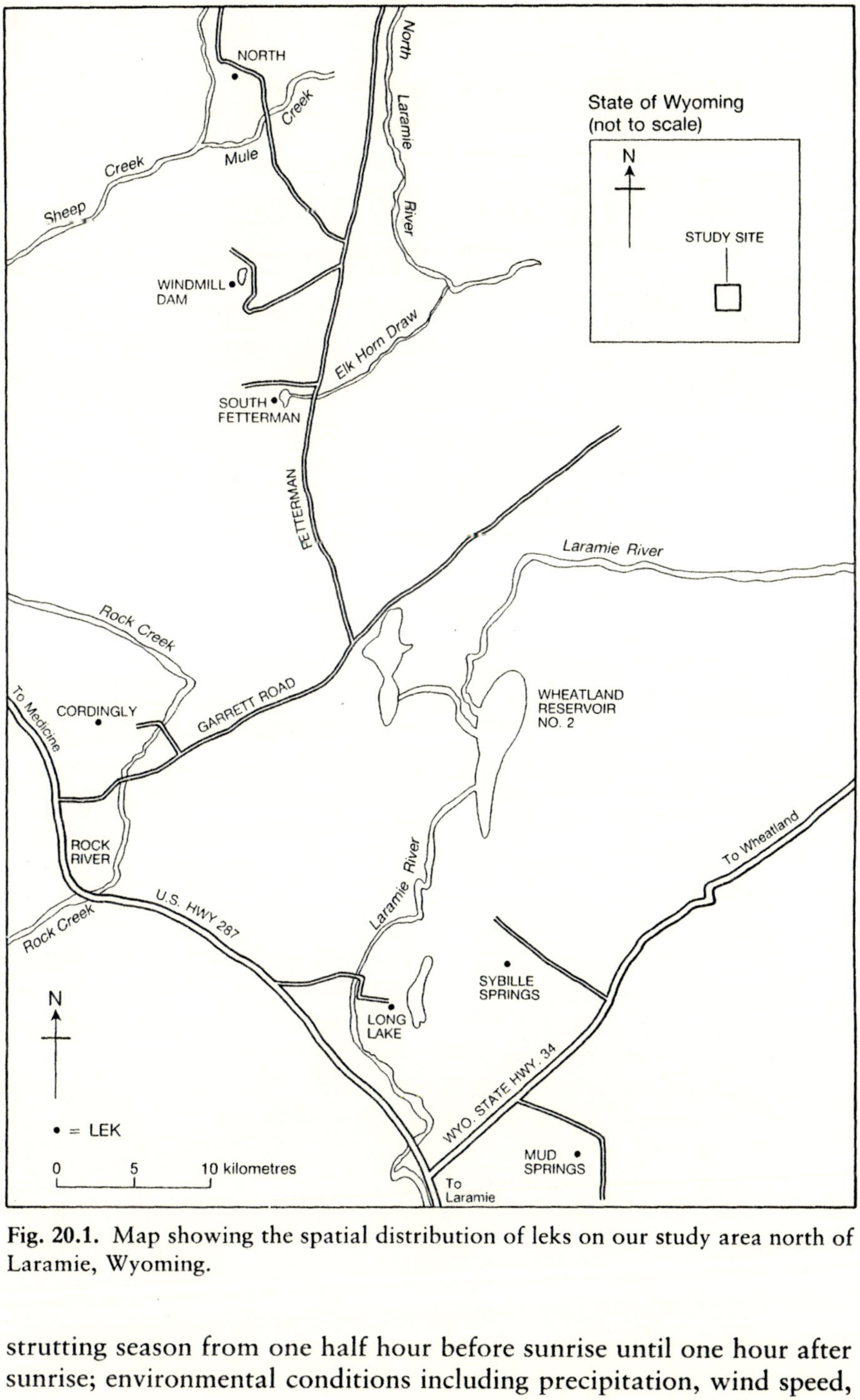

Fig. 20.1. Map showing the spatial distribution of leks on our study area north of Laramie, Wyoming.

strutting season from one half hour before sunrise until one hour after sunrise; environmental conditions including precipitation, wind speed, and lek ground condition were recorded. Disturbances, such as predators or night lighting the night prior to morning observations, and the total number of males and females attending the lek each day were also noted on observation forms.

Strutting frequencies for marked males were recorded by counting the number of struts performed during a 5-min interval. Distance to hens, time of day, and fights with other males were recorded to evaluate their effect on strutting frequency (Gibson and Bradbury 1985). Lek attendance for each bird was calculated by dividing the number of days a male was observed on the lek by the total number of days the lek was observed after capture of each individual bird. All observations of copulations were recorded.

Results

Mean strutting frequency $\bar{x}$ (mean number of struts/ 5 min) was significantly higher for breeding ($\bar{x} = 22.7, n = 27$) than for non-breeding males ($\bar{x} = 16.8$, $n = 124$; $t = 3.03$, $p < 0.005$). Strutting frequency was influenced by proximity to females and time of day as previously shown by Gibson and Bradbury (1985) and Hartzler and Jenni (1988). Therefore, we corrected for these two variables using the regression procedure described by Gibson and Bradbury (1985). Mean corrected strutting frequency was also significantly higher for breeding males ($\bar{x} = 4.16$, $n = 24$) than non-breeding males ($\bar{x} = -0.46, n = 122$) and this difference was significant ($t = 2.61, p = 0.01$). Also breeding males attended the lek more frequently (53 per cent, $n = 25$) than non-breeding males (21 per cent, $n = 183$, $t = 5.58$, $p < 0.001$).

Common parasites found include chewing lice, *Lagopoecus gibsoni* and *Goniodes centrocerci*, and the haematozoa, *Plasmodium pediocetii*, *Leucocytozoon* sp., and *Haemoproteus* sp. The prevalence of infection with *Leucocytozoon* (18.5 per cent) and *Haemoproteus* (11.8 per cent) was low. Coccidia was not detected in the faecal samples.

Lek attendance was significantly lower for males with avian malaria, *Plasmodium pediocetti* (18 per cent, $n = 75$), than for non-infected males (29 per cent, $n = 133$, $t = 3.01$, $p < 0.01$), perhaps affording a cue which females could use to distinguish males with malaria. However, there was no significant difference in mean corrected strutting frequency for malaria-infected males ($\bar{x} = -0.34$, $n = 35$) and non-infected males ($\bar{x} = 0.97$, $n = 75$; $t = 0.678$, $p = 0.5$).

Prevalence of malaria was slightly less for breeding males (24 per cent, $n = 25$) than for non-breeding males (32 per cent, $n = 177$), but not significantly so ($\chi^2 = 0.3333$, $p > 0.5$). Similarly, of breeding males, we

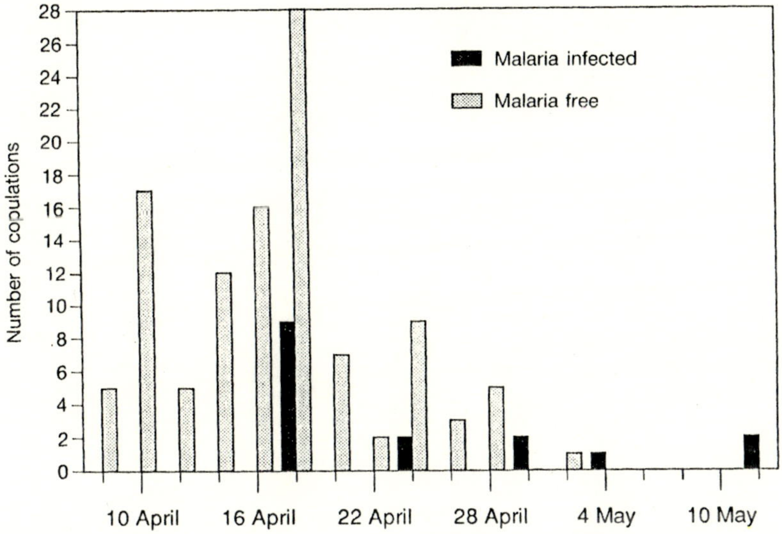

Fig. 20.2. Temporal distribution of copulations on sage grouse leks north of Laramie, Wyoming during 1987 and 1988. Copulations by malaria-infected males are significantly ($p < 0.001$) later than for males in whom malaria was not observed.

could not reject the null hypothesis that malaria occurred in the same frequency as in the population ($\chi^2 = 0.6$, $df = 1$, $p > 0.2$). However, non-infected breeding males secured a significantly higher number of copulations than expected, whereas infected males obtained fewer copulations than expected ($\chi^2 = 21.54$, $df = 1$, $p < 0.001$).

Consistent with the Darwin–Fisher hypothesis for sexual selection in monogamous birds (see Bradbury and Andersson 1987), male sage grouse infected with malaria breed significantly later in the season than malaria-free males ($\chi^2 = 24.59$, $df = 3$, $p < 0.001$) as illustrated in Fig. 20.2. Females attending the lek earlier in the season are more likely to be adults than yearlings ($\chi^2 = 18.87$, $df = 2$, $p < 0.001$). Also, condition of females, (body mass) (keel length)$^{-3}$, attending the lek is inversely correlated with date ($r_s = -0.264$, $n = 148$, $p < 0.001$).

There is significant temporal variation between years in the prevalence of malaria infection ($\chi^2 = 3.84$, $df = 1$, $p = 0.05$). Significant temporal variation occurred on two individual leks: North Lek ($\chi^2 = 6.48$, $df = 1$, $p < 0.025$) and Mud Springs Lek ($\chi^2 = 7.05$, $df = 1$, $p < 0.01$) (Table 20.1). There is significant variation in malaria prevalence among leks for males ($\chi^2 = 11.02$, $df = 5$, $p < 0.05$), but not for females ($\chi^2 = 0.99$, $df = 5$, $p > 0.9$). However, there is no difference in infection rates for males versus females ($\chi^2 = 2.35$, $df = 1$, $p > 0.1$). Between-lek differences in both males and females, for 1987 and 1988 combined, range from 8.3 per cent infected on Windmill Dam Lek to 46.2 per cent on South Fetterman Lek.

Table 20.1 Spatial and temporal distribution of avian malaria infections in sage grouse

Lek	1987		1988	
	Percentage with plasmodium	N	Percentage with plasmodium	N
Long Lake	41.0	39	41.5	65
North	14.0	22	50.0	20
Windmill Dam	0.0	7	20.0	5
South Fetterman	28.5	14	10.0	10
Cordingly	66.7	6	28.6	7
Sybille Springs	33.3	24	18.2	11
Mud Springs	14.8	27	46.2	39

Table 20.2 Spatial and temporal distribution of lice infestation among male sage grouse

Lek	1987		1988	
	Percentage with lice	N	Percentage with lice	N
Long Lake	40.7	27	23.8	21
North	0.0	11	50.0	18
Windmill Dam	83.3	6	25.0	4
South Fetterman	40.0	10	71.4	7
Cordingly	100.0	6	100.0	4
Sybille Springs	41.7	12	20.0	10
Mud Springs	33.3	21	13.0	23

Breeding males were significantly less likely to bear lice (12.1 per cent, $n=29$) than non-breeding males (42.5 per cent, $n=219$, $\chi^2=8.84$, $df=1$, $p<0.005$). Significantly fewer copulations were performed by males with lice than expected ($\chi^2=35.92$, $df=1$, $p<0.001$). Males with lice often had associated haematomas on their air sacs. Indeed, males possessing lice are substantially more likely to have haematomas on their air sacs ($\chi^2=31.65$, $df=1$, $p<0.001$). The haematomas were readily seen when the males were strutting, and these haematomas are a cue which females use to discern lousy males (Spurrier *et al.*, Chapter 21, this volume).

There is significant spatial variation among leks in lice infestation ($\chi^2=20.06$, $df=6$, $p<0.001$). As with malaria, there is significant variation between years as well ($\chi^2=5.29$, $df=1$, $p<0.025$). For example, in 1987, none of the males on North Lek were infested compared to 50 per cent in 1988 (Table 20.2). Results are reported only for males because we seldom found lice on females. This is possibly a sampling problem because lice appeared to be concentrated on the large air sacs of males and were harder to find on the smaller air sacs of females.

Finally, we evaluated whether there was independence between prevalence of malaria or lice. Males with malaria are not more likely to bear lice ($\chi^2=0.45$, $df=1$, $p=0.5$).

Discussion

On days when erythrocytes burst in malarious birds, they probably do not attend leks and thereby may lose opportunities to be selected by females. On other days when erythrocytes are not bursting, infected males attend the lek and appear able to strut as frequently as non-infected males. Lek attendance is highly correlated with mating success, apparently because females switch to other males if attendance is interrupted.

It is possible that infected birds suffer from relapse and therefore are unable to attend the lek on a regular basis. Relapse in avian species is comparable to human responses and includes fever and a return of parasites into the circulating blood (Hewitt 1940; Kitzmiller, personal communication). The duration of relapse events varies with degree of infection but short-term relapses are more common (Hewitt 1940).

In human, rodent, and avian malaria there is loss of erythrocytes because uninfected as well as infected cells are removed from circulation through an immune response in which cells are engulfed by macrophages resulting in anaemia (McGhee 1970; Cohen and Lambert 1982). Anaemia can exist even when there is an extremely low parasitaemia (McGhee 1970). Anaemia involves a decrease in haemoglobin levels which leads to decreased aerobic capacities. In malarious western fence lizards, (*Sceloporus occidentalis*) infected males are able to compete for females for a short period of time but, due to decreased aerobic capacities, stamina is reduced and males quickly become exhausted (Schall *et al.* 1982; Schall and Dearing, 1987).

Infected male sage grouse may be able to strut vigorously for a short period of time maintaining high strut frequencies. But, due to decreased aerobic capacities, they become exhausted and are unable to strut consistently while non-infected birds are able to continue strutting throughout the morning courtship period. However, there was no significant difference in strutting frequency between infected and non-infected males.

This may be a result of sampling techniques used during morning observation periods. When possible, multiple strutting frequencies for marked males were recorded on each day that the lek was observed.

However, if there were a number of marked males, it was often difficult to obtain multiple strutting counts for individual birds. When only one strutting frequency per day was recorded, it is possible that strutting frequencies for malaria-infected birds were recorded during the time when the birds were strutting vigorously. Such a bias is likely to appear because the patagial tags are most visible when the bird is strutting, making individual recognition possible.

It is still possible that, due to exhaustion, males were unable to sustain high strutting frequency throughout the morning display period. If multiple frequencies were not recorded, the strutting frequency for that day may not be representative of the bird's strutting activity. For future studies, it may be necessary to concentrate efforts on obtaining multiple strut frequencies for individual birds on a daily basis to determine if strut frequencies show greater temporal decline for infected birds.

The reproductive advantage of malaria-free males over those infected with malaria is amplified by the temporal distribution of matings. Malaria-infected males breed later and as a consequence are more likely to mate with younger females and females in poor condition. Young hens ovulate fewer follicles, lay smaller clutches of eggs, and are less successful at bringing off a brood (Wallestad and Pyrah 1974). Because age and condition effects on female breeding dates are probably non-heritable, breeding date is not likely to evolve as a consequence of this differential reproductive success (Price *et al.* 1988). Nevertheless, the pattern of hen attendance in sage grouse affords a strong selective advantage to males breeding early. This mechanism was previously proposed by Darwin (1871) and Fisher (1930) to drive sexual selection in monogamous birds.

We found significant spatial and temporal variation in prevalence of malaria, which is consistent with predictions of the Hamilton–Zuk hypothesis. In fact, we would argue that demonstration of such dynamics ought to be amongst the criteria listed by Read (1988) for studies attempting to provide empirical evidence for the Hamilton–Zuk hypothesis (see also van Riper, Chapter 8, this volume). Without such temporal variance in the selective regimen, resistance genes should become fixed and the dynamic feature of parasite–host interaction would be lost. Although we have demonstrated such temporal variation in malaria abundance, we cannot rule out the possibility that temporal fluctuations may have occurred in the abundance of the invertebrate vectors.

The existence of spatial variation in malaria prevalence for males but not for females is puzzling. Sex differences in interlek movements might be hypothesized to contribute to this since males appear to show high fidelity to leks (Wiley 1978), although Dunn and Braun (1985) claim that dispersal does not differ between male and female sage grouse in Colorado. More probable is a Type II error regarding female distribution due to an inadequate sample of females.

Read (1988) has suggested that convincing studies evaluating the Hamilton and Zuk hypothesis need to test four premises: (1) a negative relationship exists between an individual's fitness and its parasite load; (2) the expression of secondary sexual characteristics varies with parasite burdens; (3) preferred mates have fewer parasites; and (4) there is heritable variation in parasite resistance.

Regarding criterion 1, malaria can affect survival of infected birds. In malaria infections, the mild, chronic types show no signs but, as intensity of infection increases, effects on the host become more severe. In moderate, chronic forms of the disease, birds become anaemic and weak and the infection is probably a contributing factor to death under conditions of exposure or breeding exhaustion. In heavy, acute types, birds lose flesh rapidly, are unable to fly, become droopy, and some die (Barger *et al.* 1958).

Clear fitness costs are evident for males with malaria because they secure fewer copulations. Also, because they breed later in the season, the copulations that they do get contribute less to fitness.

The secondary sexual characteristics associated with malaria appears to be lek attendance. Master cocks tend to be persistent in their attendance at leks (Hartzler and Jenni 1988), and indeed, we observed significantly higher lek attendance by breeders than by non-breeders. Hens typically observe males for 3 days prior to mating. If a male misses a day, she is likely to switch to another male who is consistently in attendance.

Regarding Read's third point, females clearly show a preference for males free from both lice and malaria. However, we are unable at this time to offer support for Read's fourth criterion of heritability for resistance to either lice or malaria in sage grouse. Genetic resistance to malaria has been documented for humans (Allison 1963; Rosenstreich 1980) and, as far as has been determined, the general principles of transmission, development, and physiology of malaria are similar for all hosts (McGhee 1970; Atkinson and van Riper, Chapter 2, this volume). Genetic resistance to a large number of pathogens is known (Hamilton 1982), and genetic resistance to *Eimeria* in poultry has been well documented (Barriga 1981).

Lagopoecus gibsoni and *Goniodes centrocerci* are biting and chewing lice of the suborder Ischnocera. We hypothesized that haematomas caused by lice may be a visual cue used by females to assess the presence or absence of lice. Avoidance of infested males may offer a direct fitness advantage to the female if she is likely to contract a louse infestation through direct contact with a lousy male.

Spurrier *et al.* (Chapter 21, this volume) conducted an experiment on captive sage grouse to see if haematomas were indeed a cue used by females. In this experiment, red splotches were painted on air sacs of males and female preference was determined. Females were found to choose non-treated males, suggesting that they were avoiding males with haemotomas. However, the birds in Spurrier *et al.*'s study appeared to be infested predominantly with the louse *Goniodes centrocerçi*, whereas *Lagopoecus gibsoni* appeared most commonly on the wild birds.

Lice may destroy downy portions of the feathers reducing thermal insulation for lousy birds as Clayton (Chapter 14, this volume; see also Møller, Chapter 17) observed in rock doves (*Columbia livia*). This would thermally stress the birds during winter such that males are in poor condition while lekking, and strutting vigour may be reduced. This possibility remains to be tested.

Our observations on parasites in sage grouse add to the list of intraspecific studies supporting Hamilton and Zuk's (1982) hypothesis (see also Kennedy *et al.* 1987; Schall and Dearing 1987; Read 1988). Although the evidence is tantalizing, our results do not afford a test of the null model of Darwinian sexual selection by female choice (Kirkpatrick 1982). Yet, the Hamilton–Zuk model is not exclusive of the pure sexual selection model; indeed, both processes are thought to operate simultaneously (Bradbury and Andersson 1987). Continued studies are needed to document temporal fluctuations in parasite incidence, fluctuations in other parasites, and heritability of resistance to parasites in sage grouse.

Acknowledgements

We thank the many students, faculty, family, and friends who helped with our all-night grousing vigils, including Mike Johnson, Pat White, Connie Cole, Mark Kirkpatrick, David Burton, Margo Spurrier, Evie Merrill, Cody Boyce, Jeffry Mitton, Stan Olsen, Dan Marty, Dave Gulley, Dan Koeppen, Dan Hengel, Frank Rahel, Pat Carter, Bill Gern, Steve Buskirk, Mary Brodahl, Bryan Manly, Lyman McDonald, Maxine Lacey, and Rick Knight. Joe Bohne of the Wyoming Game and Fish Department helped us find some of our leks. Several area ranchers granted us permission to work on their land. Dale Clayton identified our lice. Beth Williams, Larry Shults, Bill Hamilton, Newt Kingston, Connie O'Neill, Nancy Kitzmiller, Nancy Stanton, Bob Lacey and Bill Jolley helped us to develop parasite sampling schemes and to identify parasites. Special thanks to Bill Lacey for supplies and equipment. Our research has been funded by the National Science Foundation under Grant No. RII-8610680.

References

Allison, A.C. (1963). An inherited factor in blood conferring resistance to protozoa. In *Immunity to protozoa* (ed. P.C.C. Garnham, A.E. Pierce and I. Roitt). Blackwell Scientific Publications, Oxford.

Avery, M.I. (1984). Lekking in birds: choice, competition, and reproductive constraints. *Ibis*, **126**, 177–87.

Barger, E.H., Card, L.E., and Pomeroy, B.S. (1958). *Diseases and parasites of poultry*. Lea and Febiger, Philadelphia.

Barriga, O.D. (1981). *The immunology of parasitic infections*. University Park Press, Baltimore, Maryland.

Borgia, G. (1979). Sexual selection and the evolution of mating systems. In *Sexual selection and reproductive competition in insects* (ed. M.S. Blum and N.A. Blum), pp. 19–50. Academic Press, New York.

Boyce, M.S., Burton, D.W., Johnson, L.L., Spurrier, M.F., White, P.A., and Mitton, J.B. (1988). Genetic variation in Sage Grouse, *Centrocercus urophasianus*. *Abstracts of the American Ornithologists' Union Annual Stated Meeting*, **106**, 9.

Bradbury, J.W. and Andersson, M.B. (1987). *Sexual selection: testing the alternatives*. John Wiley, New York.

Cohen, S. and Lambert, P.H. (1982). Malaria. In *Immunology of parasitic infections* (ed. S. Cohen and K.S. Warren), pp. 422–74. Blackwell Scientific Publications, Oxford.

Darwin, C.R. (1871). *The descent of man and selection in relation to sex*. Appleton, New York.

Dunn, P.O. and Braun, C.E. (1985). Natal dispersal and lek fidelity of Sage Grouse. *Auk*, **102**, 621–7.

Falconer, D.S. (1981). *Introduction to quantitative genetics*. Longman, London.

Fisher, R.A. (1930). *The genetical theory of natural selection*. Dover, New York.

Gibson, R.M. and Bradbury, J.W. (1985). Sexual selection in lekking grouse: phenotypic correlates of male mating success. *Behavioral Ecology and Sociobiology*, **18**, 117–23.

Gibson, R.M. and Bradbury, J.W. (1986). Male and female mating strategies on sage grouse leks. In *Ecological aspects of social evolution* (ed. D.I. Rubenstein and R.W. Wrangham), pp. 379–98. Princeton University Press, Princeton, New Jersey.

Giesen, K.M., Schoenberg, T.J., and Braun, C.E. (1982). Methods for trapping Sage Grouse in Colorado. *Wildlife Society Bulletin*, **10**, 224–31.

Hamilton, W.D. (1982). Pathogens as causes of genetic diversity in their host populations. In *Population biology of infectious diseases*, Dahlem Konferenzen (ed. R.M. Anderson and R.M. May), pp. 269–96. Springer-Verlag, New York.

Hamilton, W.D. and Zuk, M. (1982). Heritable true fitness and bright birds: a role for parasites? *Science*, **218**, 384–7.

Hartzler, J.E. and Jenni, D. (1988). Mate choice by female sage grouse. In *Adaptive strategies and population ecology of northern grouse*, Vol. 1. *Population studies* (ed. A.T. Bergerud and M.W. Gratson), pp. 240–69. University of Minnesota Press, Minneapolis.

Hewitt, R. (1940). *Bird Malaria*. Johns Hopkins Press, Baltimore, MD.

Kennedy, C.E.J., Endler, J.A., Poynton, S.L., and McMinn, H. (1987). Parasite load predicts mate choice in guppies. *Behavioral Ecology and Sociobiology*, **21**, 291–5.

Kirkpatrick, M. (1982). Sexual selection and the evolution of female choice. *Evolution*, **36**, 1–12.

McGhee, R.B. (1970). Avian malaria. In *Immunity to parasitic animals*, Vol. 2 (ed. G. J. Jackson, R. Herman, and I. Singer), pp. 331–69. Appleton-Century-Crofts, New York.

Price, T., Kirkpatrick, M., and Arnold, S.J. (1988). Directional selection and the evolution of breeding date in birds. *Science*, **240**, 798–9.

Pritchard, M.H. and Kruse, G.O.W. (1982). *The collection and preservation of animal parasites*. University of Nebraska Press, Lincoln.

Read, A.F. (1988). Sexual selection and the role of parasites. *Trends in Ecology and Evolution*, **3**, 97–101.

Rosenstreich, D.L. (1980). Genetics of resistance to infection. *Nature*, **285**, 436–7.

Schall, J.J., Bennett, A.F., and Putnam, R.W. (1982). Lizards infected with malaria: physiological and behavioral consequences. *Science*, **217**, 1057–9.

Schall, J.J. and Dearing, M.D. (1987). Malarial parasitism and male competition in the western fence lizard, *Sceloporus occidentalis*. *Oecologia*, **73**, 389–92.

Stabler, R.M. and Kitzmiller, N.J. (1976). *Plasmodium* (*Giovannolaia*) *pediocetii* from gallinaceous birds of Colorado. *Journal of Parasitology*, **62**, 539–44.

Taylor, P.D. and Williams, G.C. (1982). The lek paradox is not resolved. *Theoretical Population Biology*, **22**, 392–409.

Wallestad, R. and Pyrah, D. (1974). Movement and nesting of sage grouse hens in central Montana. *Journal of Wildlife Management*, **38**, 630–3

Wiley, R.H. Jr (1978). The lek mating system of the sage grouse. *Scientific American*, **238**, 114–25.

21 *Effects of parasites on mate choice by captive sage grouse*

MARGO F. SPURRIER, MARK S. BOYCE, and BRYAN F.J. MANLY

Introduction

A consequence of selection is reduced genetic variability (Falconer 1981, Chapter 20). Therefore, one expects low heritability for sexually-selected traits; yet lekking species maintain intense female choice with no apparent benefit to the female except male genes. Possible resolution to this 'paradox of the lek' (Taylor and Williams 1982) is offered by coevolution of parasites and host as articulated by Hamilton and Zuk (1982) and Zuk (1984; Chapter 16, this volume). Fluctuations in the abundance of parasites and the frequency of resistant hosts may play a role in maintaining variance in male fitness. Females may choose mates which are genetically resistant to disease by an examination of characters whose full expression is dependent upon health and vigour. If the genes for resistance to certain pathogens are heritable, then a female choosing the most colourful or vigorously displaying male may, in fact, be selecting the 'best genes' for her offspring.

In field studies, male sage grouse (*Centrocercus urophasianus*) bearing chewing lice (*Lagopoecus gibsoni* and *Goniodes centrocerci*; Order Mallophaga, Suborder Ischnocera) were less successful breeders than those free from lice (Johnson and Boyce, Chapter 20, this volume). Only 13.8 per cent of the males that were successful breeders were lousy, whereas 43 per cent of the non-breeders bore louse loads. Parasite load may be an expression of poor condition. Thus, individuals that are in poor condition are more likely to contract a disease than individuals in good body condition. Or, parasite load may be the cause of poor condition. Nevertheless, the end result is that parasitized males are less attractive to females.

These observations suggested experiments for our captive flock of sage grouse. Such experiments are necessary to reinforce interpretation of field studies because alternative hypotheses usually plague such correlational results. For example, without experimental manipulations of parasite loads, we cannot be certain that the lower reproductive success of parasitized males is not a consequence of unattractive males being more susceptible to parasites.

Here we present the results of three experiments to test predictions of Hamilton and Zuk's (1982) hypothesis for the role of parasites in sexual selection. First, we experimentally reduced endoparasites in male sage grouse by treating birds with oxytetracycline. In a second experiment, we further manipulated parasite loads by administering carbaryl to eliminate ectoparasites. In our third experimental treatment we applied red spots to the air sacs of males to simulate haematomas created by lice.

Methods

Sage grouse nests were found near leks north of Laramie, Wyoming and eggs were hatched in a poultry incubator. A few young chicks captured in the field were also added to our captive flock. Holding pens were constructed at the Red Buttes Environmental Research Laboratory south of Laramie, Wyoming (Johnson and Boyce 1990). Sage grouse were fed Purina Game Bird Chow supplemented weekly with fresh sagebrush (*Artemisia tridentata*) and carrots. During summer the diet was also supplemented with mealworms and alfalfa.

Two outdoor arenas were constructed for six males in pens (2.2 × 1.85 × 1.3 m) surrounding a central hexagonal arena with 2.2-m sides into which a female was released. Arenas were 4 m apart such that both could be watched by a single observer. Observations were conducted for 90 min each morning beginning one half hour before sunrise to coincide with display periods on natural leks. Experimental trials were conducted from mid-March through early May in 1987 and from 1 April 1988 through 28 April 1988.

Male pens were movable so that juxtaposition of males could be randomized without handling the males. Position of males in trials was determined by a semi-randomized block design with male locations being altered every other day in 1987 and every fourth day in 1988. The same 12 males were used in trials during the entire lekking season, in 1987, except that male no. 8 died on the 12th day of the trial and was replaced with another control male for the remainder of the season. During the 1988 experiments, two independent sets of 12 males were used in trials. Both sets were used to compare the effects of drug treatment versus no treatment. The second set was also used to examine the effects of artificial haematomas.

Females were assigned to one of the two arenas at random during the afternoon prior to each morning trial, and then rotated with the hen assigned to the other arena on the next day. This required that two females were handled on each of 2 days, but we then selected a different pair of females every other day, so that a female never saw the same arena arrangement twice. A total of 25 females participated in the 1987 trials. In 1988 there were 14 hens used in the trials with each female participating a total of four times throughout this season since there were two sets of 12 males.

Mate choice was defined by recording the position of the hen in the arena at 5-min intervals. There was a strong tendency for females to sit or stand adjacent to the male that was strutting most frequently, although the strutting frequency of neighbouring males also contributed to female choice. On natural leks, females form hen clusters surrounding the master cock where they sit and watch the male for an average of 3 days before soliciting copulation (Patterson 1952; Wiley 1973; Hartzler and Jenni 1988). In many cases, attraction of a female to a particular male was evident; although solicitations were infrequent, we are nonetheless confident that our arenas measured active female choice.

We performed three experiments to evaluate the role of parasites in mate choice by female sage grouse. In 1987, one-half of the males were treated with oxytetracycline, an antibiotic commercially available under the trademark Terramyacin. The intent here was to treat possible bacterial infections. The drug was administered by adding 3-ml 0.6 per cent oxytetracycline to 3.5 l of water, and provided *ad libitum* for 7 consecutive days each month. Controls were not given oxytetracycline for 3 months prior to the trials. Female choice was analysed for 21 days beginning on 7 April 1987 because very little strutting occurred before that date.

In 1988, we attempted to evaluate the role of lice in female choice. Since most of the experimental males visible had lice infestations (83 per cent), a random sample of 12 males was treated with carbaryl (l-napthyl-N-methylcarbamate), marketed under the trademark Sevin. Head, back, vent, and air sacs of treatment males were dusted with the powder the day before the initiation of trials.

Before placing a male in an arena pen, the male was examined for the presence of ectoparasites, coccidia, and haematozoa. Each male was handled and examined on the head, air sacs, and back for ectoparasites. On head and back feathers were separated with fingers and area was observed for 5 to 15 s. Air sacs were pulled out from the throat and examined. Each air sac examination lasted approximately 20 to 30 s.

Five cc of blood was drawn at this time from the right brachial vein for later examination. Haematocrit tubes were examined for the presence of *Trypanosoma*. Blood smears stained with Giemsa were examined microscopically for the presence of *Haemoproteus, Leucocytozoon, Plasmodium,* and *Trypanosoma* (Pritchard and Kruse 1982).

Faecal samples were also collected from each bird immediately before placement in an arena pen and three times during the trials. Samples were collected from the handling bag and the floor of the pens respectively. A floatation technique for faecal sampling was used to separate and concentrate coccidia. Slides from this preparation were then examined microscopically for oocysts (Pritchard and Kruse 1982).

An experiment to evaluate the role of haematomas on air sacs was also conducted in April 1988 on the second set of 12 males. Artificial haematomas were applied with a red felt-tipped marker to air sacs of six males, and the other six were untreated controls. We attempted to simulate haematomas observed on lousy wild males (Johnson and Boyce, Chapter 20, this volume) by applying six red splotches, 3–8 mm in diameter, to each oesophageal air sac. All birds were handled whilst examining them for lice. Birds were watched for 8 days prior to treatment, and for 6 days after treatment. One control bird, male no. 17, died after 5 days of pre-treatment sampling. We replaced this male with a control male (no. 4) from a previous experiment, and replaced the 5 days of choices from bird no. 17 with 5 days for bird no. 4 from previous observations.

In addition to standard statistics, a log-linear model was developed to account for strutting frequency of males in female choice. We assume that the probability P_i of a female choosing the *i*th male from the six available can depend upon: (1) an inherent attractiveness for the male, M_i; (2) the rate at which the male is currently strutting in struts per minute, X_i; and (3) the average rate of strutting for the two neighbouring males, Y_i. The probability function takes the form

$$P_i = \exp(M_i + AX_i + BY_i) \,/\, [\Sigma \exp(M_i + AX_i + BY_i)] \qquad (21.1)$$

where *A* and *B* are constants, and the summation is over the six males in the arena that were competing for the female's attentions. Maximum likelihood estimates were obtained for the M_is, *A*s, and *B*s. Statistical details of the approach are given by Manly (1985).

Results

Sage grouse males showed enormous variation in strutting frequency. Some males did not display during the entire mating season, whereas other males maintained display frequencies that rivalled any males that we have observed in the wild.

Female choice was significantly higher for males treated with oxytetracycline to reduce parasites ($F_{1,210} = 4.243$, $p = 0.041$). Different females were used in each arena each day; thus choices made on each day are

Table 21.1 Two-way analysis of variance for effects of oxytetracycline treatment on female choice of males

Source	Sample size	*df*	Mean-square	*F*-ratio	*p*
Date	24.37	20	1.22	0.055	1.0
Treatment	93.22	1	93.22	4.243	0.041
Date*treatment	410.7	20	20.54	0.935	0.544
Error	4613.5	210	21.96		

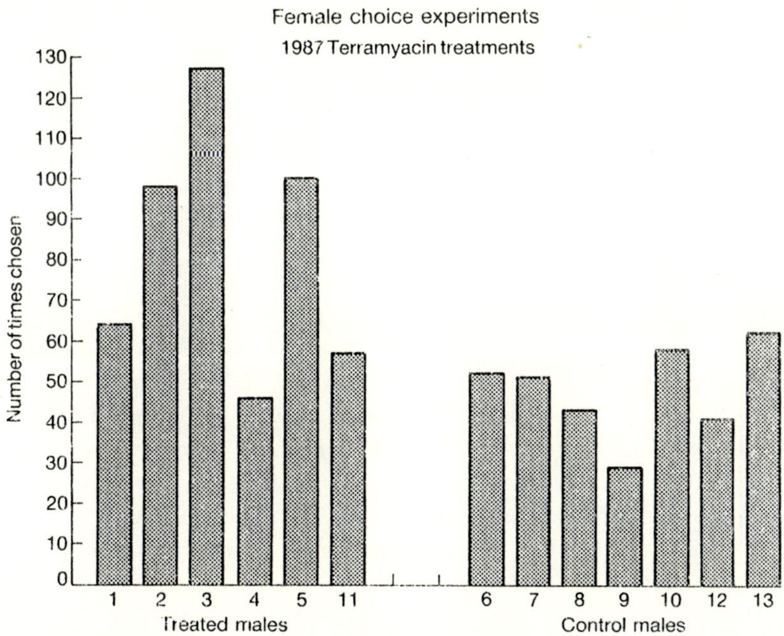

Fig. 21.1. Female choice of males treated with oxytetracycline (Terramyacin) and controls. Six treated males were chosen a total of 492 times during a 21-day trial. Seven control males were chosen a total of 336 times. Bird no. 8 died after 11 days of trials and was replaced by another control male, bird no. 13.

independent of choices made on other days. On average, females chose treated males 24 times each day, whereas control males were chosen 16 times each day. As detailed in Table 21.1, date and date*treatment effects were not statistically significant. The number of times that each male was chosen during the entire 21-day trial period is illustrated in Fig. 21.1.

We did not observe consistent differences between control and treatment birds in parasite loads, although control birds were observed with coccidia and trypanosomes, whereas no parasites were observed in treated males. One faecal sample in 1987 and one in 1988 were found to harbour a few *Eimeria* spp. oocysts. During the summer of 1987 two birds were found to be infected with *Trypanosoma*. Blood samples from 10 males were sent to the Veterinary Diagnostics Laboratory at Oregon State University in Corvallis, Oregon for immunological testing. All were found to be free from *Mycoplasma*, infectious bursal disease, avian influenza, and laryngotracheitis.

Hens showed strong preference for vigorously strutting males, and we fitted the following log-linear model for the probability, P_i, of male i being chosen by a female in the 1987 trials

$$P_i = \exp(M_i + 0.41X_i + 0.72Y_i)/[\Sigma\exp(M_j + 0.41X_j + 0.72Y_j)] \quad (21.2)$$

with variables as defined for eq (21.1). Preference for treated birds appears to be mediated by strutting frequency, because no significant treatment effect could be measured between the M_is for the treatment and control males ($t = 0.829$, 11 *df*, $p = 0.4$). This analysis may be viewed as examining female preference after removing the effect of strutting frequency and mean strutting frequency of neighbouring males.

Carbaryl treatments were highly effective at reducing louse loads, and no lice were observed on treated males. Most lice on control birds appeared to be *Goniodes centrocerci* which were commonly observed on the back and oesophageal air sacs of males. Lice were never observed on females, but this may be a sampling difficulty since lice on males are often seen on their large oesophageal air sacs. As Darwin (1871) pointed out, in sage grouse, 'The oesophagus of the female is not in any way remarkable.' We could not show any difference in female choice ($t = 0.272$, 22 *df*, $p = 0.79$) or strutting frequency ($\chi^2 = 3.23$, 1 *df*, $0.05 < p < 0.1$) for treated or control males.

For the 1988 carbaryl experiments, we found the following maximum likelihood model

$$P_i = \exp(M_i + 0.99X_i + 0.76Y_i)/[\Sigma\exp(M_j + 0.99X_j + 0.76Y_j)]. \quad (21.3)$$

Although the M_is contribute significantly to the model ($p < 0.01$), this measure of male attractiveness shows no significant difference for treated and control males ($t = 0.066$, 22 *df*, $p = 0.95$). In fact the mean M_i for treated males was actually slightly smaller ($\bar{M}_i = -0.0006$, SE $= 0.37$) than for control males with lice ($\bar{M}_i = 0.0006$, SE $= 0.2$).

Application of artificial haematomas to air sacs of treatment males in the third experiment resulted in a decrease in choice for marked males.

Since no physiological effects from marking with red ink were anticipated, the control birds were not marked. For 8 days prior to the application of artificial haematomas, pre-treatment and control males were chosen by females approximately equal numbers of times (139 and 132 respectively). For the 6 days of trials after application of artificial haematomas, 119 choices were for control males, whereas treated males were chosen only 81 times.

Analysis of the effect of haematoma application is a repeated measures problem with two trial factors, i.e. pre-and post-treatment, requiring that the first 2 days of pretreatment samples be dropped from the analysis to balance the number of days pre- and post-treatment. We found a significant haematoma-treatment effect ($F=4.9$, $p=0.05$). Associated with this increase in female choice was a 10 per cent increase in strutting frequency for controls relative to treatment males ($\chi^2=81.2$, 1 *df*, $p<0.001$), such that control males strutted significantly more often than treated males during the post-treatment period ($t=2.39$, $p=0.019$).

As for the other two experiments, we calculated a maximum likelihood model for the 6 days that trials were conducted

$$P_i=\exp(M_i+0.62X_i+0.17Y_i)/[\Sigma\exp(M_j+0.62X_j+0.17Y_j)]. \quad (21.4)$$

Although M_is were smaller for treatment males, they do not differ significantly between treatment and control groups ($t=1.47$, 10 *df*, $p=0.172$). This is not surprising since strutting frequency is correlated with female choice, and strutting frequency is a variable already accounted for in the fitted model.

Discussion

Oxytetracycline was probably influencing bacteria which we were unable to observe or measure reliably. Although the results of the oxytetracycline experiment offer support for Hamilton and Zuk's (1982) hypothesis, the results may be somewhat equivocal because we could not show a consistent difference in parasites between treatment and control groups. The only parasites observed were on control birds as should be expected, but most birds appeared to be free of parasites.

The results of the carbaryl treatments are surprising in view of the large difference in reproductive success for lousy versus louse-free males on natural leks (Johnson and Boyce, Chapter 20, this volume). However, we suspect that the timing of our treatment accounts for the lack of preference differences between control and treatment birds. We applied carbaryl powder immediately prior to the trials. Therefore, any residual consequences of lice would still affect female choice for the treated males. Such residual consequences may include haematomas on air sacs and effects on the plumage caused by lice.

Another hypothesis may account for the lack of preference for dusted males due to thermoregulatory costs to males infested with lice. Clayton (1990; Chapter 14, this volume) observed that pigeons (*Columba livia*) with feather lice lost much of the insulating capacity of their plumage because lice ate fluffy portions of the feathers. As a consequence, males with lice suffered greater thermoregulatory costs during winter and were in poorer condition during the spring mating season and were less able to secure mates than males without lice (see also Møller, Chapter 17, this volume). A similar consequence may befall male sage grouse infested with lice but, because our treatments were not applied until spring, no differences in thermal stress caused by lice should exist between treatment and control males. If our treatments had been applied throughout the preceding winter, we would have expected to see greater vigour of treated males, and consequent female preference for treated males.

A final possibility exists that *Goniodes centrocerci* rather than *Lagopoecus gibsoni* may have predominated as the lice on captive versus wild birds. We have not studied the life histories of the two louse species; thus more information is needed to clarify the effect that each of these lice are having on mate choice.

Nevertheless, carbaryl treatment was remarkably effective at eliminating lice. Therefore, if females were discriminating lousy males because they could actually see the lice, we should have seen a difference in female choice of treatment versus control males. But we did not.

The results of our artificial haematoma experiments support our hypothesis that females use haematomas to avoid lousy males. An increase in strutting frequency for the control males (that were chosen more often) is consistent with field observations that male strutting frequency is stimulated by the proximity of females (Gibson and Bradbury 1985, 1986; Hartzler and Jenni 1988; Johnson and Boyce, Chapter 20, this volume). Such a response to the presence of females appears to be particularly amplified for males that have secured copulations (M.S. Boyce, unpublished).

If hens avoid a male infested with lice by observing haematomas on his air sacs, the female is able to enhance her own fitness directly. This fitness advantage to the female does not require the passage of heritable resistance by louse-free males, but rather the female simply avoids contracting a louse infestation herself. Such a direct fitness advantage can evolve by natural selection and does not require the Hamilton–Zuk (1982) mechanism.

Nevertheless, the outcome of such selection of mates can be much the same as the Hamilton–Zuk hypothesis, creating low-frequency oscillations in parasite and resistant forms of host. Female avoidance of infested males still has the consequence of driving rapid changes in the frequency of resistance genes through sexual selection, even though we may not see the evolution of an associated secondary sexual characteristic (Read 1988). This interaction can still maintain genetic variance for resistance in the population, thereby offering resolution to the lek paradox (*sensu* Taylor and Williams 1982).

Our study design eliminates direct contact between males and therefore a component of male–male interaction that may be important in sexual selection. Indeed, it was our objective to evaluate female choice as described by Darwin (1871) with a minimum of male–male competition. Additional study is necessary to define the role of interactions amongst males in determining sexual selection in sage grouse.

Conclusions

We found support for an interaction between parasites and sage grouse which complements field observations of Johnson and Boyce (1990) and supports, but does not confirm, Hamilton and Zuk's (1982) hypothesis. Treatment with antibiotics to reduce parasite loads resulted in increased strutting frequency and increased female choice for treated males. Experimental application of carbaryl to reduce ectoparasites did not reveal any differences in female choice, suggesting that females are unable to see lice on the males. Application of artificial haematomas to resemble those created by lice resulted in female avoidance of treated males.

Acknowledgements

Our research is funded by the National Science Foundation under grant no. RII-8610680. We thank G.D. Johnson for initiating our captive flock. Dr Beth Williams performed necropsies and offered assistance in our care for the birds. We thank Dale Clayton for identifying lice, and Lyman McDonald for assistance with statistics. Beth Williams, Henry J. Harlow, and Paul J. Watson offered helpful reviews of the manuscript.

References

Clayton, D.H. (1990). Mate choice in parasitized rock doves: lousy males lose. *American Zoologist*.

Darwin, C.R. (1871). *The descent of man, and selection in relation to sex.* Appleton, New York.

Falconer, D.S. (1981). *An introduction to quantitative genetics* (2nd edn). Longman. New York.

Gibson, R.M. and Bradbury, J.W. (1985). Sexual selection in lekking grouse: phenotypic correlates of male strutting success. *Behavioral Ecology and Sociobiology*, **18**, 117–23.

Gibson, R.M. and Bradbury, J.W. (1986). Male and female mating strategies on sage grouse leks. In *Ecological aspects of social evolution* (ed. D.I. Rubenstein and R.W. Wrangham, pp. 379–98. Princeton University Press, Princeton, New Jersey.

Hamilton, W.D. and Zuk, M. (1982). Heritable true fitness and bright birds: a role for parasites? *Science*, **218**, 384–7.

Hartzler, J.E. and Jenni, D. (1988). Mate choice by female sage grouse. In *Adaptive strategies and population ecology of northern grouse*, Vol. 1. *Population studies* (ed. A.T. Bergerad and M.W. Gratson), pp. 240–69. University of Minnesota Press, Minneapolis.

Johnson, G.D. and Boyce, M.S. (1990). Rearing sage grouse in captivity. *Wildlife Society Bulletin* (in press).

Manly, B.F.J. (1985). *The statistics of natural selection.* Chapman and Hall, New York.

Patterson, R.L. (1952). *The sage grouse in Wyoming.* Sage Books, Denver.

Pritchard, M.H., and Kruse, G.O.W. (1982). *The collection and preservation of animal parasites.* University of Nebraska Press, Lincoln.

Read, A.F. (1988). Sexual selection and the role of parasites. *Trends in Ecology and Evolution*, **3**, 97–103.

Taylor, P.D. and Williams, G.C. (1982). The lek paradox is not resolved. *Theoretical Population Biology*, **22**, 392–409.

Wiley, R.H. (1973). Territoriality and non-random mating in sage grouse, *Centrocercus urophasianus. Animal Behavior Monographs*, **6**, 28–34.

Zuk, M. (1984). A charming resistance to parasites. *Natural History*, **93**(4), 28–34.

Index

Printed in the United States
795700001B